High-Frequency Electromagnetic Techniques

WILEY SERIES IN MICROWAVE AND OPTICAL ENGINEERING

KAI CHANG, Editor
Texas A & M University

FIBER-OPTIC COMMUNICATION SYSTEMS
Govind P. Agrawal
COHERENT OPTICAL COMMUNICATIONS SYSTEMS
Silvello Betti, Giancarlo De Marchis and Eugenio Iannone
HIGH-FREQUENCY ELECTROMAGNETIC TECHNIQUES: RECENT ADVANCES AND APPLICATIONS
Asoke K. Bhattacharyya
COMPUTATIONAL METHODS FOR ELECTROMAGNETICS AND MICROWAVES
Richard C. Booton, Jr.
MICROWAVE SOLID-STATE CIRCUITS AND APPLICATIONS
Kai Chang
MULTICONDUCTOR TRANSMISSION-LINE STRUCTURES: MODAL ANALYSIS TECHNIQUES
J. A. Brandão Faria
MICROSTRIP CIRCUITS
Fred Gardiol
HIGH-SPEED VLSI INTERCONNECTIONS: MODELING, ANALYSIS, AND SIMULATION
A. K. Goel
HIGH FREQUENCY ANALOG INTEGRATED CIRCUIT DESIGN
Ravender Goyal (ed.)
OPTICAL COMPUTING: AN INTRODUCTION
Mohammad A. Karim and Abdul Abad S. Awwal
MICROWAVE DEVICES, CIRCUITS AND THEIR INTERACTION
Charles A. Lee and G. Conrad Dalman
ANTENNAS FOR RADAR AND COMMUNICATIONS: A POLARIMETRIC APPROACH
Harold Mott
SOLAR CELLS AND THEIR APPLICATIONS
Larry D. Partain (ed.)
ANALYSIS OF MULTICONDUCTOR TRANSMISSION LINES
Clayton R. Paul
INTRODUCTION TO ELECTROMAGNETIC COMPATABILITY
Clayton R. Paul
OPTICAL SIGNAL PROCESSING, COMPUTING AND NEURAL NETWORKS
Francis T. S. Yu and Suganda Jutamulia

High-Frequency Electromagnetic Techniques: Recent Advances and Applications

ASOKE K. BHATTACHARYYA
Lincoln University

A WILEY-INTERSCIENCE PUBLICATION
JOHN WILEY & SONS, INC.
NEW YORK / CHICHESTER / BRISBANE / TORONTO / SINGAPORE

This text is printed on acid-free paper.

Copyright © 1995 by John Wiley & Sons, Inc.

All rights reserved. Published simultaneously in Canada.

Reproduction or translation of any part of this work beyond
that permitted by Section 107 or 108 of the 1976 United
States Copyright Act without the permission of the copyright
owner is unlawful. Requests for permission or further
information should be addressed to the Permissions Department,
John Wiley & Sons, Inc., 605 Third Avenue, New York, NY 10158-0012.

Library of Congress Cataloging in Publication Data:
Bhattacharyya, Asoke K.
   High-frequency electromagnetic techniques: recent advances and
applications / Asoke K. Bhattacharyya.
      p.   cm.—(Wiley series in microwave and optical engineering)
   Includes bibliographical references (p.).
   ISBN 0-471-55903-2
   1. Radar—Antennas. 2. Electromagnetic waves—Scattering.
3. Electromagnetic compatibility. 4. Electromagnetic interference.
I. Title. II. Series.
TK6590.A6B47 1995   94-23584
621.384'151—dc20

Printed in the United States of America

10 9 8 7 6 5 4 3 2

*To the memory of my beloved parents,
Mr. Manmothonath Bhattacharyya
and Mrs. Kironbala Bhattacharyya*

Contents

Preface	xv
List of Symbols	xvii

1 High-Frequency Electromagnetic Techniques and Recent Advances   1

1.1 Introduction	1
1.2 A Brief Historical Review of High-Frequency Techniques and Recent Advances	8
1.3 Asymptotic Solutions of Maxwell's Equations	11
1.3.1 Eikonal and Transport Equations	11
1.4 Geodesics	15
1.4.1 The Geodesic Curvature	15
1.4.2 Christoffel Symbols and Finding a Geodesic	16
1.5 Determination of Specular Point	18
1.5.1 Analytical Treatment	19
1.6 Electromagnetic Reflection Using GO	23
1.6.1 Reflection from a Convex Conducting Surface	23
1.6.2 Reflection from a Convex Dielectric Interface	27
1.7 Nonconventional Boundary Conditions	31
1.7.1 Generalized Impedance Boundary Conditions (GIBC)	32
1.7.2 Absorbing Boundary Conditions	34
1.7.3 Generalized Higher-Order Boundary Conditions	35
1.7.4 Hard and Soft Boundary Conditions	35
1.8 Physical Optics (PO) Method	36
1.8.1 Polarization-Corrected Physical Optics	39
1.9 The Diffraction Process	40

	1.9.1	A Quick Review of Diffraction by Perfectly Conducting Wedges and Half Planes	41
	1.9.2	Tip and Corner Diffraction	46
	1.9.3	Diffraction Matrix of a Discontinuity of Curvature	51
1.10	Spectral Theory of Diffraction (STD)	53
	1.10.1	Mathematical Basis of STD	53
	1.10.2	Applications of STD	54
1.11	Diffraction by an Impedance Wedge and Impedance Half Planes	57
	1.11.1	Maliuzhinets' Solution	59
	1.11.2	Uniform GTD Formulation for an Impedance Wedge	61
	1.11.3	UTD Solution for the Half Plane Two Face Impedance Problem	64
	1.11.4	UAT Solution for the Half Plane Two Face Impedance Problem	66
1.12	Diffraction from the Edges of Impedance Discontinuities on a Half Plane	70
1.13	Diffraction by Thick Half Planes	72
	1.13.1	Diffraction by a Thick PEC Half Plane	73
	1.13.2	Diffraction by a Thick Half Plane with Two Face Impedances	76
1.14	Diffraction by a Dielectric Half Plane	80
	1.14.1	Diffraction of a Plane Wave by a Thin Dielectric Half Plane	80
1.15	Diffraction by a Dielectric Wedge	86
	1.15.1	Dual Integral Equations	87
	1.15.2	Total Field at Any Point	88
	1.15.3	Geometrical Optics Field	88
	1.15.4	Physical Optics Field	89
	1.15.5	Diffraction Coefficients of the Dielectric Wedge	90
	1.15.6	Some Results and Discussions	92
1.16	Diffraction by Double and Multiple Wedges	92
	1.16.1	Uniform Double PEC Wedge Diffraction Coefficient	94
	1.16.2	Diffraction Coefficient of a Double Impedance Wedge	97
1.17	Hybrid Techniques	103
	1.17.1	Method of Moments Combined with GTD	103
	1.17.2	Diffraction by a 90° Dielectric Wedge	109
1.18	Surface Wave Diffraction	109
	1.18.1	Introduction	109
	1.18.2	A Brief Review of Analytical Models	112
	1.18.3	Source Representation	119
	1.18.4	Surface Waves on Coated Bodies	121
	1.18.5	Selected Results and Discussions on Surface Wave Propagation	123
1.19	Field Estimation at a Caustic	129

	1.19.1 Method of Equivalent Current (MEC)	130
1.20	Incremental Length Diffraction Coefficient (ILDC)	133
	1.20.1 Applications of ILDC	135
1.21	Relationship Between GTD, PTD, and MEC	135
1.22	Diffraction by Ferrite Half Planes	137
1.23	Diffraction by Anisotropic Half Planes	141
	1.23.1 Discontinuity at the Junction of Two Anisotropic Impedance Half Planes	143
1.24	Diffraction by Special Wedges	144
	1.24.1 Dielectric-Coated Metal Wedge	144
	1.24.2 Infinite Wedges with Tensor Impedance Boundary Conditions	144
	1.24.3 A Heuristic Uniform Slope Diffraction Coefficient for a Lossy Rough Wedge	145
1.25	Accuracy Testing of High-Frequency Solutions	146
1.26	Bridging the Gap Between High- and Low-Frequency Regimes	151
1.27	Time-Domain High-Frequency Techniques	153
	1.27.1 Time-Domain Physical Optics (TDPO)	153
	1.27.2 Time-Domain Uniform Theory of Diffraction (TDUTD)	153
	1.27.3 Results and Discussion	157
1.28	Summary	158
1.29	Problems	158
1.30	References	160
1.31	Additional Sources	174

2 Near and Far Fields of Electromagnetic Horn Antennas — 175

2.1	Introduction	175
2.2	The Diffraction Problem of Electromagnetic Horns	175
2.3	Pyramidal Horns: UAT Analysis	176
	2.3.1 E-Plane Diffraction Problem	176
	2.3.2 H-Plane Patterns	182
2.4	Conical Horns	185
	2.4.1 Uncorrugated Conical Horns	186
2.5	Corrugated Conical Horns	192
	2.5.1 Near-Field Patterns	194
	2.5.2 Far-Field Patterns	194
2.6	Special Horns	196
	2.6.1 Broad-Band Flared Horns with Low Sidelobes	197
2.7	Control of Sidelobes	199
2.8	Cross-Polarization Characteristics of Horns	205
2.9	Summary	207
2.10	References	210

3 Reflector Antennas and Rim Loading — 213

- 3.1 Introduction — 213
- 3.2 Reflector Analysis as an Electromagnetic Scattering Problem — 215
- 3.3 Focus-Fed Reflector — 216
  - 3.3.1 Near-Field Patterns of Paraboloids — 217
  - 3.3.2 Axially Symmetric Reflectors: GTD Analysis — 222
- 3.4 Reflectors with Offset Feed — 224
  - 3.4.1 Offset Reflector Analysis Using a PO Method — 224
  - 3.4.2 GO and Aperture Field Formulation — 226
- 3.5 Cassegrainian Reflectors — 229
  - 3.5.1 Near- and Far-Field Patterns of the Subreflector — 230
  - 3.5.2 The Subreflector–Main Reflector System — 236
- 3.6 A Typical Shaping Scheme — 243
  - 3.6.1 GO Shaping Scheme — 243
- 3.7 Effect of Rim Loading and Sidelobe Control — 247
  - 3.7.1 Introduction — 247
  - 3.7.2 Rim-Loaded Focus-Fed Parabolic and Hyperbolic Dishes — 247
- 3.8 Cross-Polarization Characteristics — 252
  - 3.8.1 Cross-Polarization Properties of Front-Fed and Offset Reflectors — 254
  - 3.8.2 Cross-Polarization Characteristics of Offset Antennas — 262
- 3.9 Reflector Antennas for High-Power Microwave (HPM) Applications — 262
- 3.10 Control of Radiation Characteristics of a Reflector Antenna with Loaded and Unloaded Shroud and Flange — 264
- 3.11 Summary — 267
- 3.12 References — 268
- 3.13 Additional Sources — 273

4 Slot Antennas — 275

- 4.1 Introduction — 275
- 4.2 Slots on a Ground Plane — 276
- 4.3 Slots on Circular Cylinders — 282
- 4.4 Slots on Elliptic Cylinders — 285
- 4.5 Slots on a Cone — 292
- 4.6 Summary — 298
- 4.7 References — 300

5 Radar Cross Sections of Complex Objects — 302

- 5.1 Introduction — 302
- 5.2 RCS of Polygonal Plates — 302
  - 5.2.1 Using ILDC — 302
  - 5.2.2 Using GTD — 304

	5.3	RCS of Strips and Plates at Grazing Incidence	306
		5.3.1 A Uniform High-Frequency Solution for Strip Scattering at Grazing Incidence	306
	5.4	RCS of Cylindrical Plates	311
		5.4.1 Physical Optics Current on Curved Smooth Conducting Surfaces	311
	5.5	RCS of Open-Ended Cavities	313
		5.5.1 Aperture Integration (AI) and the GO Shooting and Bouncing Ray (SBR) Method	314
		5.5.2 The Gaussian Beam (GB) Method	316
		5.5.3 Results and Discussion on Scattering from Open-Ended Cavities	322
	5.6	Bistatic RCS of Radar Targets	329
	5.7	Control of Scattering by Rim Loading	331
	5.8	Real-Time Prediction of RCS of Complex Objects Using High-Frequency Techniques	345
		5.8.1 Graphical Electromagnetic Computing (GRECO)	346
	5.9	Summary	348
	5.10	References	351
6	**High-Frequency Treatment of Antennas in Complex Environments**		**355**
	6.1	Introduction	355
	6.2	Monopole on a Finite PEC and Composite Ground Planes	355
	6.3	Microstrip Patch Antenna on a Finite PEC Ground Plane	358
		6.3.1 Introduction	358
		6.3.2 Radiation Pattern Calculations Using Slot Theory and GTD	359
		6.3.3 Modal Expansion and GTD	363
	6.4	Monopole on a Finite Cylinder	365
	6.5	Monopole on a Cone	369
	6.6	Monopole on a Rocket-Shaped Body	369
	6.7	Antennas on Aircraft	369
		6.7.1 Near-Field Patterns	371
		6.7.2 Far-Field Patterns	380
	6.8	Summary	385
	6.9	References	385
	6.10	Additional Sources	387
7	**Estimation of Mutual Coupling Between Antennas on Structures: EMC and EMI Studies**		**388**
	7.1	Introduction	388
	7.2	Mutual Coupling Between Antennas on a Cylinder	389
		7.2.1 Using Uniform Theory of Curved Surface Diffraction	389
		7.2.2 Using Hybrid Techniques	391

7.3	Shading Loss Calculations	394
7.4	Mutual Coupling Between Antennas on a Cone	395
7.5	Limitations of Current Antenna Coupling Models	396
7.6	Some Experimental Results	401
7.7	Cross-Polarization Components in Interantenna Coupling Calculations	401
7.8	Conclusions	404
7.9	Summary	405
7.10	References	405
7.11	Additional Sources	406

8 Terrain Scattering and Propagation Modeling Using High-Frequency Techniques — 407

8.1	Introduction	407
8.2	A Summary of Longley–Rice and GTD Propagation Models	408
	8.2.1 Longley–Rice Model	408
	8.2.2 GTD Model	409
	8.2.3 A Comparison Between Longley–Rice and GTD Propagation Models	410
8.3	Knife-Edge Diffraction and an Attenuation Function for Multiple Knife-Edge Diffraction	412
8.4	Finite Conductivity GTD	415
	8.4.1 Comparison of Finite Conductivity Uniform GTD and Knife-Edge Diffraction in Path Loss Prediction	417
8.5	Hilly Terrain Modeling	420
8.6	Impedance Wedge Modeling	424
8.7	Some Experimental Results	426
8.8	A Three-Dimensional Polarimetric Terrain Propagation Model	427
8.9	Propagation Along Buildings for Low-Power Radio Systems	430
8.10	Summary	435
8.11	References	435
8.12	Additional Sources	436

9 Radar Clutter Modeling Using High-Frequency Techniques — 437

9.1	Introduction	437
9.2	Sources of Radar Clutter	439
9.3	Physical Optics Method	439
9.4	Radar Clutter from Different Important Clutter Sources	441
	9.4.1 Roads and Roadside Materials	441
	9.4.2 Snow-Covered Terrain	441
	9.4.3 Sea Surface Scattering Using High-Frequency Techniques	450
	9.4.4 Airport Humped Runways	454
9.5	Bistatic Clutter Measurements	456

| | | CONTENTS | xiii |

	9.6 A Typical Radar Clutter Model and Simulation Program	457
	9.7 Summary	460
	9.8 References	461

10 High-Frequency Electromagnetic Computer Codes — 463

    10.1 Introduction — 463
    10.2 Steps to Develop Computer Codes — 463
    10.3 Characteristics of a Good Code — 464
    10.4 Some of the Existing High-Frequency Computer Codes — 464
        10.4.1 Intrasystem Electromagnetic Compatibility Analysis Program (IEMCAP) — 464
        10.4.2 Aircraft Interantenna Propagation with Graphics (AAPG) — 465
        10.4.3 GTD-Based Codes — 465
        10.4.4 General Electromagnetic Model for Analysis of Complex Systems (GEMACS) — 468
        10.4.5 G3F-TUD1 — 468
        10.4.6 GENSCAT — 468
    10.5 Physical Optics (PO) and Physical Theory of Diffraction (PTD) Codes — 469
    10.6 Scope for Further Investigations — 472
    10.7 Summary — 473
    10.8 References — 473

Appendix A: A Summary of UTD and UAT — 475

Appendix B: Time-Saving Sampling Methods for Evaluation of Time-Consuming PO Integral — 479

Appendix C: Evaluation of the Function $M_N(\phi, \phi_0, \theta_0, \theta_N)$ — 481

Author Index — 483

Subject Index — 484

Preface

This book addresses state-of-the-art high-frequency techniques and their many practical applications in radiation, scattering, antennas on structures, mutual coupling, electromagnetic compatibility (EMC), electromagnetic interference (EMI), terrain propagation, and terrain modeling. Some repetition of existing material cannot be avoided, but every attempt has been made to include fresh topics and fresh treatments. For example, high-frequency diffraction by a half plane and by a curved perfectly electrically conducting (PEC) wedge, adequately discussed in other books, are covered in an appendix, not in the text.

Significantly modified and expanded, this book is based on a graduate course on high-frequency techniques and applications that I taught in the Department of Electrical and Computer Engineering at New Mexico State University, Las Cruces.

This book will be of interest to researchers in electromagnetic radiation, scattering, interference, propagation and clutter modeling as well as to instructors teaching graduate courses on the above topics. Some of these topics also have applications in the environmental area, such as modeling buried metallic and dielectric objects (referred to as "unexploded ordnance" by the U.S. Army), which is of great importance in environmental clean up operations.

There are ten chapters in the book. Chapter 1 provides the theoretical prerequisites for the text and also describes some of the most recent high-frequency problems. Chapter 2 goes beyond existing textbook analyses of near- and far-field patterns for horns and corrugated horns, and looks at their design. Chapter 3 deals with the analysis and design of different reflector systems, including the effect of aperture, rim loading and shrouds. This chapter also discusses the estimation of the effect of rim loading and use of shrouds on the reflector performance, such as reduced wide-angle sidelobes and high front-to-back lobe ratio. Chapter 4 is on slot antennas, their HF formulations and design. In Chapter 5, the radar cross sections of some objects are analyzed using the methods of Chapter 1. This chapter also touches upon how to reduce the radar cross section (RCS) with the use of radar-absorbing material (RAM) by just coating around the discontinuities on a target. Chapter 6 deals with the HF treatment of antennas on finite bodies and aircraft. The estimation of mutual coupling between antennas on structures and the EMC and EMI problems of aircraft antennas, antennas on structures and shading loss calculations are described in Chapter 7. Chapter 8 deals with terrain propagation modeling and prediction of the role of terrain geometry and terrain properties on electromagnetic wave propagation, particularly the attenuation and positions of transmitting and receiving antennas in a given terrain of ground-based radar systems. Chapter 9 is about modeling of ground, sea, and atmospheric clutter and the choice of radar parameters to obtain a clutter reduction under given scenarios. Chapter 10 discusses some key computer codes used for high-frequency calculations within the confines of the availability of information. Whenever possible, an attempt has been made to discuss their advantages and disadvantages.

I apologize to those whose work I have not cited. The references and credits are by no means exhaustive. Readers are requested to bring any incompleteness, mistake, or error to my attention.

It is impossible to compile a book without assistance from many individuals and organizations. I wish to thank my colleague and friend, Dr. Steven Castillo of New Mexico State University, for suggesting that I teach a graduate course on this subject. My students, in particular, Russ Jedlicka, Gary Halligan, Scott Hutchinson and David Pearson, helped me develop the course through many useful discussions. Many researchers and friends, in particular John Volakis, A. H. Serbest, Prabhakar Pathak, Ray Luebbers, Andy Lee, and Arthur Yaghjian, helped by providing their work on the subject. I acknowledge the support provided by the Physical Science Laboratory and the Department of Electrical and Computer Engineering at New Mexico State University.

My special thanks to Dr. Kai Chang, Editor, John Wiley Series in Microwave and Optical Engineering, for including my book in his series, Mr. George Telecki and Ms. Angioline Loredo, without whose active support and expert advice, this book could never have come to print. I am indebted to Dr. Marshall Holman of Lincoln University, Jefferson City, for his encouragement.

Last but not the least, I must thank my wife Shibani and daughter Anasuya for their patience and understanding during the preparation of this book.

ASOKE K. BHATTACHARYYA

Jefferson City, Missouri

List of Symbols

Any symbol not listed here has been defined locally.

A	physical area	(m²)
	dimension of one side of a flat plate	
A	magnetic vector potential	
AbBC	absorbing boundary conditions	
ApBC	approximate boundary conditions	
a_e	effective aperture of an antenna/scatterer	(m²)
a	radius of a loop	(m)
	radius of a disk	(m)
	dimension of one side of a plate	(m)
B	magnetic flux density	(Wb/m² = T)
b	length	(m)
	dimension of one side of a flat plate	
BC	boundary conditions	
c	velocity of light in free space	(m/s)
D	electric flux density vector	(C/m²)
$D^{s,h}$	diffraction coefficients in soft/hard cases	
d	length	(m)
dB	decibel	
dl	scalar length element	(m)
$d\mathbf{l}$	vector length element	(m)
ds	scalar area element	(m²)
$d\mathbf{s}$	vector area element	(m²)
dv	volume element	(m³)
E	electric field intensity	(V/m)
$\mathbf{E}^i, \mathbf{E}^r, \mathbf{E}^s$	incident, reflected, and scattered electric field intensity vectors	(V/m)

Symbol	Description	Units
$\mathbf{E}^G(\cdot)$	geometrical optics electric field intensity vector	(V/m)
$\mathbf{E}^d(\cdot)$	diffracted optics electric field intensity vector	(V/m)
E_x, E_y, E_z	components of electric field intensity in rectangular system	(x, y, z)
E_ρ, E_ϕ, E_z	components of electric field intensity in cylindrical system	(ρ, ϕ, z)
E_r, E_θ, E_ϕ	components of electric field intensity in spherical polar coordinate system	(r, θ, ϕ)
$E_{\|,\perp}$	parallel ($\|$) and perpendicular ($\perp$) components of electric field intensity	
$F(x)$	Fresnel integral function	
$\hat{F}(x)$	Dominant term of $F(x)$	
f	frequency	(Hz)
G	Green's function	
	gain of an antenna	
$\bar{\bar{G}}$	dyadic Green's function	
$\bar{\bar{g}}$	incremental dyadic Green's function	
GIBC	Generalized impedance boundary conditions	
GTD	geometric theory of diffraction	
$\mathbf{H}$	magnetic field intensity vector	(A/m)
$\mathbf{H}^{i,r,d,s}$	incident/reflected/diffracted/scattered magnetic field intensity	(A/m)
$H(\cdot)$	Heaviside unit step function	
$H_n^{(1)}(x), H_n^{(2)}(x)$	ordinary Hankel functions of first and second kind of order n	
$h_n^{(1)}(x), h_n^{(2)}(x)$	spherical Hankel functions of first and second kind of order n	
$i = \sqrt{-1}$		
$\hat{i}$	unit vector in the direction of incidence	
$\mathbf{J}$	volume electric current density	(A/m^2)
ISB	incident shadow boundary	
J_s	surface current density	(A/m)
$J_n(x)$	ordinary Bessel function of first kind of order n	
$j_n(n)$	spherical Bessel function of first kind of order n	
k_0	wave number $(2\pi/\lambda_0)$ in free space	(1/m)
k	wave number $(2\pi/\lambda)$ in a medium	
L	distance	(m)
	length	(m)
M	Magnetic surface current density	(V/m)
$M(u)$	Maliuzhinets' function	

MEC	method of equivalent current
μ	refractive index
	magnetic permeability of a medium
n	an integer or noninteger related to the wedge angle of a straight/curved wedge, $n = (2\pi - \alpha)/\pi$, where α is the inteior wedge angle
$\hat{n}$	unit normal, positive if outwardly drawn
PEC	perfectly electrically conducting
PMC	perfectly magnetically conducting
PTD	physical theory of diffraction
R_1 and R_2	the two principal radii of curvature at the point of incidence
R	distance of observation point
RB	reflection boundary
$R_\parallel, R_\perp$	complex reflection coefficient for parallel/perpendicular polarizations
σ	conductivity
	radar cross section
	arc length along a geodesic
ψ	a scalar function
	scalar potential
	a phase factor
$T_\parallel, T_\perp$	complex transmission coefficient for parallel/perpendicular polarization
VSWR	voltage standing wave ratio

CHAPTER ONE

High-Frequency Electromagnetic Techniques and Recent Advances

1.1 INTRODUCTION

An electromagnetic engineer has to deal with a variety of problems while not much can be accomplished with the exact solution of wave equations. This is because often in practical problems the surfaces of the geometries under study do not conform to those of the 11 coordinate systems in which the wave equations are separable. Hence, one turns to approximate methods where the solutions approximately satisfy the wave equations. Among these approximations are *high-frequency asymptotic* techniques. An asymptotic method is one whose accuracy increases the higher the value of an independent parameter. This parameter in electromagnetics and optics is the frequency f in Hz or wavenumber $k = 2\pi/\lambda$. Such methods also come under ray-optical methods, where the energy travels in a straight line in the form of a ray–the limiting case of a tube of rays. The ray-optical field is best represented by a Luneberg–Kline (LK) series [1] which aymptotically satisfies Maxwell's electromagnetic field equations [2]. The LK series in a space domain where the permittivity of the medium does not change with position is given by [1]

$$\mathbf{E}(\mathbf{r}, \omega) = e^{ik\psi(\mathbf{r})} \sum_{n=0}^{\infty} (i\omega)^{-n} \mathbf{E}_n(\mathbf{r}) \qquad (1.1)$$

where $\mathbf{E}(\mathbf{r}, \omega)$ is the electric field at the observation point at distance $\mathbf{r}$ at an angular frequency of ω, $\psi(\mathbf{r})$ is the phase factor and $\mathbf{E}_n(\mathbf{r})$, the nth complex term.

The leading term $E_0(\mathbf{r})$ of the series with $n = 0$ is the geometrical optics (GO) term. The terms for higher values of n can be thought of as correction terms for lower frequencies, e.g. in millimeter and microwave bands. Recently, a physical interpretation has been given [3] for the second term of the LK series on the role of the geometric parameters based upon the examination of the reflected field from two-dimensional surfaces. For further details of the numerical and hybrid results and discussions the reader is referred to the source [3].

A large number of analytical and numerical techniques (Tables 1.1 through 1.3 [4]) have been developed over the decades to obtain accurate solutions to handle the ever increasing complexity of practical problems. The high-frequency methods which are the subject of this book can be classified as follows:

1. Geometrical optics (GO)
2. Physical optics (PO)
3. Geometrical theory of diffraction (GTD)
4. Uniform asymptotic theory (UAT)
5. Uniform theory of diffraction (UTD)
6. Physical theory of diffraction (PTD)
7. Spectral theory of diffraction (STD)
8. Method of equivalent current (MEC)
9. Hybrid methods

Hybrid methods are any combination of two or more from (1) through (8); or one technique from (1) through (8) and one or more of the techniques in Tables 1.1, 1.2, and 1.3 [4]. Each approach has its own advantages, limitations, and difficulties. High-frequency techniques have three principal advantages: the memory requirements for even electrically large problems are sometimes not very high; recent advances allow such techniques to handle many new classes of problems; and they provide adequate physical insight into the problems under investigation. The chief disadvantages are that ray tracing may be involved for complex objects, and they are not recommended for electrically small objects and objects without canonical boundaries.

Numerical and iterative approaches have been useful and popular with the rapid increase in computer storage and speed and the increasing use of supercomputers. Computer simulations of complicated electromagnetic problems have been increasingly important and popular because of complexity, high cost, and inconvenience of in situ measurements. But it turns out that with the present rate of increase of computer storage and speed it may be difficult to meet future storage requirements. Figures 1.1 and 1.2 respectively describe the projected mainframe computer performance in flops and the growth of electromagnetic modeling capability versus calender year. An examination of these two graphs leads to two important conclusions: (1) even if continued order-of-magnitude increases in computer size and speed could be anticipated, the maximum object size that might be routinely handled would not increase in the very same proportion; and (2) given that such computer advances are unlikely to be realized, the past growth rate in EM problem-solving capability,

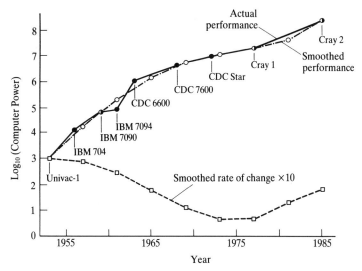

FIGURE 1.1 Projected mainframe computer performance to early 1990s.

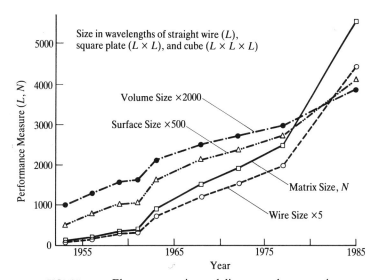

FIGURE 1.2 Electromagnetic modeling growth versus time.

attributable to improved computer technology, is also unlikely to continue. Parallel/vector processing, use of special algorithms to handle large matrices, and replacement of semiconductor memory and magnetic disks by optical storage significantly help users in dealing with electrically large problems. Tables 1.1 through 1.3 briefly project electromagnetic problem-solving capability as regards maximum dimension, materials, and shape for three-dimensional models.

TABLE 1.1 Structure Size Table (3D Volumetric Models)

	$<\lambda/10$	$\to \lambda$	$\to 10\lambda$	$\to 100\lambda$	$>100\lambda$
Method of Moments	✓	✓	?		
MOM/Iterative	✓	✓	✓		
Finite Elements	✓	✓	?		
Spectral Iterative		✓	✓	?	
High-Frequency Techniques			✓	✓	✓
FD–TD	?	✓	✓	✓	?
Space–Time Integral Equation	✓	✓	?		

Blank = not applicable
? = not sure if applicable
✓ = applicable

Recent advances in high-frequency techniques have been necessitated by various civilian and military requirements. Understanding the effect of the properties of non perfectly electrically conducting non-PEC composite materials on diffraction phenomenon is an important and interesting investigation and has numerous civilian and military applications. Some practical examples are interactions of the electromagnetic wave with property discontinuities of the terrain and the composite structures of aircraft, satellites, and missiles. In some practical situations one needs to consider more than one wedge as multiple or joint, where the contribution from a higher-order ray and/or the surface waves may be the only thing to consider. This is very important in terrain modeling and in electromagnetic interference (EMI) and electromagnetic compatibility (EMC). Typical of this are the mutual interactions between antennas on a surface with arbitrary radii of curvature and material composition. Electromagnetic waves diffracted by the edge of a curved surface and the consequent generation of surface wave modes on both the convex and concave sides—the modes on the concave side are known as whispering gallery (WG) modes—is yet another area of interest. A good example is the accurate estimation of the characteristics of a reflector antenna or a slot array antenna. The diffraction processes at every step with relative contributions must be known explicitly to design a reflector antenna with ultralow sidelobe and large front-to-back ratio to maintain secrecy and simultaneously as an antijamming measure.

In my view the principal advances in HF techniques over the last decade have been along the following lines and are still growing.

1. Electromagnetic wave interactions with a half plane with arbitrary impedances on two sides
2. Wave interactions with a straight wedge with arbitrary impedance on two sides
3. Scattering from the edges of impedance discontinuities on a plane surface

TABLE 1.2 Materials Table (3D Volumetric Models)

	Perfect Electrical Conductors	Perfect Magnetic Conductors	Homogeneous Dielectrics	Inhomogeneous Dielectrics	Lossy Media	Anisotropic Media Diag.	Anisotropic Media Gen.
Method of Moments	✓	✓	✓	✓	✓	*	
MOM/Iterative	✓	✓	✓	✓	✓	?	
Finite Elements	✓	✓	✓	✓	✓	?	
Spectral Iterative	✓	✓	✓	✓	✓	?	
High-Frequency Techniques	✓	✓					
FD–TD	✓	✓	✓	✓	✓	✓	*
Space–Time Integral Equation	✓	✓	✓	✓	✓	?	

Blank = not applicable
? = not sure if applicable
✓ = applicable
* = theory and program developed

TABLE 1.3 Shapes Table (3D Volumetric Models)

	Close to Spherical	With Edges	With Corners	With Corner Reflectors	With Arbitrary Unloaded Cavities	With Arbitrary Loaded Cavities
Method of Moments	✓	✓	✓	✓	?	?
MOM/Iterative	✓	✓	✓	✓	?	?
Finite Elements	✓	?	?	?	?	?
Spectral Iterative	✓	?	?	?	?	?
High-Frequency Techniques	✓	✓	✓	✓		
FD–TD	✓	✓	✓	✓	✓	✓
Space–Time Integral Equation	✓	✓	✓	✓	?	?

Blank = not applicable
? = not sure if applicable
✓ = applicable

4. Scattering by a half plane with various resistance/impedance tapers
5. Scattering by dielectric half planes and wedges
6. Scattering by double and multiple wedges
7. Improved models of curved surface diffraction: radiation, scattering, and coupling
8. Improved methods of fast sampling and efficient evaluation of otherwise time-consuming PO integrals for large surfaces, and fast algorithms for ray tracing in complicated objects
9. Scattering by electrically thick PEC and dielectric half planes
10. Development of hybrid methods combining the advantages of two or more methods to widen the scope of the HF technique in resonance regions and for efficient solution of objects of intermediate electrical size
11. Extensive applications in antenna scattering, wave propagation, coupling, EMC and EMI, terrain propagation, clutter modeling, and so on
12. Generation of various user-friendly computer codes for users with different requirements

In its ten chapters this book addresses state-of-the-art methods of high-frequency analysis and their numerous practical applications in radiation, scattering, antennas on structures, mutual coupling, EMC, EMI, terrain propagation, and terrain modeling. Some repetition is unavoidable but every attempt has been made to cover topics not previously dealt with or only briefly dealt with in existing books.

The first chapter establishes a theoretical base, the analytical tools, and the relevant details for using and understanding the subsequent nine chapters. It also describes the current state of research in high-frequency problems. Chapter 2 analyses near- and far-field patterns of horns—corrugated conical horns and low-sidelobe horns—and their design. Chapter 3 deals with analysis and design of different reflector systems, including the effect of aperture blocking. It also discusses how to estimate the effects on reflector performance of rim loading and shrouds, effects such as reduced wide-angle sidelobes and high front-to-back lobe ratio. Chapter 4 is on slot antennas and their HF formulations and design. The radar cross sections of some simple and complex objects are analyzed in Chapter 5, which also indicates how to cut down on radar-absorbing material (RAM) by coating only around the discontinuities of a target. Chapter 6 contains a high-frequency (HF) treatment of antennas on finite bodies and aircraft. The estimation of mutual coupling between antennas on structures and the electromagnetic compatibility (EMC) and electromagnetic interference (EMI) problems of aircraft antennas, antennas on structures are described in Chapter 7 along with shading loss calculations. Chapter 8 deals with terrain propagation modeling and prediction; it looks at the way terrain geometry and eight other properties affect electromagnetic wave propagation. And for ground-based systems it discusses the positions of transmitting and receiving antennas to achieve best performance in given terrain scenarios. Chapter 9 covers the modeling of ground and sea clutter using the techniques and the choice of radar parameters to obtain a clutter reduction under given scenarios. Chapter 10 compares the advantages and disadvantages of key computer codes for high-frequency calculations. It concludes by outlining the scope for further investigations.

In this chapter we develop the foundations of the high-frequency techniques listed in Section 1.1 for use in a variety of applications in later chapters. The aim of this discussion is to get the reader acquainted with the latest advances in a form which can be directly applied to various problems of practical interest by practising engineers and researchers with full knowledge of the scope, limitations, and applicability. We omit detailed discussions whenever felt necessary and merely summarize the earlier advances since they have been adequately treated in a number of available texts [5–7]. Necessary references have always been made to the source where the missing details, if any, can be readily obtained. The material in each section is supplemented with results and discussions. Some relevant examples are worked out and suitably chosen problems to be solved by the reader are suggested depending upon the topics discussed. An extensive reference list is provided for each topic discussed so that the reader can easily continue his or her quest for knowledge on any desired topic in the book. No attempt has been made to make the list of references complete. This book also provides an additional list of recent books, review papers, codes, and so on, in the area of high-frequency electromagnetic techniques and applications.

The time convention used throughout the book is $e^{i\omega t}$ with $i = \sqrt{-1}$ and ω is the angular frequency in radians/second. The SI system of units are used. The medium properties are assumed to be linear with the electric and magnetic field intensities.

1.2 A BRIEF HISTORICAL REVIEW OF HIGH-FREQUENCY TECHNIQUES AND RECENT ADVANCES

The aim of this section is to present a quick historical review of developments of the high-frequency techniques to date. Key contributions have been summarized chronologically; this review is not exhaustive.

Classical geometrical optics [1] can describe incidence, reflection, and refraction phenomena. But it cannot deal with diffraction and consequently "bending of beams." A major contribution from Keller extended classical GO to include diffracted rays and the geometrical theory of diffraction (GTD) [8, 9] was originated. He presented a diffraction coefficient of a PEC straight wedge by asymptotic evaluation of the Sommerfelds' diffraction integral. The major drawback of Keller's GTD, namely, the singularity of the field at the incident and reflected shadow boundaries, was removed using two different approaches. One is known as the uniform theory of diffraction (UTD) [10] by Kouyoumjian and Pathak for the PEC curved wedge and the other, by Ahluwalia, Lewis, and Boersma, is called uniform asymptotic theory (UAT) [11] for a PEC half plane. Ahluwalia [12] extended UAT to diffraction by an edge on a three-dimensional PEC body. While UTD presents a singularity-free dyadic diffraction coefficient, UAT expresses incident, reflected, and diffracted fields in asymptotic forms and the total field satisfies Maxwell's electromagnetic field equations asymptotically. As expected with the ray techniques, none of the two can predict finite fields at the caustic. Separate treatment is hence needed to predict caustic fields. Physical optics (PO) [13] and the method of equivalent current (MEC) [14] are used to treat caustics correctly. In 1976, UAT was extended to the problem of diffraction by a curved wedge by Lee and Deschamps [15]. Physical optics has also been a powerful method to estimate fields but is inaccurate, in its classical form, in situations where the fringing field and current near the edges contribute significantly to the field at the observation point. The fringe effect has been incorporated in what is known as physical theory of diffraction (PTD), essentially an extension of the PO method, by Ufimtsev [16, 17]. Another diffraction theory, called the spectral theory of diffraction (STD) was formulated by Mittra, Rahmat-Samii, and Ko [18]. This is based on the concept that the scattered field is related to the Fourier transform of the induced current on the scatterer, and the ray description of the scattered field is derivable from the spectral representation via the asymptotic evaluation of an integral.

Before 1980, very little of high-frequency research and applications involved non-PEC objects. With the possibility of fabrication of composites and their marketing, an urgent need was felt to systematically investigate canonical problems with arbitrary surface impedances with immediate practical applications in mind. In this case also, the solutions were often obtained by asymptotic evaluation of the pertinent diffraction integral in the canonical cases and in UTD (Pauli–Clemmow approach [19]) and UAT (Van der Waeden's approach [20]) formats. The problem of scattering by a half plane with two face impedances was formulated by Maliuzhinets in 1975 [21] in integral form. A UTD diffraction coefficient was extracted by Tiberio et al. [22] in 1985 for the straight wedge with arbitrary impedances on the two faces. Volakis [23] reported a UTD coefficient for an imperfectly conducting half plane based on exact Weiner–Hopf solution. Tiberio and Pelosi [24] also formulated the problem of scattering from a surface impedance discontinuity on a flat surface in UTD format. Earlier, considerable investigations were reported by Senior [25–27] on the integral equation formulations involving surfaces with imperfect conductivity and presented what they call diffraction tensors. Also, Senior [28, 29] presented several investigations on imperfect wedges at skew incidence. Several investigators have dealt with the problem of the strips and half planes with resistive tapers. Resistive tapers can be used to control induced current and hence RCS. Senior [30] treated backscattering from resistive strips. Haupt and Liepa [31] systematically carried out synthesis of resistive tapers. Exact solution was obtained by Young, Ra, and Senior [32] for E-polarized scattering from resistive half plane with linearly varying resistivity. Solutions in the UAT format for arbitrary incidence for scattering by a half plane with two face impedances have been obtained by Sanyal and Bhattacharyya [33], and for an imperfect right-angled wedge with one face having imperfect impedance and one face PEC by Senior and Volakis [34]. The former solution [33] also includes estimation of surface wave contributions on both sides of the half plane. GTD solutions for wave interactions with thick PEC and imperfect half planes were treated by Volakis and Ricoy [35] and Volakis [36]. A generalized version of Maliuzhinets' method was used by Volakis and Senior [37] to apply the generalized boundary conditions to the case of scattering by a metal-backed dielectric half plane. Simultaneously, high-frequency solutions for dielectric half planes and wedges were also progressing. Scattering by dielectric half planes were presented by Anderson [38], Chakravorty [39], and Senior and Volakis [40, 41]. Dielectric wedge problems have been treated by using PO [42], by corrections to the PO solution [43], and by hybrid techniques [44–50]. The diffraction problem of a high-frequency wave by the edge of a PEC curved sheet and estimation of the surface wave contributions on the convex as well as the whispering gallery modes on the concave side have been extensively treated by Ideman [51], Serbest [52, 53], Ideman and Felsen [54, 55], and Topuz, Niver, and Felsen [56]. The problem of high-frequency fields of point and line sources located on PEC and impedance surfaces was discussed by Felsen et al. [57–59]. The high-frequency line source field in the presence of a curved dielectric interface has been treated in [60] by Heyman and Felsen. The high-frequency treatment of curved convex surface diffraction in general on a smooth curved object has also progressed steadily. It was initiated by Keller's presentation [61, 62] of the surface wave diffraction coefficients and attenuation constants for hard and soft cases from canonical solutions. These coefficients and attenuation constants were obtained also from an asymptotic evaluation of the exact formulation. Hesserjian and Ishimaru [63] obtained an approximate expression for the magnetic field induced by slots on a circular cylinder of infinite length and infinite radius. A similar analysis was reported earlier by Wait [64] to obtain the surface field of a large conducting sphere excited by a short thin dipole. However, these investigations did not consider torsion, which complicates the field distribution on the surface. Hwang and Kouyoumjian [65] first included torsion in such an analysis. Pathak and Wang [66] proposed a formulation of the surface wave phenomenon of elementary electric and magnetic dipoles on a convex surface whose one radius of curvature is large compared to the other and is therefore valid for cylindrical structures.

In some practical scenarios, more than one edge is involved. From practical and academic points of view the investigations involving multiple wedges [67–72] were performed. Kaminetsky and Keller [67] obtained diffraction coefficients for higher-order edges and vertices. Multiple knife-edge diffraction was treated by Deygott [68]. Vogler [69] presented an attenuation function for multiple edge diffraction. Jones [70] and Herman and Volakis [71] also treated double knife-edge diffraction problems. Very recently, a general double edge diffraction coefficient for impedance wedges [72], based on extended spectral rays, and for PEC wedges [73] was derived in terms of double Fresnel integrals. Another important contribution is the corner diffraction coefficient [74, 75]. A more accurate UTD corner diffraction coefficient is being worked out [76].

All the above formulations have been applied to a variety of situations, including very complex environments. Chapters 2 through 9 describe many such applications. Merits and demerits of some HF codes are discussed in Chapter 10.

1.3 ASYMPTOTIC SOLUTIONS OF MAXWELL'S EQUATIONS

A good account of the theory of geometrical optics is available in [73]. We therefore discuss the key aspects for use in later sections of this chapter and in subsequent chapters. An asymptotic solution is one which becomes more and more exact as a given parameter increases; in our case, frequency f or propagation constant k in the medium. Mathematically, an asymptotic expansion of a given function $f(z)$ is a power series $P(z)$ given by

$$P(z) = a_0 + a_1/z + a_2/z^2 + \cdots + a_n/z^n + \cdots \quad (1.2)$$

$P(z)$ tends to $f(z)$ as $|z|$ tends to ∞. The series $P(z)$ may be divergent for small values of z but it rapidly converges for higher values of z. In electromagnetics, the advantage of using such an expansion is that the first few terms of $P(z)$ often make an accurate approximation of $f(z)$.

1.3.1 Eikonal and Transport Equations

In order that the LK series in eqn. (1.1) be a ray-field solution, it must satisfy Maxwell's electromagnetic field equations or the vector wave equation. So if eqn. (1.1) is substituted in the scalar wave equation, eqn. (1.3), the resulting equations obtained by equating like powers of ω would represent the wave phenomena completely. Substituting eqn. (1.1) in the wave equation below

$$\nabla^2 \psi + k^2 \psi = 0 \quad (1.3)$$

one obtains on equating terms with the same power of ω

$$|\nabla \psi| = \mu \quad (1.4a)$$

$$\nabla \psi \cdot \nabla \mathbf{E}_0 + \tfrac{1}{2}(\nabla^2 \psi)\mathbf{E}_0 = 0 \quad \text{for} \quad n = 0 \quad (1.4b)$$

$$\nabla \psi \cdot \nabla \mathbf{E}_n + \tfrac{1}{2}(\nabla^2 \psi)\mathbf{E}_n = \tfrac{1}{2}\nabla^2 \mathbf{E}_{n-1} \quad \text{for} \quad n = 1, 2, 3, \ldots \quad (1.4c)$$

$$\nabla \psi \cdot \mathbf{E}_n = -\nabla \cdot \mathbf{E}_{n-1} \quad \text{for} \quad n = 1, 2, 3, \ldots \quad (1.4d)$$

where, μ is the refractive index of the medium and ψ is a phase factor. Equation (1.4a) is the Eikonal equation, eqn. (1.4b, c) are transport equations, and eqn. (1.4d) is a representation of Gauss's law. For very high frequency ($f \to \infty$) and zero wavelength ($\lambda \to 0$) limits, $\Delta E_n/E_n$ is negligible for $n = 0, 1, 2, 3, \ldots$ and the equations reduce to their GO limit.

We now try to discuss the implications of eqn. (1.4a–d). Since ψ is a phase factor, eqn. (1.4a) means that the surfaces of constant phase are perpendicular to the direction of propagation. As long as the refractive index μ does not change with position, the energy travels in a straight line. Equations (1.4b, c) are transport equations and express the transport of the electric field from one point to another. For the case $n = 0$ (infinite frequency or zero wavelength limit) using eqn. (1.1) and eqn. (1.4b) for a homogeneous medium it turns out that the electric field intensity at an observation point can be expressed as

$$\mathbf{E}(\mathbf{r}) \approx \mathbf{E}_0(\mathbf{R}_0) \exp(ik_0\psi_0 + k_0|\mathbf{R} - \mathbf{R}_0|) \exp\left(\frac{-1}{2}\int_{r_0}^{r} \nabla^2\psi \, dl'\right) \quad (1.5)$$

where 0 represents reference values of the parameters and R_0 is a reference point where the field is known a priori. In isotropic and homogeneous media, it turns out that

$$\exp\left(-\int_{R_0}^{R} \nabla^2\psi \, dl'\right) = \frac{G(R)}{G(R_0)} \quad (1.6)$$

where G is the Gaussian curvature of the wavefront at the point under consideration and is equal to $1/R_1R_2$ where R_1 and R_2 are the two principal radii of curvature of the surface at that point. The Gaussian curvature G is different for wavefronts of different shapes. A simple problem involving geometrical optics is discussed below.

Example 1.1 An infinite electric line source oriented in the z-direction and carrying a constant current I_0 is held symmetrically at a height h above an infinitely long PEC strip (Fig. 1.3). Find the GO incident, GO reflected, and total GO field at any point (ρ, ϕ). Assume the field in absence of the strip is given by

$$E_z^i = -\frac{\beta^2 I_0}{4\omega\varepsilon_0} H_n^{(1)}(\beta\rho_i)$$

Solution Assuming the electric line source is infinitely thin and extends to infinity we use only the zeroth-order Hankel function. Therefore, the field in absence of the strip is given by

$$E_z^i = \frac{\beta^2 I_0}{4\omega\varepsilon_0} H_0^{(1)}(\beta\rho_i)$$

1.3 ASYMPTOTIC SOLUTIONS OF MAXWELL'S EQUATIONS

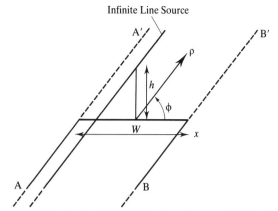

FIGURE 1.3 An infinitely thin wire above a PEC strip.

Since the z-direction is tangential to the PEC strip, the reflected field E_z^r from the PEC surface is simply

$$E_z^r = -\frac{\beta^2 I_0}{4\omega\varepsilon_0} H_0^{(2)}(\beta\rho_r)$$

Therefore, the total field ($\mathbf{E}^t = \mathbf{E}^i + \mathbf{E}^r$) at any point is given by

$$\mathbf{E}^t = \begin{cases} -\hat{a}_z \dfrac{\beta^2 I_0}{4\omega\varepsilon} [H_0^{(2)}(\beta\rho_i) - H_0^{(2)}(\beta\rho_r)] & \text{above the strip} \\ 0 & \text{below the strip} \end{cases}$$

For short distances from the strip we can make a low-argument approximation for the special function. This gives

$$J_0(X) = 1$$

$$Y_0(X) \simeq \frac{2}{\pi} \ln\left(\frac{\gamma X}{2}\right) \quad \text{with} \quad \gamma = 1.781 \quad X \to 0$$

$$J_p(X) \simeq \frac{1}{p!}\left(\frac{X}{2^p}\right)$$

With a large-argument approximation, the incident and reflected fields are given by

$$\mathbf{E}^i = \hat{a}_z \eta I_0 \exp(-j\beta\rho_i)/\sqrt{\rho_i}$$
$$\mathbf{E}^r = -\hat{a}_z \eta I_0 \exp(-j\beta\rho_r)/\sqrt{\rho_r}$$

The total field is then

$$\mathbf{E}^t = \mathbf{E}^i + \mathbf{E}^r = \begin{cases} \hat{a}_z \eta I_0 \left[\dfrac{\exp(-j\beta\rho_i)}{\sqrt{\rho_i}} - \dfrac{\exp(-j\beta\rho_r)}{\sqrt{\rho_r}} \right] & \text{above the infinite plane} \\ 0 & \text{below the infinite plane} \end{cases}$$

For field points at large distances ($\rho \gg h$) we have

For phase considerations

$$\rho_i \approx \rho - h\cos\left(\frac{\pi}{2} - \phi\right) = \rho - h\sin\phi$$

$$\rho_r \approx \rho + h\cos\left(\frac{\pi}{2} - \phi\right) = \rho + h\sin\phi$$

For amplitude considerations

$$\rho_i \approx \rho_r \approx \rho$$

With a low-argument approximation it turns out that near- and far-field expressions for the incident, reflected and total fields are given by

Near-field expressions

$$H_0^{(2)}(X) \approx 1 + j\frac{2}{\pi}\ln\left(\frac{\gamma X}{2}\right) \quad \text{as} \quad X \to 0$$

$$E_z^i = \frac{\beta^2 I_0}{4\omega\varepsilon_0}\left[1 + j\frac{2}{\pi}\ln\left(\frac{\gamma\beta\rho_i}{2}\right)\right]$$

$$E_z^r = \frac{\beta^2 I_0}{4\omega\varepsilon_0}\left[1 + j\frac{2}{\pi}\ln\left(\frac{\gamma\beta\rho_r}{2}\right)\right]$$

$$E_t = E_z^i + E_z^r = \frac{\beta^2 I_0}{4\omega\varepsilon_0}\ln\left(\frac{\rho_i}{\rho_r}\right)$$

Far-field expressions

$$\mathbf{E}^i = \hat{a}_z \eta I_0 \frac{\exp(-j\beta\rho_i)}{\sqrt{\rho_i}}$$

$$\mathbf{E}^r = -\hat{a}_z \eta I_0 \frac{\exp(-j\beta\rho_r)}{\sqrt{\rho_r}}$$

$$\mathbf{E}^t = \begin{cases} -\hat{a}_z j\eta_0 I_0 \sqrt{\dfrac{j\beta}{2\pi}} \sin(\beta h \sin\phi) \dfrac{\exp(-j\beta\rho)}{\sqrt{\rho}} & \text{above the infinite plane} \\ 0 & \text{below the infinite plane} \end{cases}$$

1.4 GEODESICS

A geodesic is the shortest path between two points that lies in a given surface. Determination of geodesics, to identify the scattering centers of first and higher orders on a given object for specified source and field positions, is very important in antenna, scattering, and coupling problems. This may get quite involved when one deals with complex structures. In such cases, one may resort to an efficient numerical approach, perhaps a solution of a nonlinear differential equation. In the next sections we will discuss how to find a geodesic and validate the theory with some simple examples. Determination of geodesics is a problem of differential geometry [77–79].

1.4.1 The Geodesic Curvature

The curvature vector **K** of a surface is the vector sum of normal and tangential curvature vectors $\mathbf{K}_n$ and $\mathbf{K}_g$ respectively and is given by

$$\mathbf{K} = \mathbf{K}_n + \mathbf{K}_g \tag{1.7}$$

The tangential curvature vector $\mathbf{K}_g$ is also called the geodesic curvature vector. Figure 1.4 shows the three orthogonal vectors at a point P on a surface. $\mathbf{K}_g = k_g \hat{u}$ where $\hat{u}$ is a unit vector in the direction of the geodesic curvature and k_g is the magnitude of the geodesic curvature vector. The tangential curvature vector of a curve at P is the (ordinary) curvature vector of the projection on the tangent plane of the surface at P. In the language of differential geometry [77–79], geodesics are the curves of zero geodesic curvature ($k_g = 0$). When k_g is zero, $\hat{n} = +\hat{N}$ and the normal on a geodesic is nothing but the surface normal in the same (or, opposite) directions.

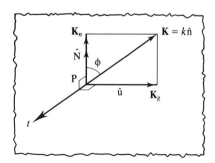

FIGURE 1.4 The orthogonal vectors at a point P on a given surface.

1.4.2 Christoffel Symbols and Finding a Geodesic

Geodesic calculations [79] use the Christoffel symbols [79]:

$$\Gamma^1_{11} = \frac{GE_n - 2FF_u + FF_v}{\Delta}; \qquad \Gamma^2_{11} = \frac{2EF_u - EE_v - FF_u}{\Delta}$$

$$\Gamma^1_{12} = \frac{GE_v - FG_u}{\Delta}; \qquad \Gamma^2_{11} = \frac{EG_u - FE_v}{\Delta} \qquad (1.8)$$

$$\Gamma^1_{22} = \frac{2GF_v - GG_u - FG_u}{\Delta}; \qquad \Gamma^2_{22} = \frac{EG_v - 2FF_v + FG_v}{\Delta}$$

where $\Delta = EG - F^2$ and subscripts indicate partial derivatives given by

$$E_u = \frac{\partial E}{\partial u} = 2\mathbf{r}_u \cdot \mathbf{r}_{uu}; \quad E_v = 2\mathbf{r}_u \cdot \mathbf{r}_{uv}$$

$$F_u = \mathbf{r}_u \cdot \mathbf{r}_{uv} + \mathbf{r}_v \cdot \mathbf{r}_{uu}; \quad F_v = \mathbf{r}_u \cdot \mathbf{r}_{vv} + \mathbf{r}_v \cdot \mathbf{r}_{uv} \qquad (1.9)$$

$$G_u = 2\mathbf{r} \cdot \mathbf{r}_{uv}; \quad G_v = 2\mathbf{r}_v \cdot \mathbf{r}_{vv}$$

$$e = \mathbf{r}_{uu} \cdot \hat{N}; \quad f = \mathbf{r}_{uv} \cdot \hat{N}; \quad g = \mathbf{r}_{vv} \cdot \hat{N}$$

where

$$\hat{N} = \mu \frac{\mathbf{r}_u \times \mathbf{r}_v}{|\mathbf{r}_u \times \mathbf{r}_v|}$$

with $\mu = \pm 1$ depending upon the choice of normal.

From the relation $k_g = 0$ two differential equations of the geodesic can be derived and are given by

$$\frac{d^2 u}{d\sigma^2} + \Gamma^1_{11}\left(\frac{du}{d\sigma}\right)^2 + 2\Gamma^1_{12}\frac{du}{d\sigma}\frac{dv}{d\sigma} + \Gamma^1_{22}\left(\frac{dv}{d\sigma}\right)^2 = 0 \qquad (1.10)$$

$$\frac{d^2 v}{d\sigma^2} + \Gamma^2_{11}\left(\frac{du}{d\sigma}\right)^2 + 2\Gamma^2_{12}\frac{du}{d\sigma}\frac{dv}{d\sigma} + \Gamma^2_{22}\left(\frac{dv}{d\sigma}\right)^2 = 0 \qquad (1.11)$$

where σ is the arc length along the curve (geodesic).

A solution of eqns. (1.10) and (1.11) can be written in the form

$$u = f_1(\sigma) \qquad (1.12a)$$

$$v = f_2(\sigma) \qquad (1.12b)$$

u and v would describe the geodesic. If the eqn. (1.12a, b) are not independent, then they are related through the first fundamental form given by the relation

$$d\sigma^2 = E(du)^2 + 2F\, du\, dv + G(dv)^2 \qquad (1.13)$$

1.4 GEODESICS

Eliminating σ from eqns. (1.10) and (1.11), the differential equation for the geodesic is given by

$$\frac{d^2v}{du^2} = \Gamma^1_{22}\left(\frac{dv}{du}\right)^3 + (2\Gamma^1_{12} - \Gamma^2_{22})\left(\frac{du}{du}\right)^2 + (\Gamma^1_{11} - 2\Gamma^1_{12})\frac{dv}{du} - \Gamma^2_{11} \quad (1.14)$$

where the Γ^1_{ij} are given in eqn. (1.8).

At a given point (u, v), which may be the starting point for the tracing program on a given surface, once dv/du is specified, d^2y/du^2 is unique and the way in which the geodesic curve can proceed is also determined. Hence, a geodesic is uniquely determined once a tangential direction is specified. For unconventional and complicated cases, a numerical solution of eqn. (1.14) is to be tried perhaps by using a standard code. The details of the numerical solutions will not be described here. We now examine a few simple examples to validate the method described above.

Example 1.2 Show that the geodesic on a plane surface is a straight line.

Solution Let us assume the surface to be coincident with the (x, y) plane. A point on the (x, y) plane is described by

$$\mathbf{r}(x, y) = (x, y, z = \text{constant})$$

Let $u = x$, $v = y$, then from eqn. (1.8)

$$\Gamma^1_{11} = \Gamma^2_{11} = \Gamma^1_{12} = \Gamma^2_{12} = \Gamma^1_{22} = \Gamma^2_{22} = 0 \quad (1.15)$$

Therefore the differential equation (1.14) reduces to

$$\frac{d^2y}{dx^2} = 0$$

whose solution is $y = Ax + B$. This is an equation of a straight line, which is the shortest path between two points on a plane surface.

Example 1.3 Find the geodesic for a cylindrical surface of radius r with the initial ray at angle ϕ.

Solution A cylindrical surface can be described by

$$\mathbf{r}(u, v) = (a \cos u, a \sin u, v)$$

The r-vectors are given by

$$r_u = (-a \sin u, a \cos u, 0); \quad r_v = (0, 0, 1); \quad r_{uu} = (-a \cos u, -a \sin u, 0);$$

$$r_{uv} = r_{vv} = 0;$$

$$\hat{N} = \mu \frac{r_u \times r_v}{|\hat{r}_u \times \hat{r}_v|} \quad \text{with} \quad \mu = +1 = (\cos u, \sin u, 0)$$

$$E = r_u \cdot \hat{r}_u = a^2; \quad F = r_u \cdot \hat{r}_v = 0; \quad G = v_v \cdot v_v = 0;$$

$$e = r_{uu} \cdot \hat{N} = -a; \quad f = 0; \quad g = 0$$

$$E_u = E_v = F_u = F_v = G_u = G_v = 0$$

$$\Gamma^1_{11} = \Gamma^2_{11} = \Gamma^1_{12} = \Gamma^2_{12} = \Gamma^1_{22} = \Gamma^2_{22} = 0$$

Substitution of the above into the differential equation (1.14) produces

$$\frac{d^2 z}{d\phi^2} = 0 \tag{1.16}$$

and the solution is

$$z = C_1 \phi + C_2 \tag{1.17}$$

where C_1 and C_2 are constants. It is easy to note that this curved geodesic on the cylinder will be a straight line when the cylindrical surface is unfolded into a plane.

1.5 DETERMINATION OF SPECULAR POINT

A specular point on the smooth surface of the object is one where the first law of reflection is satisfied; in other words, the outward normal drawn at the point of incidence bisects the incidence and observation directions. The use of GO to calculate the reflected field, from a surface finds many applications. For given positions of transmitter and receiver, specular points on a surface must be found in order to apply GO to calculate the reflected field. And specular points need to be found efficiently by time-saving fast algorithms as an exhaustive search may not be cost-effective. This is true of large surfaces like those encountered in large reflectors. Algorithms that lead to fast answers in the search for specular points have been developed for curved conductors and dielectric surfaces. W. V. T. Rusch and O. Sorensen [80] presented a procedure based on Fermat's principle for a general convex surface of revolution. They considered the case where the source and the field points are coplanar with the symmetry axis of the surface. In many situations, only a numerical description of the surface is available. R. Mittra and A. Rushdi [81] obtained a procedure for fast determination of specular points from such a numerically specified surface. S. W. Lee and P. W. Cramer [82] developed an algorithm which reduces a conventional two-dimensional search to a one-dimensional search. For certain classes of surface, in which the source and observation point geometry makes it difficult to perform a two-dimensional search, this one-dimensional search may be globally economically feasible.

1.5.1 Analytical Treatment

The geometry and coordinate system of the reflection of an electromagnetic wave from the surface of a conducting body of revolution is shown in Figure 1.5. We present below the method [84] which reduces the conventional two-dimensional search to a one-dimensional search. This technique has the potential to significantly reduce processing times.

Let the surface of revolution Σ be described by

$$z = f(\rho) \tag{1.18}$$

for any ρ and ϕ in domain D. Let the coordinates of the source and observation points be (x_1, y_1, z_1) and (x_2, y_2, z_2). Let the test specular point be P_0 on the surface and let its coordinates be $(x_{P_0}, y_{P_0}, z_{P_0} = f(\rho))$ in rectangular and $(\rho, \phi, z = f(\rho))$ in cylindrical systems. It can be shown that by using Snell's law

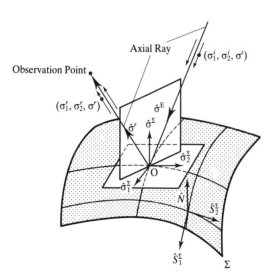

FIGURE 1.5 The geometry and coordinate system of the reflection from a surface of a conducting body of revolution [85].

of reflection the coordinates of the test specular point must satisfy the two nonlinear equations

$$-\frac{d_2}{d_1} = \frac{(x_{P_0} - x_2) + (f - z_2)f_x}{(z_{P_0} - x_1) + (f - z_1)f_x} \quad (1.19a)$$

$$= \frac{(y_{P_0} - y_2) + (f - z_2)f_y}{(y_{P_0} - b_1) + (f - z_1)f_y} \quad (1.19b)$$

where $f_{x,y}$ are the partial derivatives of the function f with respect to x, y and

$$d_n = \sqrt{(x_{P_0} - a_n)^2 + (y_{P_0} - b_n)^2 + (f - c_n)^2} \quad \text{with} \quad n = 1, 2, \ldots .$$

The next task is to solve for (x, y) or (ρ, ϕ) in the desired form of eqn. (1.19) by trial and error in two dimensions completely in the desired domain. The unknown $\phi = \tan^{-1}(y/x)$ can be solved from eqn. (1.19).

The angle ϕ can be expressed as [84]

$$\phi = \cos^{-1}[(A^2 + B^2)^{-1}(-AC + \tau B\sqrt{\Delta})]$$
$$= \sin^{-1}[(A^2 + B^2)^{-1}(-BC - \tau A\sqrt{\Delta})] \quad (1.20)$$

where

$$A = (b_2 - b_1)(\rho + ff_\rho) + (b_1 c_2 - b_2 c_1)f_\rho \quad (1.21a)$$

$$B = (a_1 - a_2)(\rho + ff_\rho) + (a_2 c_1 - a_1 c_2)f_\rho \quad (1.21b)$$

$$C = a_2 b_1 - a_1 b_2; \quad \Delta = A^2 + B^2 - C^2 \quad (1.21c)$$

$$f_\rho = \partial f/\partial \rho \quad \text{and} \quad \tau = \pm 1 \quad (1.21d)$$

There are two solutions of ϕ for each given ρ since τ has two values.

It turns out [84] that a given value of ρ must satisfy the following two equations simultaneously:

$$G_x(\rho) = d_1 a_2 - d_1[f_\rho(f - c_2) + \rho]\cos\phi$$
$$+ d_2 a_1 - d_2[f_\rho(f - c_1) + \rho]\cos\phi = 0 \quad (1.22a)$$
$$G_y(\rho) = d_1 b_2 - [f_\rho(f - c_2) + \rho]\sin\phi + d_2 b_1 - [f_\rho(f - c_1) + \rho]\sin\phi = 0 \quad (1.22b)$$

The iteration process can be performed using two steps.

1. Choose a trial value of ρ, say, ρ_{P_0}. Using ϕ from eqn. (1.20) calculate $G_x(\rho)$ at $\rho = \rho_{P_0}$. If it does not satisfy eqn. (1.22) select another ρ till the eqn (1.22a, b) are satisfield simultaneously within a specified tolerence ε, a small quantity, say, $|\varepsilon| = 1.0 \times 10^{-4}$.

2. Substitute the solution of eqn. (1.22a) in eqn. (1.22b). If it is satisfied then the corresponding coordinates are those of the specular point or points. If the answer is no, then there is no specular point under the given conditions.

As discussed in [82], a few tips may be helpful. It may be noted that a specular point does not exist if the trial value of ρ is such that $\Delta < 0$, $\cos \phi > 1$, or $\sin \phi > 1$. If the source and observation points are in the plane containing the z-axis, then the specular point will also lie in the plane. The following example illustrates the method.

Example 1.4 The problem of reflection from a paraboloid is shown in Fig. 1.6. The paraboloid is described by the equation

$$z = f(\rho) = F - \frac{\rho}{4F} \tag{1.23}$$

where F is the focal length, D_1 and D_2 are the projections of the intersection of the paraboloid in the (x, y) plane on the x-axis. Find the specular point in the restricted range $0 \leq \rho \leq D_2$. Let $F = 19.2$; $D_1 = 6$ and $D_2 = 34$. Find numerically the specular point(s).

Solution Let us define a sharper search range for specular point(s). Let this be $\rho_0 \leq \rho \leq D_2$ where ρ_0 is the real root of the equation

$$\alpha_3 X^3 + \alpha_2 X^2 + \alpha_1 X + \alpha_0 = 0 \tag{1.24}$$

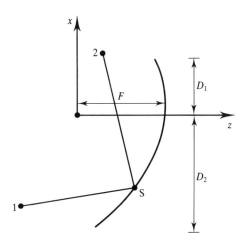

FIGURE 1.6 The problem of reflection from a paraboloid.

where

$$X = (\rho_0/2F)^2, \quad \alpha_3 = F^{-2}[(a_2 - a_1)^2 + (b_2 - b_1)^2] \tag{1.25a}$$

$$\alpha_2 = 2F^{-2}[(a_2 - a_1)^2 + (b_2 - b_1)^2 + F^{-1}(a_2 - a_1)(a_2 c_1 - a_1 c_2)$$
$$+ F^{-1}(b_2 - b_1)(b_2 c_1 - b_1 c_2)] \tag{1.25b}$$

$$\alpha_1 = F^{-2}\{[(a_2 - a_1) + F^{-1}(a_2 c_1 - a_1 c_2)^2 + (b_2 - b_1)$$
$$+ F^{-1}(b_2 c_1 - b_1 c_2)]^2\} \tag{1.25c}$$

$$\alpha_0 = -F^{-4}(a_2 b_1 - a_1 b_2)^2 \tag{1.25d}$$

In general, there might be serveral roots of eqn. (1.24). The correct ranges will be those for which $\Delta > 0$.

For the numerical example, the values of $a_1, b_1, c_1, a_2, b_2, c_2$ are

$$a_1 = -14; \quad b_1 = -10; \quad c_1 = -6; \quad a_2 = 34.645;$$
$$b_2 = 35.355; \quad \text{and} \quad c_2 = -89.619$$

Substituting eqn. (1.25a–d) in eqn. (1.24) we find $\rho_0 = 27.25$. The specular points must lie in the range

$$27.25 < \rho < 34$$

The four roots of $G_x = 0$ are tabulated below.

τ	Root	ρ	ϕ(deg)
+1	A	27.4797	208.10
−1	B	28.8875	219.35
−1	C	28.8875	140.65
−1	D	30.3630	116.80

Substituting the above four roots into $G_y = 0$, in it can be shown that only A and D satisfy eqn. (1.22b). Hence we have two specular points on the paraboloid. Their rectangular coordinates are tabulated below.

Root	x	y	z
A	−24.2396	−12.945	9.33675
D	−13.6922	27.1004	7.1959

The time taken for a typical run to solve this problem on a Univac 1108 was 16 msec.

1.6 ELECTROMAGNETIC REFLECTION USING GO

In electromagnetic problems it is often necessary to determine high-frequency fields reflected from a PEC surface and from a dielectric interface with arbitrary principal radii of curvature. The starting point is to express the incident and reflected fields in LK series (see Section 1.1). To study the effect of curvature of the interface, higher-order terms need to be included in the LK representation. Keller, Lewis, and Seckler [83] formulated reflection of point and line sources from a number of second degree surfaces, namely, spheres, cylinders, paraboloids, and hyperboloids. But only scalar fields were considered. A similar analysis using vector fields was carried out by Schensted [84]. A systematic procedure using ray techniques and a vector field formulation for reflection from an arbitrary conducting surface was presented in [85, 86, 87] by S. W. Lee. A study of reflection from curved dielectric interfaces has been presented by Lee et al in [88] based on ray methods. For dielectric–dielectric interfaces, plane or curved, it is well known that one must distinguish between two cases of transmission: (1) from rarer to denser and (2) from denser to rarer. The electromagnetic tunneling phenomenon in the problem of reflection at a curved dielectric interface (rarer to denser medium propagation) is presented in [89].

Ray techniques fail when the observation point is at the caustic (focal plane or point) or at the incident and reflected shadow boundaries. In such cases, a modified ray technique is required. The problems of high-frequency fields in the presence of a curved dielectric—total internal reflection and whispering gallery (WG) modes—have been extensively treated in ray format by Felsen and Marcuvitz [90] and Felsen et al [91].

A good discussion on different aspects of GO has already been presented in a text on high-frequency techniques [92]. The aim of this section is to systematically present the reflection phenomenon and other associated effects in the ray format for curved dielectric–conductor and dielectric–dielectric interfaces.

1.6.1 Reflection from a Convex Conducting Surface

The geometry and coordinate system for the problem of reflection from a curved conducting surface is shown in Figure 1.5. R_1 and R_2 are the two principal radii of curvature of the surface at the point of incidence. It is worthwhile to mention here that even if the surface is curved, plane wave reflection ($R_\parallel$ and $R_\perp$) and transmission ($T_\parallel$ and $T_\perp$) coefficients are used with a tangent plane approximation. The basic assumption is that the reflection is occuring from a plane tangent to the surface at the point of incidence. The incident field is expressed in a ray-optical form given by eqn. (1.1). The steps [85] required to calculate the reflected field from the given incident field are summarized below.

Before we proceed to determine the reflected field, let us introduce some notation. $(\hat{u}, \hat{v}, \hat{w})$ are the three mutually perpendicular unit base vectors describing a pencil of rays. A pencil is constituted of an axial ray and paraxial rays. $\hat{w}$ is in the direction perpendicular to the tangent plane in which $(\hat{u}, \hat{v})$ lie.

The phase function ψ at any point $(\hat{u}, \hat{v}, \hat{w})$ on the ray is related to the initial phase $\psi(0, 0, 0)$ and curvature matrix $\mathbf{Q}$ and is given by

$$\psi(\hat{u}, \hat{v}, \hat{w}) = \psi(0, 0, 0) + w + \frac{1}{2}\begin{bmatrix} u \\ v \end{bmatrix} \cdot \mathbf{Q}(w) \begin{bmatrix} u \\ v \end{bmatrix} + O(u^2) + O(v^2) \quad (1.24)$$

where $\mathbf{Q}(w)$ is the curvature matrix of the wavefront at the point $(0, 0, w)$ on the axial ray. Along the axial ray, the phase function is

$$\psi(0, 0, w) = \psi(0, 0, 0) + w \quad (1.25)$$

$\mathbf{Q}(w)$ is diagonal when the directions u and w are the same as the principal directions of the wavefront. $\mathbf{Q}(w)$ is then given by

$$\mathbf{Q}(w) = \begin{bmatrix} \dfrac{1}{R_1 + w} & 0 \\ 0 & \dfrac{1}{R_2 + w} \end{bmatrix} \quad (1.26)$$

when the plane containing $(\hat{u}, \hat{v})$ is at angle Θ with the principal directions, the curvature matrix $\mathbf{Q}$ at $w = w$ is no longer diagonal and can be shown to be related to $\mathbf{Q}(w = 0)$ as follows [85]:

$$[\mathbf{Q}(w)]^{-1} = [\mathbf{Q}(w = 0)]^{-1} + w\begin{bmatrix} 1 & 0 \\ 0 & 1 \end{bmatrix} \quad (1.27)$$

$$\mathbf{Q}(w = 0) = \begin{bmatrix} \dfrac{\cos^2 \Theta}{R_1} + \dfrac{\sin^2 \Theta}{R_2} & \dfrac{1}{2}\left(\dfrac{1}{R_1} - \dfrac{1}{R_2}\right)\sin 2\Theta \\ \dfrac{1}{2}\left(\dfrac{1}{R_1} - \dfrac{1}{R_2}\right)\sin 2\Theta & \dfrac{\sin^2 \Theta}{R_1} + \cos^2 R_2 \end{bmatrix} \quad (1.28)$$

The phase function is calculated from eqn. (1.25). The formula for amplitude continuation is already presented in eqn. (1.5).

The incident and reflected fields are expressed in LK series and are given by

$$E^i(\mathbf{r}, w) = e^{ik\psi^i(\mathbf{r})} \sum_{n=0}^{\infty} (iw)^{-n} E_n^i(\mathbf{r}) \quad (1.29)$$

$$E^r(\mathbf{r}, w) = e^{ik\psi^r(\mathbf{r})} \sum_{n=0}^{\infty} (iw)^{-n} E_n^r(\mathbf{r}) \quad (1.30)$$

Let us express the field $E_n^i(\mathbf{r})$ in three mutually perpendicular directions. To do that three local orthonormal base vectors are needed [85] and are defined as follows:

$$\hat{S}_1 = \hat{u} + \hat{w}\frac{-\hat{u}}{R_1 + w} + O(u^2) \tag{1.31a}$$

$$\hat{S}_2 = \hat{v} + \hat{w}\frac{-\hat{v}}{R_2 + w} + O(v^2) \tag{1.31b}$$

$$\hat{S}_3 = \nabla\psi = \hat{u}\frac{u}{r_1 + u} + \hat{v}\frac{v}{r_2 + v} + w + O(u^2) + O(w^2) \tag{1.31c}$$

It turns out that $E_n(u, v, w)$ can be expressed as

$$E_n(u, v, w) = \hat{u}\left(E_{n1} + \frac{u}{r_1 + wE_{n2}}\right) + \hat{v}\left(E_{n2} + \frac{v}{r_2 + w}E_{n2}\right)$$
$$+ \hat{w}\left(E_{n2} - \frac{u}{R_1 + u}E_{n1} - \frac{v}{r_2 + w}E_{n2}\right) + O(u^2) + O(v^2) \tag{1.32}$$

In essence, at the starting point one needs to know three things to find the reflected field: (1) the initial value of ψ^r; (2) the initial value of the reflected field $E_n^r(\mathbf{r})$ on the surface of reflection; and (3) the curvature matrix of the reflected field. The initial value of $E_n^r(\mathbf{r})$ may be obtained from boundary conditions at the reflecting surface.

Gaussian Curvature of the Reflected Wavefront. An excellent discussion on the derivation of Gaussian curvature and its role in the GO solution of reflection from a PEC arbitrarily curved surface is available in [85]. We briefly summarize the key features of the problem here. The reflected field at any point P can be expressed as a product of the amplitude of the reflected wave $E(R_0)$ at a reference point R_0, which may be chosen at the point of reflection on the surface, the reflection coefficient, the divergence or spreading factor Γ_R, which accounts for the divergence when the wavefront is nonplanar, and the associated phase $\exp(-ikR)$. Mathematically,

$$E(P) = R_\perp E_\perp(R_0)\Gamma_R \exp(-ikR) \quad \text{for perpendicular polarization} \tag{1.33a}$$

$$H(P) = R_\parallel H_\parallel(R_0)\Gamma_R \exp(-ikR) \quad \text{for parallel polarization} \tag{1.33b}$$

where the reflection coefficients are given by

$$R_\| = \frac{\sqrt{\varepsilon_r} \cos\theta_i - \cos\theta_i}{\sqrt{\varepsilon_r} \cos\theta_i + \cos\theta_i} \tag{1.34a}$$

$$R_\perp = \frac{\cos\theta_i - \sqrt{\varepsilon_r} \cos\theta_t}{\cos\theta_i + \sqrt{\varepsilon_r} \cos\theta_t} \tag{1.34b}$$

The expression for divergence, or spreading factor, can be derived using the law of conversation of energy along a divergent tube of rays and on the shape of the wavefront. The derivation is available in [8].

It is given by

$$\Gamma = \sqrt{\frac{R_1 R_2}{(R_1 + R)(R_2 + R)}} \tag{1.35}$$

For a spherical wave ($R_1 = R_2 = R$),

$$\Gamma = \frac{1}{R} \tag{1.36a}$$

For a cylindrical wave, $R_1 = \infty$, $R_2 = \rho$

$$\Gamma = \frac{\rho}{\sqrt{\rho + R}} \tag{1.36b}$$

For a plane wave, $R_1 = R_2 = \infty$

$$\Gamma = 1 \text{ (no spreading)} \tag{1.36c}$$

The reflected field at the PEC surface is given by

$$\mathbf{E}(\mathbf{R}_0) = -\mathbf{E}^i(\mathbf{R}_0) + 2[\mathbf{E}^i(\mathbf{R}_0) \cdot \hat{\mathbf{n}}]\hat{\mathbf{n}} \tag{1.37}$$

where $\hat{n}$ and $\mathbf{E}_i(\mathbf{R}_0)$ are respectively the unit normal and the incident field at the point of incidence.

The Gaussian curvature of the reflected and transmitted wavefronts depends on the principal radii of curvature of the incident wavefront, the principal radii of curvature of the reflecting surface, and the orientation of the direction of incidence with the principal directions. The principal radii of curvature R_1 and R_2 of the reflected wavefront are determined by the familiar lens formulae

$$\frac{1}{R_1} = \frac{1}{2}\left\{\frac{1}{R_1^i} + \frac{1}{R_2^i}\right\} + \frac{1}{f_1} \tag{1.38a}$$

$$\frac{1}{R_2} = \frac{1}{2}\left\{\frac{1}{R_1^i} + \frac{1}{R_2^i}\right\} + \frac{1}{f_2} \tag{1.38b}$$

where R_1^i and R_2^i are the principal radii of curvature of the incident wavefront. For a spherical wave, $R_1^i = r_2^i = s'$; for a cylindrical wave, $R_1^i = \infty$, $R_2^i = \rho'$; for a plane wave $R_1^i = R_2^i = \infty$ and f_1 and f_2 can be obtained from the relations [87]

$$1/f_1 = T_1 + T_2 \tag{1.39a}$$

$$1/f_2 = T_1 - T_2 \tag{1.39b}$$

$$T_1 = \frac{1}{\cos\theta_i}\frac{\sin^2\theta_2}{R_{s1}} + \frac{\sin^2\theta_1}{R_2} \tag{1.39c}$$

$$T_2 = \frac{\sin^2\theta_1}{R_{s2}} + \sqrt{\frac{1}{\cos^2\theta_i}\frac{\sin^2\theta_2}{R_{s1}} + \frac{\sin^2\theta_i}{R_{s2}} - \frac{4}{R_{s1}R_{s2}}} \tag{1.39d}$$

where

R_{s1}, R_{s2} = principal radii of curvature of the reflecting surface
$\hat{u}_1$ = unit vector in the principal direction of the surface at Q_R with principal radius of curvature R_{1s}
$\hat{u}_2$ = unit vector in the principal direction of the surface at Q_R with principal radius of curvature R_{2s}
θ_1, θ_2 = angles between the direction of incident ray and the unit vectors $\hat{u}_1, \hat{u}_2$.

1.6.2 Reflection from a Convex Dielectric Interface

It is well known that when an electromagnetic wave is incident on a dielectric surface, a part of the energy is reflected and a part is transmitted. The reflected field can be found in the same way as discussed earlier with a new reflection coefficient which depends on the incident angle, the relative permittivities of the two media, frequency, and polarization. The transmitted wave can be written as a product of the incident field at the point of incidence, the transmission coefficient, the divergent vector and the transmission phase. The transmitted divergence factor depends on the shape of the incident wavefront and the principal radii of curvature of the interface. The transmitted field at any point is given by [93]

$$E_T(R) = T_\perp E^i(R_0)\Gamma \exp(-ikR) \tag{1.40a}$$

$$H_T(R) = T_\parallel H^i(R_0)\Gamma \exp(-ikR) \tag{1.40b}$$

where $T_\perp$ and $T_\parallel$ are the transmission coefficients for perpendicular and parallel polarizations. Γ is the divergence factor for the transmitted field and is given in (1.36a–c). The transmission coefficients are given by

$$T_\parallel = \frac{2\cos\theta_i}{\sqrt{\varepsilon_r} + \cos\theta_t} \tag{1.41a}$$

$$T_\perp = \frac{2\cos\theta_i}{\cos\theta_i + \sqrt{\varepsilon_r}\cos\theta_t} \tag{1.41b}$$

where θ_i and θ_t are angles of incidence and transmission and ε_r is the relative dielectric constant of the second medium. The incident medium in the above formulae is free space.

Assuming the dielectric interface to be infinite and electrically smooth the reflected and transmitted fields can be expressed in the Luneberg–Kline series in the form given by eqn. (1.1). Likewise the incident field is also expressed in the same form. It is worthwhile to mention here that the geometrical optics solutions [89] fail to predict the cross-polarization effects, if any, and in order to study the effect of the curvature of the interface, higher-order terms are needed in the GO expansions. The geometry and coordinate system of the problem of reflection and transmission from a dielectric–dielectric interface is shown in Figure 1.7. The incident field is expressed as

$$E^i(r) = \frac{\exp(-jkr)}{r} [\hat{\theta} A_\theta(\theta, \phi) + \hat{\phi} A_\phi(\theta, \phi)] \quad (1.42)$$

The time convention used is $\exp(j\omega t)$. The scalar field u^i is given by

$$u^i = \begin{cases} E^i_\perp & \text{for perpendicular polarization} \\ H^i_\| & \text{for parallel polarization} \end{cases}$$

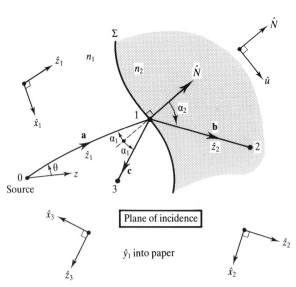

FIGURE 1.7 Geometry and coordinate system of the problem of reflection from a dielectric–dielectric interface.

The GO solution for the reflected and transmitted fields is expressed as

$$u^r = u^i R \Gamma_r e^{-jk_1 c} \tag{1.43a}$$

$$u^t = u^i T \Gamma_t e^{-jk_2 b} \tag{1.43b}$$

where R and T are complex Fresnel reflection and transmission coefficients, Γ_r and Γ_t are the reflection and transmission divergence (spreading) factors. The divergence factors Γ_r and Γ_t are given by

$$\Gamma_r = \frac{1}{\sqrt{1 + (c/R_{31})}\sqrt{1 + (c/R_{32})}} \tag{1.44a}$$

$$\Gamma_t = \frac{1}{\sqrt{1 + (b/R_{21})}\sqrt{1 + (b/R_{22})}} \tag{1.44b}$$

where R_{21} and R_{22} are the principal radii of curvature for the transmitted waves, R_{31} and R_{32} are the principal radii of curvature for reflected wavefronts, and c is the distance of the observation point along the reflected ray.

The expressions for R_{21}, R_{22}, R_{31}, and R_{32} are now needed for calculation of the spreading factors. It is quite involved to calculate the four radii of curvature but their expressions are available in [85] and we present them below.

$$R_{21} = (n \cos^2 \alpha_2) \left[\frac{1}{a} \cos^2 \alpha_1 + \frac{1}{\rho}(n \cos \alpha_2 - \cos \alpha_1) \right]^{-1} \tag{1.45a}$$

$$R_{22} = \left[\frac{1}{na} + \frac{1}{n\rho}(n \cos \alpha_2 - \cos \alpha_1) \right]^{-1} \tag{1.45b}$$

$$R_{31} = \left[\frac{1}{a} - \frac{2}{\rho \cos \alpha_1} \right]^{-1} \tag{1.45c}$$

$$R_{32} = \left[\frac{1}{a} - \frac{2 \cos \alpha_1}{\rho} \right]^{-1} \tag{1.45d}$$

Snyder and Love [89] investigated the reflection and transmission of a locally plane wave from a curved lossless dielectric–dielectric interface and the results from Lee et al. [88] do not quite agree. For the normal incidence and central ray the divergence factors are missing in the field expressions; this makes Snyder's field expressions less accurate. Also, the power transmission coefficient as defined by Snyder and Love [89] seems to be different from the conventional power reflection coefficient. The following example illustrates the applications of the treatment developed in this section.

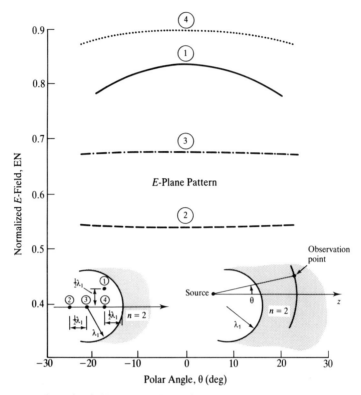

FIGURE 1.8 E-plane far-field pattern through a concave spherical interface. The isotropic source is located at 1, 2, 3, and 4 [88].

Example 1.5 Find an expression for the normalized far field for observation point 2 in medium 2 (Fig. 1.8) for (i) a concave spherical interface and (ii) a concave hyperboloidal interface.

Solution The spherical interface is given by

$$(z/\lambda_1) = 1 - \tfrac{1}{2}[1 - (x^2 + y^2)/\lambda_1^2] \tag{1.46}$$

The hyperboloidal interface is described by

$$(z/\lambda_1) = \tfrac{1}{2}[1 + 2(x^2 + y^2)/\lambda_1^2]^{1/2} - \tfrac{1}{2} \tag{1.47}$$

The normalized far field is defined as

$$E_N = \left|\frac{E^t(2)}{E^i(2)}\right| = \frac{|E^t - \text{field when } n_1 \neq n_2|}{|E^i - \text{field when } n_1 \neq n_2|} - \frac{1}{2} \tag{1.48}$$

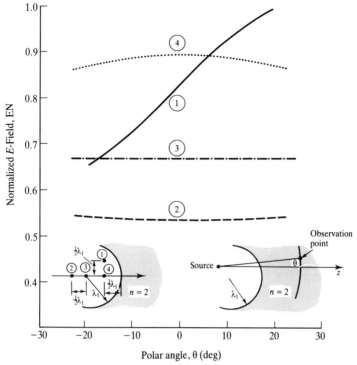

FIGURE 1.9 Same as in Figure 1.8 except that it is H-plane [88].

Using eqns. (1.43b) and (1.42) and letting $b \to \infty$,

$$E_N \sim \frac{1}{a} T\sqrt{R_{21}R_{22}}, \quad b \to \infty \tag{1.49}$$

where a is the source to interface distance. The factor $\sqrt{R_{21}R_{22}}$ is the Gaussian curvature.

Figures 1.8 and 1.9 show the E- and H-plane far-field patterns E_N as a function of θ for a concave spherical surface.

1.7 NONCONVENTIONAL BOUNDARY CONDITIONS

Different types of boundary conditions are used in the analytical investigations according to the nature of the electromagnetic problems. The exact boundary conditions [2] are very common and too elementary to discuss in this text. Instead we discuss here the generalized boundary conditions, the absorbing boundary conditions, and the generalized higher-order boundary conditions.

1.7.1 Generalized Impedance Boundary Conditions (GIBC)

Approximate boundary conditions (ApBC) [95] are extremely useful in complex situations in all areas of work, including electromagnetics, to simplify the analytical and/or numerical solutions of electromagnetic radiation, scattering, and coupling problems. Some examples are the analysis of reflection from a periodically varying substrate beneath a coal layer and design of radar absorbing materials for treating edges.

One way of treating [94, 95] a surface of finite conductivity is to simulate the properties using a boundary condition which involves only the external fields. An excellent discussion is presented in [95] on different aspects of approximate boundary conditions. The approximate impedance boundary condition is described by the equation

$$\mathbf{E} - (\hat{n} \times \mathbf{e})\hat{n} = \eta_0 \Gamma(\hat{n} \times \mathbf{H}) \qquad \text{on} \quad S \qquad (1.50)$$

where $\mathbf{E}$ and $\mathbf{H}$ are the electric and magnetic fields outside the metal surface. The factor Γ and the refractive index N are given by

$$\Gamma = \frac{\mu}{\mu_0 N} \qquad (1.51)$$

$$N = \left[\frac{\mu}{\mu_0}\left(\varepsilon/\varepsilon_0 - j\frac{\sigma}{\omega\varepsilon_0}\right)\right]^{1/2} \qquad (1.52)$$

Equation (1.50) holds when the refractive index N and the pertinent radius of curvature of the surface are both large and also when the intrinsic impedance of the medium is independent of direction. The regions of validity of the impedance boundary conditions have been described in [96, 97] and will not be repeated here.

A generalization of the above is to extend the idea to an inhomogeneous medium [99]. An inhomogeneous medium can be considered as a superposition of a good number of homogeneous layers stratified perpendicular to the z-direction. Then the intrinsic impedance is the equivalent impedance seen by the electromagnetic wave. As a simple case of a metal-based single layer of thickness t, the intrinsic impedance is given by

$$\eta = iZ \tan(k_0 t \sqrt{\varepsilon\mu}) \qquad (1.53)$$

where Z, ε and μ are respectively the intrinsic impedance, permittivity, and permeability of the material as a medium.

If the material primary constants ε and/or μ vary continuously with position then the first-order approximation suggested by Rytov [98] is to substitute

$$\eta = Z(1 + O(k_0^{-1} \partial Z/\partial z)) \qquad (1.54)$$

1.7 NONCONVENTIONAL BOUNDARY CONDITIONS

If the medium happens to be anisotropic, eqn. (1.50) is still applicable except that ε and/or μ are now tensors. A small departure from the planar surface can also be taken care of by the impedance boundary condition.

Yet another possibility is the lateral variation of the primary constants. It has been established [95] that this variation of μ and/or ε does not get involved explicitly. Hence, to first order, the impedance boundary condition can be treated as a local boundary condition. Apparently, it is not known what should be the limit of the rate of change of η for valid ApBC.

Minor departures, perturbations, from the planeness of the surface can also be incorporated in ApBC. Assuming the perturbations are statistically uniform and isotropic, have rms height h, and have correlation length l, the field satisfies eqn. (1.50) at the mean surface at $z = 0$ with the condition $k_0 h \ll \sqrt{k_0 l}$ and the surface slopes are small. Similar arguments are valid for corrugated surfaces.

Impedance boundary conditions can be used to do a sheet simulation, i.e. to simulate a layer of electrically small thickness by actually using a sheet of infinitely small thickness.

Karp and Karal [100] proposed a generalized condition in the form

$$\prod_{m=1}^{M}\left(\frac{\partial}{\partial z} - ik_0 \Gamma_m\right) u = 0 \tag{1.55}$$

The constant Γ_m can be found by using two or more terms in eqn. (1.55) to improve the solution incorporating homogeneous multilayers. The reflection coefficient still remains independent of the angle of incidence.

It is perhaps worthwhile here to present Maxwell's equation on ApBC surfaces. These are given by [101]

$$E^i(\mathbf{r})|_{\tan} - \eta_0 Z_s(\phi) \hat{n} \times \hat{\mathbf{H}}^i(\mathbf{r}) = L\mathbf{J}(\mathbf{r})|_{\tan} + K[\eta_0 Z_s(\phi') \hat{n}' \times \mathbf{J}(\mathbf{r})]|_{\tan}$$
$$- \eta_0 Z_s(\phi) \hat{n} \times K\mathbf{J}(\mathbf{r}) + Z_s(\phi) \hat{n}$$
$$\times L[\hat{n} \times Z_s(\phi') \mathbf{J}(\mathbf{r})] \tag{1.56}$$

where L and K are integrodifferential operators defined as

$$L\mathbf{X}(\mathbf{r}) = jk\eta \int_{\partial R}\left[\mathbf{X}(r') - \frac{1}{k^2}\nabla\nabla\mathbf{X}(r')\right]\phi(\mathbf{r} - \mathbf{r}')\,ds' \tag{1.57}$$

and

$$K\hat{\mathbf{X}}(r') = \int_{\partial R} \hat{\mathbf{X}}(r') \times \nabla\Phi(\mathbf{r} - \hat{\mathbf{r}}')\,ds' \tag{1.58}$$

where $\mathbf{E}^i$, $\mathbf{H}^i$ are incident fields, ∂R is the boundary of the equivalent ApBC surface, and Z_s is the surface impedance.

For further details and applications, the reader may refer to [101].

One problem with ApBC is that accuracy is generally poor for oblique incidence, particularly for cases with low-loss dielectrics. In such situations one needs higher-order boundary conditions to ensure accuracy. This is discussed in a later section.

1.7.2 Absorbing Boundary Conditions

A major difficulty often encountered in electromagnetic scattering problems is that the boundary conditions need to be specified at infinite distances. To get a numerical solution, these infinite distances must be truncated to finite distances. Depending upon the computer power at one's disposal, the shorter the distance the less time will be required to get to the solution. We illustrate here the foundations of absorbing boundary conditions (AbBC) with a one-dimensional model of reflection and transmission of an electromagnetic wave from a dielectric slab, as shown in Figure 1.10.

The scalar wave equations in the three different regions are

$$(\nabla^2 + k^2)u = 0 \quad x < 0 \quad (1.59a)$$

$$[\nabla^2 + k^2 n^2(x)]u = 0 \quad 0 < x < L \quad (1.59b)$$

$$(\nabla^2 + k^2 n_0^2)u = 0 \quad x > L \quad (1.59c)$$

where, $n(x)$ and n_0 are the refractive index at any point x and n_0 is the refractive index of the space to the right of the slab.

The equation needs to be solved at the artificial boundary at L'. The solution of the problem for x_0 for the incident wave in Figure 1.10 is given by

$$u(x) = e^{ikx} + R(k)e^{-ikx} \quad (1.60)$$

where $R(k)$ is the reflection coefficient at the first boundary. The solution in the transmitted region can be written as

$$u(x) = T(k)e^{ikn_0 x} \quad (1.61)$$

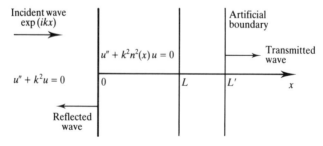

FIGURE 1.10 One-dimensional reflection and transmission model and the absorbing boundary condition (AbBC).

It turns out that eqn. (1.61) takes the form

$$u'(L') - ikn_0 u(L') = 0 \qquad (1.62)$$

The boundary condition (1.62) is the exact AbBC for the problem. Though in a practical problem, things get complicated when the medium has intrinsic resonances and particularly when the volume is of arbitrary shape.

Recently [102], a numerical study has been made for vector absorbing boundary conditions for the finite element solution of Maxwell's equations. It is found that the second-order condition gives smaller errors that decrease further as the outer boundary moves away. The computational cost of the two orders are comparable. An absorbing boundary condition for an arbitrary outer boundary in a large elongated scatterer has been described in [103]. This requires a smaller matrix size, unlike a circular outer boundary.

1.7.3 Generalized Higher-Order Boundary Conditions

We have already mentioned the need of higher-order boundary conditions in practical problems. A very good treatment of this aspect and its applications has been presented by Senior and Volakis. It is shown in [36, 37] that, as already described by Karp and Karal [100], higher-order boundary conditions can correctly predict the oblique incidence scattering, including grazing incidence. These higher-order conditions involve higher-order derivatives of the surface field, which take care of currents normal to the layer and hence provide corrections to the standard impedance boundary conditions.

1.7.4 Hard and Soft Boundary Conditions

In electromagnetics, optics, and acoustics it is sometimes necessary to design reflecting surfaces that produce no cross-polarization components or weak cross-polarization components. This may be possible by fabricating a surface which is artificially soft or artificially hard. For geometrical optics fields, it is established [104] that both the hard and soft surfaces have a polarization-independent reflection coefficient. This gives rise to an interesting feature of the surface that a circularly polarized wave is returned in the same sense after reflection.

The definitions of hard and soft boundary conditions are available in [104]. The longitudinal and transverse surface impedances Z_l and Z_t are given by

$$Z_l = -E_l/H_t \quad \text{and} \quad Z_t = E_t/H_l \qquad (1.63)$$

where l and t are longitudinal and transverse components of the E- and H-field and $\hat{l} \times \hat{t} = \hat{n}$ where $\hat{n}$ is surface normal.

A soft surface is defined by

$$Z_l = \infty \quad Z_t = 0 \qquad (1.64a)$$

A hard surface is defined by

$$Z_l = 0 \quad Z_t = \infty \qquad (1.64b)$$

A soft surface may be realized by cutting transverse grooves on a metal surface; a hard surface may be realized by cutting longitudinal grooves on a metal surface.

1.8 PHYSICAL OPTICS (PO) METHOD

The physical optics method has already been described in a number of references [105–107]. This technique has the following main features: (1) the fields and surface currents on only the illuminated portion of the target are considered; (2) the fields and currents are considered to be zero in the nonilluminated portions of the target. This necessarily means that in practical applications the surface and creeping waves in the shadow regions are not taken into account.

The PO currents are given by Schelkunoff's principle [108].

In the illuminated region

$$\mathbf{J} = 2(\hat{n} \times \mathbf{H}) \qquad (1.65)$$

$$\mathbf{M} = \mathbf{E} \times \hat{n} \qquad (1.66)$$

In the shadow region

$$\mathbf{J} = 0 \quad \mathbf{M} = 0 \qquad (1.67)$$

The radiated fields can be found using the currents in eqns. (1.65) through (1.67) and the well-known vector potential method. The electric and magnetic vector potentials $\mathbf{F}$ and $\mathbf{A}$ are expressed as

$$\mathbf{A}(\mathbf{r}) = \mu \int \mathbf{J}(\mathbf{r}') \frac{e^{ik|\mathbf{r}-\mathbf{r}'|}}{4\pi|\mathbf{r}-\mathbf{r}'|} dV' \qquad (1.68a)$$

$$\mathbf{F}(\mathbf{r}) = \varepsilon \int \mathbf{K}(\mathbf{r}') \frac{e^{ik|\mathbf{r}-\mathbf{r}'|}}{4\pi|\mathbf{r}-\mathbf{r}'|} dV' \qquad (1.68b)$$

The electric and magnetic scalar potentials are given by

$$\phi_e(\mathbf{r}) = \frac{1}{\varepsilon} \int \rho_e(\mathbf{r}') \frac{e^{ik|\mathbf{r}-\mathbf{r}'|}}{4\pi|\mathbf{r}-\mathbf{r}'|} dV' \qquad (1.68c)$$

$$\phi_m(\mathbf{r}) = \frac{1}{\mu} \int \rho_m(\mathbf{r}') \frac{e^{ik|\mathbf{r}-\mathbf{r}'|}}{4\pi|\mathbf{r}-\mathbf{r}'|} dV' \qquad (1.68d)$$

The electric and magnetic fields can be computed from the following relations:

$$\mathbf{E} = i\omega \left[\mathbf{A} + \frac{1}{k^2} \nabla(\nabla \cdot \mathbf{A}) \right] - \frac{1}{\varepsilon} \times \mathbf{F} \quad (1.69)$$

$$\mathbf{H} = \frac{1}{\mu} \nabla \times \mathbf{A} + i\omega \left[\mathbf{F} + \frac{1}{k^2} \nabla(\nabla \times \mathbf{F}) \right] \quad (1.70)$$

Since in many situations the currents and fields at the observation point cannot be determined easily by directly applying eqns. (1.65) through (1.70), the physical optics current is embedded in an integral equation. The solution of this integral equation for unknown current gives the current distribution, then eqns. (1.69) and (1.70) may be used to calculate the field at any observation point. The generalized formulation of the integral equations for the unknown current is given by

$$\mathbf{E}^s = + \frac{i e^{-i k_0 r}}{r} \int \hat{s} \times (\mathbf{M} + \eta_0 \hat{s} \times \hat{\mathbf{J}}) \exp[i k_0 \mathbf{r}' \cdot (\hat{i} - \hat{s})] \, ds' \quad (1.71)$$

$$\mathbf{H}^s = -i \frac{e^{-k_0 r}}{r} \int \hat{s} \times \left\{ \mathbf{J} \times \left[\mathbf{J} \times \left(\mathbf{J} - \frac{1}{\eta_0} \hat{s} \times \mathbf{M} \right) \right] \right\} \exp[-i k_0 \mathbf{r}' \times (\hat{i} - \hat{s})] \, ds' \quad (1.72)$$

where S is the illuminated area of the surface, $\hat{i}$ and $\hat{s}$ are unit vectors in the direction of incidence and observation. Primed quantities are local qualities. Equations (1.71) and (1.72) may be solved by the method of moments [111]. For a dielectric object both $\mathbf{J}$ and $\mathbf{M}$ are nonzero. For electrically perfectly conducting (PEC) targets, $\mathbf{J} = 2(\hat{n} \times \mathbf{H})$ and $\mathbf{M} = 0$.

The physical optics method has gained importance in recent years due to three reasons: (1) the PO method does not require any special treatment of shadow boundaries and caustics; (2) due to increased computer power and improved fast sampling algorithms the PO integral evaluation is not a serious problem anymore, even for large surfaces; and (3) the accuracy of the estimation of fringe currents can be increased by using moment and finite difference methods.

The version of PO described in eqns. (1.65) through (1.67) is only one of the 12 possible versions [109] of the PO method. One can generate other versions by combining each of the following three options:

1. Define the lit region according to either the transmitter or the receiver.
2. The three possible current source representations are

$$\text{E-H formulation: } \mathbf{J} = \hat{n} \times (\mathbf{H}^i + \mathbf{H}^s); \quad \mathbf{M} = (\mathbf{E}_i + \mathbf{E}^s) \times \hat{n} \quad (1.73)$$

$$\text{H formulation: } \mathbf{J} = 2\hat{n} \times \mathbf{H}^s, \quad \mathbf{M} = 0 \quad (1.74)$$

$$\text{E formulation: } \mathbf{J} = 0; \quad \mathbf{M} = 2(\mathbf{E} \times \hat{n}) \quad (1.75)$$

3. Since the fields calculated by PO generally do not satisfy reciprocity, one may solve the problem twice. Firstly with the original transmitter and receiver positions and secondly with the transmitter and receiver positions interchanged. The average of this solution is found to satisfy reciprocity and is designated as the reciprocal solution (EH_{avg}).

We can combine (1) to (3) above to get 12 versions of PO; none is exact.

Figure 1.11 shows the RCS characteristics of a coated cone ellipsoid with observation angles. It is found that the $E-H$ and EH_{avg} formulation differ around aspect angles of 60° and 140°.

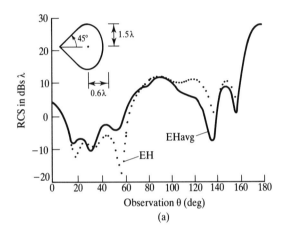

(a)

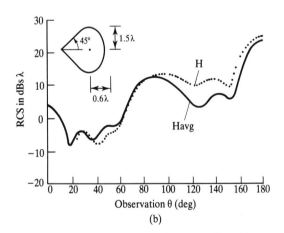

(b)

FIGURE 1.11 Radar cross section of a coated cone ellipsoid for axial incidence. Uniformly coated with a dielectric of $\varepsilon_r = 93.75 + j50.0$; $\mu_r = 1.00$; thickness = 0.318λ: (a) EH = $E-H$ formulation of PO; EHavg = average of original positions of transmitter and receiver and their positions interchanged using the E–H formulation; (b) same as (a) except the E-formulation is replaced by the H-formulation of PO [109].

1.8 PHYSICAL OPTICS (PO) METHOD

Some special problems associated with successful applications have been discussed in [110]. The reader is referred to this reference for more information on the PO method.

1.8.1 Polarization-Corrected Physical Optics

The PO current in eqns. (1.65) through (1.67), which is the current that would have flowed if the area were infinitely extending, must be corrected by the contributions coming from various discontinuities on the object. The total electric current density on an open surface is given by

$$\mathbf{J}_t = \mathbf{J}_0 + \mathbf{J}_1 \tag{1.76}$$

where $\mathbf{J}_0$ is the uniform part computed from eqns. (1.66) and (1.67) with the assumption of infinite surface and $\mathbf{J}_1$ is the nonuniform part of the current density, which is an additional contribution due to the presence of discontinuities.

There are two extreme cases of the effect of these currents on the PO result when the excitation is a pulse. If the incident pulsewidth is short compared to the size of the object then the correction terms will have an insignificant effect. If the size of the object is comparable to the pulsewidth the correction terms have a strong effect on the solution, the PO current given by eqn. (1.65) is an inadequate representation, and needs correction [112].

As discussed earlier, the GO field is the leading term of the LK series of inverse powers of frequency. Likewise, the physical optics fields can also be obtained from the leading term of LK type of expansion for induced current. The series expansion of the current can be generated through the treatment of the magnetic field integral equation (MFIE). The polarization correction to the PO field can thus be obtained through the higher-order terms. We here present the first-order correction of the PO-induced current.

Let a plane wave $\mathbf{H}^i$ propagating in the $+z$-direction be incident on a perfectly electrically conducting smooth surface S. The source of the incident field is at $z = z_0$ and the phase reference is at the source. The induced surface current satisfies the magnetic field integral equation given by [111, 112]

$$\mathbf{J}(\mathbf{r}') = 2(\hat{n}(\mathbf{r}') \times \mathbf{H}^i) + \frac{1}{2\pi}\left[\hat{n}(\mathbf{r}') \times P \oint_S d^2\mathbf{r}\, L\, \mathbf{J}(\mathbf{r}) \times \hat{r}(\mathbf{r}', \mathbf{r})\right] \tag{1.77}$$

where

$$L \equiv \frac{1 - ikR}{R^2} e^{ikR}; \quad \mathbf{R} \equiv \mathbf{r}' - \mathbf{r}, \quad R \equiv |\mathbf{R}|; \quad \hat{R} \equiv \frac{\mathbf{R}}{R} \tag{1.78}$$

and P stands for principal value sense.

The classical PO approximation, $\mathbf{J}^{po}$, to $\mathbf{J}$ is the first term of the right-hand side of eqn. (1.77), on the illuminated portion S of the scatterer surface, and equal to zero in the shadow region L'. The first-order correction term, $\mathbf{J}_1$, to $\mathbf{J}^{po}$, is given by

$$\mathbf{J}_1(\mathbf{r}') = \frac{1}{2\pi} \hat{n}(\mathbf{r}') \times P\left[\oint_S d^2\mathbf{r} \, L \mathbf{J}^{po}(\mathbf{r}) \times \hat{R}(\mathbf{r}', \mathbf{r})\right] \qquad (1.79)$$

The recursive relationship to obtain higher-order correction $\mathbf{J}_i$ is given below.

$$\mathbf{J}_i(\mathbf{r}') = \frac{1}{2\pi} \hat{n}(\mathbf{r}') \times P\left[\oint d^2\mathbf{r} \, L \mathbf{J}_{i-1}(\mathbf{r}) \times \hat{R}(\mathbf{r}', \mathbf{r})\right], \qquad i = 2, 3, \ldots \quad (1.80)$$

1.9 THE DIFFRACTION PROCESS

When an electromagnetic wave is incident upon an obstacle a part of the energy is reflected, a part is scattered, and a part is dissipated in the body. The space can be divided into three regions, namely, the illuminated region, the transition region, and the shadow region (Fig. 1.13). Even if the obstacle is opaque, the transition region near the ISB and the shadow region are partially illuminated. This is exactly what happens in the penumbra of eclipses. This existence of nonzero field in the shadow is due to diffraction. This might be an edge diffraction at the edge of a wedge or a surface diffraction as the wave travels down a smooth surface.

Classical geometrical optics (GO) approximations (Section 1.3) have the serious drawback that they do not take into account the diffraction often encountered at edges, vertices, and corners in electromagnetic problems. A major contribution to surmount the shortcomings of GO approximation came from J. B. Keller [8, 9]. Keller formulated a geometrical theory of diffraction (GTD) to take into account polarization, phase, and the diffracted field from a straight PEC wedge. From the analogy of reflection and transmission coefficients, he defined a diffraction coefficient [8, 9] for the diffracted rays and obtained an explicit expression for diffraction coefficients for the general case of an edge of a straight PEC wedge. This was done by comparing the field obtained from an exact solution using asymptotic evaluation of integrals with the product of incident field, diffraction coefficient, and phase.

Keller's theory has given good results for backscattering of a right circular cone [113] and a cylinder [114] and the agreement of the theoretical results with the experiment was much better than the PO formula. Nevertheless, it suffers from a number of limitations: (1) the diffracted field becomes infinite at both incident and reflected shadow boundaries; (2) the diffraction coefficient becomes infinite at an edge, thus violating the edge condition; (3) the higher-order terms in the expressions for a diffracted wave cannot be determined; (4) the solution becomes singular at a caustic; and (5) the expression for the diffraction coefficient has been obtained only by comparison with the exact solution of the canonical problem and does not follow as an integral part of the proposed theory.

For the last 15 years, a lot of research effort has been directed to obtain uniformity of diffraction solutions at and near the shadow boundaries and to remove the limitations of Keller's solutions. Basically, two approaches have been used. An approach developed by Ahluwalia, Lewis, and Boersma [11] has been termed the uniform asymptotic theory of diffraction (UAT). This is based on the assumption that the field solution in an edge diffraction problem can be expanded in a particular asymptotic series involving a Fresnel integral. They describe a systematic procedure to determine the coefficients of the asymptotic series for all orders of k, the wave number. Their solution is uniformly valid close to and away from the shadow boundaries. UAT has also been applied to diffraction by a curved wedge [115] and recently to three-dimensional half-plane diffraction [116]. An alternative approach has been reported by Kouyoumjian and Pathak [10]. In their work, Keller's diffraction coefficient is modified by a multiplication factor involving a Fresnel integral. This multiplication factor has the property that it approaches zero when the field point approaches a shadow boundary. This compensates for the singularity in Keller's diffraction coefficient leading to a finite field at a shadow boundary. It is known in the literature as the uniform theory of diffraction (UTD). Yet another diffraction theory is the spectral theory of diffraction (STD).

1.9.1 A Quick Review of Diffraction by Perfectly Conducting Wedges and Half Planes

Diffraction by PEC wedges and half planes has been dealt with in many references [5–7]. We present here the key features for our use in Chapters 2 through 9.

Keller [8, 9] presented a diffraction coefficient by asymptotic evaluation of a Sommerfeld scattering integral for a straight wedge diffraction problem shown in Figure 1.12. The diffraction coefficient is given by [8, 9]

$$D^{s,h} \sim \frac{e^{-j\pi/4}}{2n\sqrt{2\pi\beta}\sin\beta_0}\left\{\left[\cos\left(\frac{\pi}{n}\right) - \cos\left(\frac{\phi - \phi'}{n}\right)\right]^{-1} \right.$$
$$\left. - \left[\cos\left(\frac{\pi}{n}\right) - \cos\left(\frac{\phi + \phi'}{n}\right)\right]^{-1}\right\} \quad (1.81)$$

The different symbols have the following meanings. The superscripts s and h stand for soft ($E_{tan} = 0$) and hard ($\delta E/\delta n = 0$) cases. $+$ and $-$ refer to soft and hard cases. β_0 is the smaller angle between the direction of incidence and the tangent to the edge of the wedge at the point of incidence. In many practical problems, $\beta_0 = \pi/2$. ϕ' and ϕ are the angles of incidence and observation respectively.

42 HIGH-FREQUENCY ELECTROMAGNETIC TECHNIQUES AND RECENT ADVANCES

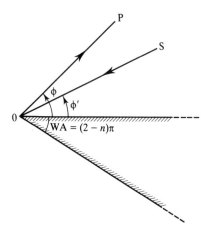

FIGURE 1.12 Diffraction by a straight PEC wedge.

A summary of the UTD and UAT for PEC cases is given in Appendix A. Though this section is concerned with edge diffraction, the general body of revolution may have edges as well as smooth surfaces. This will cause reflected, diffracted, creeping wave, and transition region fields. The different components and regions are shown in Figure 1.13.

Example 1.6 Figure 1.14 shows the geometry of TM scattering by a flat plate of size $(2A \times 2B)$ meters. Find an expression for the monostatic cross section

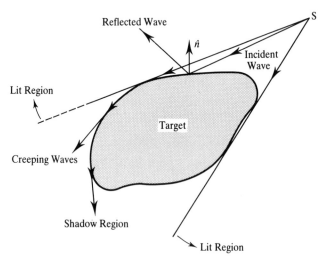

FIGURE 1.13 Scattering by an obstacle: different types of contributions to the field at any point and for different regions.

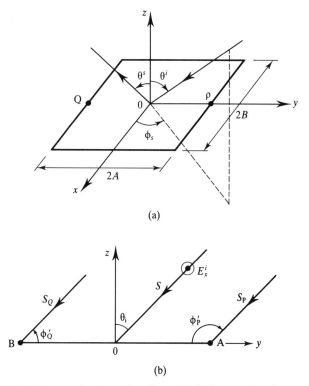

FIGURE 1.14 (a) TM scattering by a flat plate; (b) TM scattering for monostatic case.

of the plate in the $\phi = \pi/2$ plane using Keller's diffraction coefficient (consider only the singly diffracted rays neglecting corner effects). Indicate the steps to take into account the higher-order diffraction contributions.

Compare the monostatic radar cross section (RCS) of the plate of size (10 cm × 10 cm) at $f = 10$ GHz for soft polarization predicted from the above formula and the PO formula. Use aspect angle values equal to 5°, 20° 80°, and 120°. Comment on the comparison.

Solution For the monostatic case, $\theta_s = \theta_i$ and $\phi_s = \phi_i = \phi = \pi/2$. As shown in Figure 1.14(a) the scattering centers A and B are the dominant scintillation centers. Let us consider only the single diffracted rays first. We compare the two-dimensional (2D) result assuming the strip is uniform along the x-axis and compute the three-dimensional (3D) result from the cross section per unit length (along x) using the relation between σ_{3D} and σ_{2D}. Referring to Figure 1.14(b) for monostatic situations the angles at A and B are given by $\phi_A = \phi'_B$; $\phi_Q = Q'_Q$.

The definitions of RCS in the far field with plane wave illumination are

$$\sigma_{2D} = \lim_{\rho \to \infty} \left(2\pi\rho \left| \frac{E_\rho}{E_i} \right|^2 \right) \tag{1.82a}$$

$$\sigma_{2D} = \lim_{r \to \infty} \left(4\pi r^2 \left| \frac{E_r}{E_i} \right|^2 \right) \tag{1.82b}$$

Using Siegel's [117] work the two fields are related as

$$E_r = E_\rho \, 2B \, \frac{\exp(i\pi/4)}{\sqrt{\lambda\rho}} \tag{1.83}$$

where $2B$ is the dimension of the edge in the other direction.
Therefore,

$$\sigma_{3D} = \lim_{r \to \infty} 4\pi r^2 \left| \frac{2B \exp(i\pi/4)}{\sqrt{\lambda r}} \right|^2 \left| \frac{E_\rho}{E_i} \right|^2$$

$$= \frac{2(2B)^2}{\lambda} \lim_{r \to \infty} 2\pi r \left| \frac{E_\rho}{E_i} \right|^2 \tag{1.84}$$

Hence,

$$\sigma_{3D} = \frac{2(2B)^2}{\lambda} \sigma_{2D} \tag{1.85}$$

In the far field we have the following relations

For phase terms
$$S_P = S - A \sin \theta_i$$
$$S_Q = S + A \sin \theta_i$$

For amplitude terms
$$S_P = S_Q = S$$

The incident field at the origin (center of the plate) is given by

$$E_x^i = C e^{iks \cos(\phi - \phi')}$$

The single diffracted fields from the two scattering centers A and B are given by

$$E_{x_P}^d(S) = E_x^i(Q_{e_P}) D_S^k(\phi_P, \phi_P', \beta_0, n) \frac{e^{-ikS_P}}{\sqrt{S_P}} \tag{1.86a}$$

$$E_{x_Q}^d(S) = E_x^i(Q_{e_Q}) D_S^k(\phi_Q, \phi_Q', \beta_0, n) \frac{e^{-ikS_P}}{\sqrt{S_Q}} \tag{1.86b}$$

1.9 THE DIFFRACTION PROCESS

Considering the phase and far-field approximations the single diffracted fields are

$$E^d_{x_P}(s) = C e^{iks\cos(\phi-\phi')} \frac{e^{-iks}}{\sqrt{s}} D_{S_P} e^{i2kA\sin\theta_i} \tag{1.87a}$$

$$E^d_{x_Q}(s) = C e^{iks\cos(\phi-\phi')} \frac{e^{-iks}}{\sqrt{s}} D_{S_P} e^{-i2kA\sin\theta_i} \tag{1.87b}$$

where

$$D_{S_P} = \frac{e^{-i\pi/4}\sin\frac{\pi}{n}}{n\sqrt{2\pi k}\sin\beta_0}\left[\frac{1}{\cos\frac{\pi}{n}-\cos\frac{(\phi_P-\phi'_P)}{n}} - \frac{1}{\cos\frac{\pi}{n}-\cos\frac{(\phi_P+\phi'_P)}{n}}\right] \tag{1.88}$$

$$D_{S_Q} = \frac{e^{-i\pi/4}\sin\frac{\pi}{n}}{n\sqrt{2\pi k}\sin\beta_0}\left[\frac{1}{\cos\frac{\pi}{n}-\cos\frac{(\phi_Q-\phi'_Q)}{n}} - \frac{1}{\cos\frac{\pi}{n}-\cos\frac{(\phi_Q+\phi'_Q)}{n}}\right] \tag{1.89}$$

The edges are half planes (interior wedge angle $\alpha = 0$) hence $n = (2\pi - \alpha)/\pi = 2$.

The 2D RCS is given by

$$\sigma_{2D} = 2\pi|\sqrt{\lambda}D_{S_P} e^{i2kA\sin\theta_i} + \sqrt{\lambda}D_{S_Q} e^{-i2kA\sin\theta_i}|^2$$

The PO result for the plate is

$$\sigma_{3D} = \left[4\pi\frac{(2A)(2B)}{\lambda}\right]^2 \cos^2\theta_i \left[\frac{\sin(kA\sin\theta_i)}{(kA\sin\theta_i)}\right]^2 \tag{1.90}$$

The table below compares the results of the PO and the Keller GTD formulae. It may be concluded that the comparison is good out to about 24° excluding the null depths.

Comparison of Monostatic RCS of a Flat Plate Using PO and Keller's GTD

Aspect Angle (deg)	PO σ_{3D} (dBsm)	Keller's GTD σ_{3D} (dBsm)
5	−4.1	−4.1
20	−18.5	−17.6
80	−40.7	−24.8
120	−33.4	−24.4

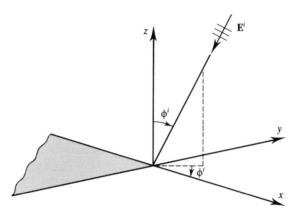

FIGURE 1.15 Tip diffraction by the cone.

The higher-order diffractions can be taken by considering the first-order fields scattered from A/B and then scattered by B/A to the observation point. Normally the second-order effect is considerable for electrically small and medium-sized plates and still higher-order effects can be neglected unless one is estimating a really weak field.

1.9.2 Tip and Corner Diffraction

Tip Diffraction. A tip diffraction (Fig. 1.15) is necessarily associated with the diffraction of an electromagnetic wave from the tip of a cone or a conelike structure. The diffraction coefficient of such a tip can be derived by an asymptotic expansion of the diffraction integral for diffraction by such a cone [118–122]. The tip contribution is considered to be strong for angles of observation less than or equal to the half angle of the cone. The key expressions are available in [105].

In cases of narrow-angle cones, where the half angle, $\alpha \ll \pi/2$, Felsen's expressions are

$$E_\perp^s \approx \frac{ie^{ik_0 r}}{k_0 r} \left(\frac{\alpha}{2}\right)^2 \left[\frac{3 + \cos^2 \theta}{4 \cos^3 \theta}\right] [\hat{x} \sin \phi - \hat{y} \cos \phi] \qquad (1.91)$$

The PO formula for the same case is given by

$$E^s \approx \frac{e^{ik_0 r}}{r} \left[\frac{i \tan^2 \alpha}{4k_0 \cos^3 \theta (1 - \tan^2 \alpha \tan^2 \alpha \tan^2 \theta)}\right]^{3/2} [\hat{x} \sin \phi - \hat{y} \cos \phi] \text{ for } \theta < \alpha$$

$$(1.92)$$

For paraxial cases (both θ and α small) the two expressions agree with each other very well. But Felsen's result is a little simpler. The axial backscattering cross section of the tip is given by

$$\frac{\sigma}{\lambda^2} = \frac{\alpha^4}{16\pi} \qquad (1.93)$$

Since the axial cross section is proportional to the fourth power of the half angle in radians, for small-angle cones the contributions are mostly insignificant. For nonaxial cases, the monostatic cross section is given by

$$\frac{\sigma}{\lambda^2} = \frac{\alpha^4}{16\pi} \left[\frac{3 + \cos^2\theta}{4\cos^3\theta}\right]^2 \qquad (1.94)$$

For wide angle cones, $0 < \alpha \leq \pi/2$ the monostatic cross section is given by

$$\frac{\sigma}{\lambda^2} \sim \frac{1 - 2\cos^2(\pi - \alpha)}{16\pi \cos^4(\pi - \alpha)} \qquad (1.95)$$

The tip diffraction contributions to the far field for antennas radiating in the presence of a cone or, an antenna on a nose cone is relatively small, particularly away from the paraxial region. But in some cases this little tip-diffracted field may cause electromagnetic interference on the on-board radar systems.

Corner Diffraction. Corners where two edges meet are commonly encountered in many practical situations. Neglecting corner effects can cause significant error in the resonance frequency region where the plate size is of the order of one wavelength squared. Sometime ago an attempt was made to solve for the problem of scattering from flat plate structures with corners [123–126] but a corner diffraction coefficient was first presented in [127]. Figure 1.16 shows the geometry of corner diffraction problems. An empirical corner diffraction coefficient was proposed in [128] using the asymptotic evaluation of the scattering integral for the flat plate structures. The corner-diffracted field is given by [127]

$$\begin{bmatrix} E^C_{\beta_0} \\ E^C_{\phi} \end{bmatrix} = -\begin{cases} E^i_{\beta_0}(Q_C) \ D^C_s(L, L_C, \phi, \phi', \beta_0, \beta_C, \beta_{0C}) \cdot \sqrt{\frac{s'}{s''(s' + s'')}} \sqrt{\frac{s(s + s_C)}{s_C}} \frac{e^{-jks}}{s} \\ E^i_{\phi'}(Q_C) \ D^C_h(L, L_C, \phi, \phi', \beta_0, \beta_C, \beta_{0C}) \cdot \sqrt{\frac{s'}{s''(s' + s'')}} \sqrt{\frac{s(s + s_C)}{s_C}} \frac{e^{-jks}}{s} \end{cases}$$

$$(1.96a)$$

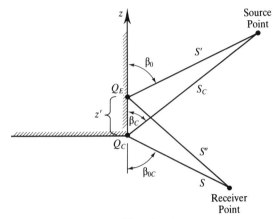

FIGURE 1.16 Diffraction by a corner.

where

$$\begin{bmatrix} D_s^C \\ D_h^C \end{bmatrix} = \begin{bmatrix} C_s(Q_E) \\ C_h(Q_E) \end{bmatrix} \frac{\sqrt{\sin \beta_C \sin \beta_{0C}}}{(\cos \beta_{0C} - \cos \beta_C)} \cdot F[kl_C a(\pi + \beta_{0C} - \beta_C)] \frac{e^{-j\pi/4}}{\sqrt{2\pi k}} \quad (1.95b)$$

and

$$C_{s,h}(Q_E) = \frac{e^{-j\pi/4}}{2\sqrt{2\pi k} \sin \beta_0} \left[\frac{F(kLa(\beta^-))}{\cos \beta^-/2} |F(X^-)| \pm \frac{F(kLa(\beta^+))}{\cos \beta^+/2} |F(X^+)| \right] \quad (1.96c)$$

where

$$X^{\pm} = \frac{La(\beta^{\mp})/\lambda}{kL_C a(\pi + \beta_{0C} - \beta_C)} \quad (1.96d)$$

$$F(x) = 2j|\sqrt{x}|e^{jx} \int_{|\sqrt{x}|}^{\infty} e^{-j\tau^2} d\tau \quad (1.96e)$$

$$a^{\pm}(\beta) = 2\cos^2\left(\frac{2n\pi N^{\pm} - (\beta)}{2}\right) \quad (1.96f)$$

where N^+ are the integers which most nearly satisfy the equations:

$$2\pi n N^+ - (\beta) = +\pi; \quad \beta = \beta^+ = \phi + \phi' \quad (1.96g)$$

$$L = \frac{s's''}{(s' + s'')} \sin^2 \beta_0 \quad \text{for cylindrical incidence} \quad (1.96h)$$

$$L_C = \frac{s_C s}{s_C + s} \quad \text{for spherical incidence} \quad (1.96i)$$

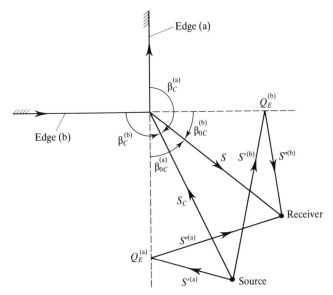

FIGURE 1.17 Corner diffraction: a case where there is only one edge-diffracted field contribution [129].

The function $C_{s,h}(Q_E)$ is a heuristic modification of the UTD half-plane diffraction coefficient. The half plane ($n = 2$) diffraction coefficient is modified by a factor

$$|F| \frac{La(\beta/\lambda)}{kL_c a(\pi + \beta_{oc} - \beta_c)} \tag{1.96j}$$

This factor ensures that the edge diffracted field is continuous at and around the shadow boundaries. Discussions on several possible cases are in order. Since two edges consitute a corner each edge has a corner diffracted field. Two situations are of our interest: (1) In the case shown in Figure 1.17 were only one edge-diffracted field component contributes; and (2) in Figure 1.18 where no edge-diffracted field component is present but two corner-diffracted field components are present.

A corner diffraction formulation can be used as an alternative to the equivalent current method (Section 1.19). The first-order field diffracted by the corner is given by

$$\begin{bmatrix} E_{\beta_0}^C \\ E_\phi^C \end{bmatrix} = \begin{bmatrix} E_{\beta_0(Q_C)}^I \\ E_{\phi'(Q_C)}^i \end{bmatrix} \begin{bmatrix} D_s^C \\ D_h^C \end{bmatrix} \frac{e^{-jks}}{s} \tag{1.97}$$

where

$$D_{s,h}^C = C_{s,h}(Q_E) \tan \beta_{oc} \frac{e^{-jks}}{\sqrt{8\pi k}} \tag{1.98}$$

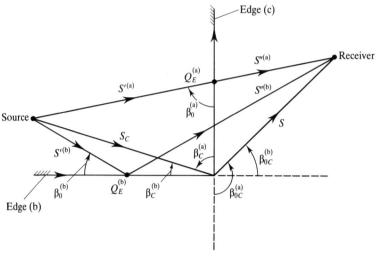

FIGURE 1.18 Corner diffraction: a case where no edge-diffracted field component is present but two corner-diffracted field components are present [129].

and

$$C_{s,h}(Q_E) = \frac{e^{-i\pi/4}}{2\sqrt{2\pi k}} \left| F\left(\frac{1/\lambda}{k \cos^2 \beta_{OC}}\right) \mp \frac{1}{\cos \phi} F\left(\frac{\cos^2 \phi/\lambda}{k \cos^2 \beta_{OC}}\right) \right| \quad (1.99)$$

Using eqn. (1.99) in eqn. (1.96) the diffraction coefficient is obtained as

$$D_{s,h}^C = \frac{j \tan \beta_{OC}}{8\pi} \left| F\left(\frac{1/\lambda}{k \cos^2 \beta_{OC}}\right) \mp \frac{1}{\cos \phi} F\left(\frac{\cos^2 \phi/\lambda}{k \cos^2 \beta_{OC}}\right) \right| \quad (1.100)$$

The values of different quantities used in the preceding discussion are

$$L = \infty, \quad L_C = \infty; \quad \beta_C = \pi - \beta_{OC}; \quad \beta_0 = \pi/2 \quad \text{and} \quad \phi = \phi'$$

T. B. Hansen [128] obtained a corner diffraction coefficient for the quarter plane based on an exact numerical solution to a canonical scattering problem. This is of course valid for forward directions of incidence. The current near a right-angled corner on a PEC flat plate scatterer consists of three parts: (1) physical optics (PO) current on the surface; (2) the fringe current due to the edges on the half planes on the structure; and (3) the corner current due to the presence of the corner. This can be obtained from the E-field integral equation solution for the square plate. The corner contribution for the structure

(Fig. 1.17) can be expressed as [128]

$$\mathbf{E}^c(\mathbf{r}) = \frac{-1}{4\pi} \left\{ \frac{\mathbf{E}^i(\mathbf{0}) \cdot \hat{\theta}^i I_\theta(\theta^i, \phi^i) + \mathbf{E}^i(\mathbf{0}) \cdot \hat{\phi}^i I_\phi(\theta^i, \phi^i)}{1 + \sin\theta \cos\phi} \cdot (\cos\theta\cos\phi\hat{\theta} - \sin\phi\hat{\phi}) \right.$$
$$+ \frac{\mathbf{E}^i(\mathbf{0}) \cdot \hat{\theta}^i I_\theta(\theta^i, 90° - \phi^i) - \mathbf{E}^i(\mathbf{0}) \cdot \hat{\phi}^i I_\phi(\theta^i, 90° - \phi^i)}{1 + \sin\theta\sin\phi}$$
$$\left. \cdot (\cos\theta\sin\phi\hat{\alpha} - \cos\phi\hat{\phi}) \right\} \frac{e^{-jkr}}{r}. \quad (1.101)$$

where

$$I_\theta = \lambda e^{j2.2}[A(\theta^i)(70° - \phi^i)^2 + B(\theta^i)(70° - \phi^i) + C(\theta^i)] \quad (1.102a)$$
$$I_\phi(\theta_i, \phi^i) = \lambda e^{-j1.0}[A(\theta^i, \phi^i)(\theta^i - 45°)^2 + B(\phi^i)(\theta^i - 45°) + C_0] \quad (1.102b)$$

where

$$A(\theta^i) = A_0 \cdot (\theta^i - 25°), \quad \theta^i \leq 75°$$
$$A(\theta^i) = A_0 \cdot (\theta^i - 25°) \left[\frac{\cos(\theta^i)}{\cos(75°)} \right]^{1.65}, \quad \theta^i > 75°$$
$$B(\theta^i) = B_0 \cdot (90° - \theta^i) \quad (1.103)$$
$$A_0 = 1.26 \times 10^{-6}, \quad B_0 = 3.48 \times 10^{-5}$$

Recently, a more rigorous UTD corner (vertex) diffraction coefficient is being presented [129] which is believed to be an improved version of the earlier corner diffraction coefficent described in this section.

1.9.3 Diffraction Matrix of a Discontinuity of Curvature

One source of scattering from a target is from the discontinuity or discontinuities in curvature on the target. A diffraction curvature matrix has been suggested in [25]. As in [25] we consider here a case where a plane wave is incident on a cylindrical surface consisting of two smoothly joined parabolic sections with different curvatures at their junction. The basic formulation was given by Weston [130] and this was later extended by Senior [25] to obtain the diffraction curvature matrix of such a discontinuity.

The geometry of the problem is shown in Figure 1.19. Without entering into the details of the derivation, the diffraction matrix is given by

$$\hat{e}^d = \Delta \hat{e}^i \quad (1.104)$$

where $\hat{u}^i$ and $\hat{u}^d$ are incident and diffracted column vectors. Δ the (3 × 3) curvature discontinuity matrix is given by

$$\Delta = \begin{bmatrix} 2i(F-G) & 0 & 0 \\ -2i(F-G)\cot\beta\sin\theta & -2i(F+G)\cos\theta\cos\alpha & -2i(F+G)\cos\theta\sin\alpha \\ 2i(F-G)\cot\beta\cos\theta & -2i(F+G)\sin\theta\cos\alpha & -2i(F+G)\sin\theta\sin\alpha \end{bmatrix}$$

(1.105)

where

$$F = \frac{a_2 - a_1}{2k} \frac{1 + \cos(\alpha + \theta)}{(\cos\alpha + \cos\theta)^3} + O(k^{-2}) \qquad (1.106)$$

$$G = -\frac{a_2 - a_1}{2k} \frac{1 + \cos\alpha_\theta}{(\cos\alpha + \cos\theta)^3} + O(k^{-2}) \qquad (1.107)$$

α and θ are shown in the Figure 1.19. The column vectors $\hat{u}^i$ and $\hat{u}^d$ have components in the directions of three base vectors $\hat{T}$, $\hat{N}$ and $\hat{B}$ and the incident and diffracted directions $\hat{i}$ and $\hat{s}$ are given by

$$\hat{i} = \hat{T}\cos\beta - \hat{N}\sin\beta\sin\alpha + \hat{B}\sin\beta\cos\alpha \qquad (1.108)$$

$$\hat{s} = \hat{T}\cos\beta + \hat{N}\sin\beta\sin\theta - \hat{B}\sin\beta\cos\theta \qquad (1.109)$$

where β is the angle lying between 0 and π.

A practical application using this (3 × 3) curvature diffraction matrix is not known to the author.

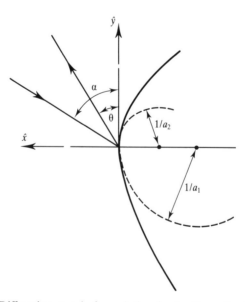

FIGURE 1.19 Diffraction matrix formulation for he discontinuity n curvature.

1.10 SPECTRAL THEORY OF DIFFRACTION (STD)

Unlike PO or GTD methods, the spectral theory of diffraction (STD) [18, 131, 132] starts with the Fourier-transformed version of the diffraction integral. This integral expresses the scattered field due to a target as the convolution of the induced surface current and the free-space Green's function. There are several advantages of using such an approach. (1) The scattered field extracted from such representation is always free from singularities at the incident and reflection shadow boundaries and the caustic. Some comments do not apply for Keller's GTD. STD is also valid for regions where fields are rapidly varying but GTD and UTD require a slope diffraction correction [10]. (2) The transform $\tilde{J}$ of the current is necessarily bounded for all spectral angles. (3) An evaluation of the STD integral gives results identical to GTD in the region away from the shadow boundaries. (4) The STD method lends itself to verification of high-frequency results obtained by other high-frequency techniques. But STD has a disadvantage. The STD integral representation is more involved except in simple cases. STD recovers the GTD results if and only if the assumed current distribution is the same as for GTD.

1.10.1 Mathematical Basis of STD

To describe the basic concepts of STD, we take the example of half-plane scalar diffraction. More complex problems will be discussed in a following section. Let a half plane be illuminated by an excitation field u^i incident at an angle ϕ' and is given by

$$u^i = \exp(ik^i \cdot \rho) = \exp[-k\rho \cos(\phi - \phi')] = \exp[-ik(x \cos \phi' + y \sin \phi')]$$
(1.110)

A Wiener–Hopf technique [133] can be used to construct the solution. The procedure of STD is summarized [134, 135] as: (1) obtain the STD diffraction coefficient $\tilde{J}$ which is the transform of the induced current; (2) insert $\tilde{J}$ in the spectral integral representation for the problem under consideration; (3) evaluate the integral asymptotically for large $k\rho$ using saddle-point integration or any other method taking care of the singularities. The final result is a diffracted field which is continuous everywhere.

The scattered field is expressed in terms of an inverse Fourier transform as

$$u^s(x, y) = \frac{1}{2\pi} \int_{-\infty + i\tau}^{\infty + i\tau} J(\alpha) \frac{e^{-i\alpha x - \gamma |y|}}{2\gamma} d\alpha, \qquad (1.111)$$

where $\gamma = \sqrt{\alpha^2 - k^2}$; τ is a real number; α is the transform variable. $\tilde{J}(\alpha)$, the

Fourier transform of the induced current on the PEC half plane, may be called the spectral diffraction coefficient. This is given by

$$\tilde{J}(\alpha) = \frac{2i(k + k\cos\phi')^{1/2}\sqrt{\alpha - k}}{\alpha - k\cos\phi'} \quad (1.112)$$

A saddle-point integration evaluation [18] of (1.111) leads to the following expression of scattered field

$$u^s(\rho, \phi) = \tilde{J}(\phi', \phi)g(k\rho) + O[(k\rho)^{-3/2}] \quad k\rho \to \infty, \quad (1.113)$$

where

$$g(k\rho) = \frac{e^{i(k\rho + \pi/4)}}{2\sqrt{2\pi k\rho}} \quad (1.114)$$

The spectral diffraction coefficient $\tilde{J}(\phi', \phi)$ is identified as Keller's diffraction coefficient. The expression in eqn. (1.111) has singularities at the incident shadow boundary (ISB) for $\phi = \pi + \phi'$ and at the reflection boundary (RB) for $\phi = \pi - \phi'$. The integral can be evaluated asymptotically so that the field remains finite at all observation angles.

1.10.2 Applications of STD

Diffraction by a Circular Aperture. The geometry and coordinate system of the diffraction by a circular aperture is shown in Figure 1.20. The STD coefficient $\tilde{J}$ is given by

$$\tilde{J} = \tilde{J}_0 + \tilde{J}_f \quad (1.115)$$

where $\tilde{J}_0$ and $\tilde{J}_f$ are the Fourier transforms of J_0, the PO current density in the area occupied by the aperture and J_f is the fringe current density due to the edges. The scattered field u is given by the integral transform

$$u = \frac{1}{4\pi^2} \int_{-\infty}^{\infty} \tilde{J}(\alpha, \beta) \frac{e^{-i(\alpha x + \beta y) - \gamma|z|}}{2\gamma} \, d\alpha \, d\beta \quad (1.116)$$

where $\gamma = \sqrt{\alpha^2 + \beta^2 - k^2}$.

We will not go through the mathematics of the evaluation of the transformed fringe current. This is available in [133]. We give the final results of the scattered field along the caustic and off-axis directions.

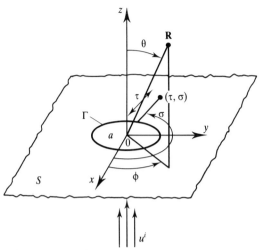

FIGURE 1.20 Diffraction by a circular aperture: spectral theory of diffraction formulation.

1. Caustic Region ($\theta \approx 0$)

$$u_f = -2\pi a + \left(-\frac{\pi a}{8} + \frac{i\pi}{8} ka^2 + \frac{\pi}{2} k^2 a^3\right) \sin^2 \theta \frac{e^{ikR}}{4\pi R} \quad (1.117a)$$

$$u^0 = -2i\pi ka^2 + \frac{i\pi}{4} k^3 a^4 \sin^4 \theta \frac{e^{ikR}}{4\pi R} \quad (1.117b)$$

$$u^i = u^0 + u^f \quad (1.117c)$$

The above are finite fields at the caustic. Equation (1.117) contains terms of order k^0, k^1, k^2, and k^3. Here STD coefficients have a distinct advantage over GTD coefficients, whose order is $k^{-1/2}$, incapable of correct representation of fields at the caustic.

2. Off-Axis Regions (Away from Caustic)
   The total field is given by

$$u^t = \frac{1}{\sqrt{k\pi}} a^{1/2} \sin^{-3/2} \theta - e^{-ika\sin\theta - i\pi/4} \cos\frac{1}{2}(\pi/2 - \theta)$$

$$+ e^{-ika\sin\theta + i\pi/4} \cos\frac{1}{2}(\pi/2 + \theta) \frac{e^{ikR}}{4\pi R} \quad (1.118)$$

This is the same as given by GTD.

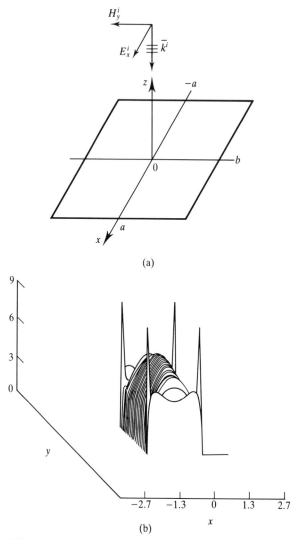

FIGURE 1.21 Diffraction by a thin flat plate: (a) geometry and coordinate system of the flat plate; (b) magnitude of the dominant x-component of the current density on a $1\lambda \times 1\lambda$ plate ($ka = 3.14$). Normal Incidence; plate region: $x \in (-1, 1), y \in (-1, 1)$ [132].

Some Results of Applications of STD. Figure 1.21 shows the geometry of the flat plate diffraction and the dominant x-component of the current density for a plane wave and normal incidence. The GTD solution has the complications of the caustic at normal incidence and also the corner effects have to be taken care of by using a corner diffraction coefficient as described in Section 1.9.2. STD does not have these problems. The STD solution makes use of GTD as a zeroth-order approximation and

1.11 DIFFRACTION BY AN IMPEDANCE WEDGE AND IMPEDANCE HALF PLANES

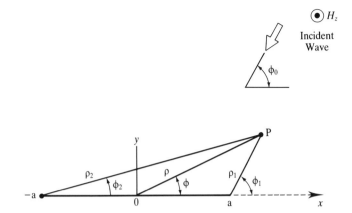

FIGURE 1.22 Geometry and coordinate system of infinite strip scattering.

its Fourier transform of the current distribution as the spectral diffraction coefficient. As seen from Figure 1.21(b), the rise in current density is predicted at the corners. So STD can take care of the effect of a corner in, for example, the case of an antenna mounted close to it. This type of application will be discussed in Chapter 6. We close this discussion on STD and applications by comparing the results of electromagnetic scattering by a strip using STD and UAT. The theory is available in [134]. It is worthwhile to compare the STD results of a typical scattering problem of a 2λ wide strip with UAT and UTD results. The geometry and coordinate system of the strip scattering problem is shown in Figure 1.22. Figure 1.23 shows a comparison of STD ($|H^b|$) and UAT ($|H^u|$) for amplitude and phase of the scattered field. Figure 1.24 shows the same case but for a comparison with UTD.

1.11 DIFFRACTION BY AN IMPEDANCE WEDGE AND IMPEDANCE HALF PLANES

Many electromagnetic scenarios include radiation, scattering, and propagation involving an impedance wedge and an impedance half plane. By *impedance* wedges and half planes we mean those wedges and half planes whose two surfaces can assume any arbitrary value of surface impedance. We only treat here the canonical problem of a straight impedance wedge. The solution of the problem of wave interaction with a curved wedge (a wedge is curved when at least one surface is curved, e.g. the ends of a cylinder with finite length) is, to my knowledge, not yet available in the literature.

An exact solution of the straight impedance wedge scattering was presented by Maliuzhinets [21] more than three decades ago but in terms of infinite integrals. Various high-frequency approximations to this solution were since then suggested by several workers and applied to different canonical problems,

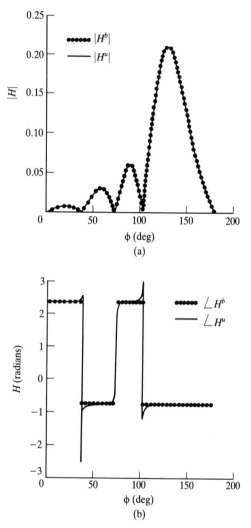

FIGURE 1.23 Comparison of STD and UAT for a 2λ-wide infinite strip: (a) comparison of magnitude; (b) comparison of phase of the scattered field [135].

e.g. half plane with two face impedances by Bucci and Franseschetti [136], for a strip with two face impedances [22] and edge-on incidence and edges of impedance discontinuities on a plane [24] by Tiberio and his associates, imperfectly conducting half plane by Volakis [23], UAT expansion for the half plane with two face impedances by Sanyal and Bhattacharyya [33]. Later we discuss the cases of the impedance wedge and impedance half plane separately. First, we present very briefly the exact analytical form of Maliuzhinets' solution.

1.11 DIFFRACTION BY AN IMPEDANCE WEDGE AND IMPEDANCE HALF PLANES 59

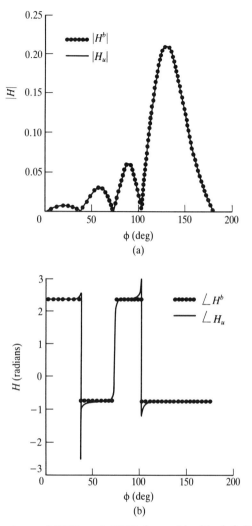

FIGURE 1.24 Comparison of STD and UTD for a 2λ-wide infinite strip: (a) comparison of magnitude; (b) comparison of phase of the scattered field [135].

1.11.1 Maliuzhinets' Solution

The geometry and coordinate system of the problem is shown in Figure 1.25. 0 and n faces of the wedge respectively have the surface impedances Z_0 and Z_n. The angles $\theta_{0,n}^{e,h}$ for an incident plane wave with the electric field either parallel (TM, e) or perpendicular (TE, h) to the edge:

$$\sin \theta_{0,n}^e = \eta_0/Z_{0,n} \quad \text{and} \quad \sin \theta_{0,n}^h = Z_{0,n}/\eta_0 \qquad (1.119)$$

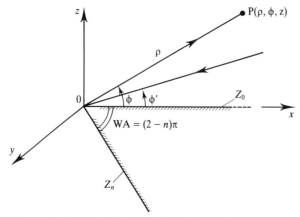

FIGURE 1.25 Scattering by a wedge with two face impedances.

An exact integral representation for the diffracted field is given in [21] as

$$u^d(P) = \frac{1}{4n\pi i}\int_{SDP(\pi)} \exp(jk\rho \cos \xi)\, \frac{\tilde{\Psi}(\xi + n\pi/2 - \phi)}{\tilde{\Psi}(n\pi/2 - \phi')} \cdot \cot\left(\frac{\xi - \beta^-}{2n}\right)$$

$$+ \frac{\tilde{\Psi}(\zeta + n\pi/2 - \phi)}{\tilde{\Psi}(n\pi/2 + \phi')} \cdot \frac{\sin \phi' - \sin \theta_0}{\sin \phi' + \sin \theta_0} \cot\left(\frac{\xi - \beta^+}{2n}\right) d\xi$$

$$+ \frac{1}{4n\pi i}\int_{SDP(-\pi)} \exp(jk\rho \cos \xi)\cdot \frac{\tilde{\Psi}(\xi + n\pi/2 - \phi)}{\tilde{\Psi}(-3n\pi/2 + \phi')}$$

$$\cdot \cot\left(\frac{\xi - \beta^-}{2n}\right) + \frac{\tilde{\Psi}(\xi + n\pi/2 - \phi)\sin(n\pi - \phi') - \sin \theta_n}{\tilde{\Psi}(-3n\pi/2 + \phi')\sin(n\pi - \phi' + \sin \theta_n)}$$

$$\cdot \cot\left(\frac{\xi - \beta^+}{2n}\right) d\xi \tag{1.120a}$$

where k is the propagation constant, $\beta^\pm = \phi \pm \phi'$ and SDP($\pm\pi$) are steepest descent paths through the saddle points $\pm\pi$. $\tilde{\Psi}(\alpha)$ is related to the auxiliary functions suggested by Maliuzhinets

$$\tilde{\Psi}(\alpha) = 2\Psi(\alpha)/[\cos(\alpha/n) + \sin(\theta_0/n)] \tag{1.120b}$$

$$\Psi(\alpha) = \Psi_n(\alpha + n\pi/2 + \pi/2 - \theta_0)\Psi_n(\alpha + n\pi/2 - \pi/2 + \theta_0)$$
$$\times \Psi_n(\alpha - n\pi/2 - \pi/2 + \theta_n)\cdot \Psi_n(\alpha - n\pi/2 + \pi/2 - \theta_n) \tag{1.120c}$$

where Ψ_n is a special meromorphic function given by

$$\Psi(z) = \exp\left[\int f_n(t)\, dt\right] \tag{1.120d}$$

1.11 DIFFRACTION BY AN IMPEDANCE WEDGE AND IMPEDANCE HALF PLANES

with

$$f_n(t) = -\frac{1}{4n\pi j} \int_{-j\infty}^{j\infty} \text{tg}\left(\frac{v}{n}\right) \frac{dv}{\cos(v-t)} \quad (1.120e)$$

Closed form expressions [22] for f_n and Ψ_n are available for several values of n, namely $n = 0.5, 1.5,$ and 2.0.

1.11.2 Uniform GTD Formulation for an Impedance Wedge

The asymptotic evaluation of eqn. (1.120a) for large $k\rho$ can be done by two methods of asymptotic evaluation. They are the Pauli–Clemmow and Van der Waerdens methods of the steepest descent path (SDP). The first one leads to the uniform theory of diffraction (UTD) and the second to the uniform asymptotic theory (UAT) of diffraction for the problem of electromagnetic scattering by a straight wedge with two face impedances.

A high-frequency asymptotic solution for scattering by an impedance wedge can be written in the form

$$u^d(P) \sim u^i(Q) \mathscr{D}(\phi, \phi'; \theta_0, \theta_n; k\rho) \exp(-jk\rho)/\sqrt{\rho} \quad (1.121)$$

where $u^i(Q)$ is the incident field at point Q.

The UTD coefficient $\mathscr{D}$ is given by

$$\mathscr{D}(\phi, \phi'; \theta_0, \theta_n; k\rho) = M_n(\phi, \phi'; \theta_0, \theta_n) C_n(\phi, \phi', \theta_0, \theta_n) D^-(\beta^-; k\rho)$$
$$+ \Gamma_n(\phi', \theta_0) D^-(\beta^+; k\rho) + C_n(n\pi - \phi, n\pi - \phi'; \theta_n, \theta_0)$$
$$\cdot [D^+(\beta^-; k\rho) + \Gamma_n(\phi', \theta_0) D^+(\beta^+; k\rho)] \quad (1.122)$$

$$M_n(\phi, \phi'; \theta_0, \theta_n) = \exp[(M_n(\phi, \phi_t); \theta_0) + M_n(n\pi - \phi, n\pi - \phi'; \theta_n)], \quad (1.123)$$

where

$$M_n(\alpha, \alpha'; \theta) = \int_0^{\pi + (\alpha - \alpha')} [f_n(t + n\pi - \alpha - \pi/2 + \theta) - f_n(t + n\pi - \alpha - \pi/2 - \theta)] \, dt$$

(1.124a)

f_n is given by

$$f_n(t) = -\frac{1}{4n} \int_{-j\infty}^{+j\infty} \text{tg}\left(\frac{v}{n}\right) \frac{dv}{\cos(v-t)} \quad (1.124b)$$

and the expressions for specific values of n are [22]:

For WA $= \pi$ and $n = 1$ (full plane)

$$f_1(X) = \frac{2X - \pi \sin X}{4\pi \cos X} \quad (1.125a)$$

For WA $= 2\pi$ and $n = 2$ (half plane)

$$f_2(X) = -\frac{\pi \sin X - 2\sqrt{2\pi} \sin(X/2) + 2X}{8\pi \cos X} \tag{1.125b}$$

For WA $= \pi/2$ and $n = \frac{3}{2}$ (90° wedge)

$$f_{3/2}(X) = -\frac{1}{6} tg(X/6) \frac{2\cos(X/3) + 3}{2\cos(x/3) + 1} \tag{1.125c}$$

and

$$C_n(\alpha, \alpha'; \theta_1, \theta_2) = \frac{S_n(\alpha; \theta_1, \theta_1)}{S_n(\theta'; \theta_1, \theta_1)} \frac{S_n(n\pi - \theta'; -\theta_1, -\theta_2)}{S_n(n\pi - \alpha + \pi; -\theta_1, -\theta_2)} \tag{1.126a}$$

where

$$S_n(x; x_1, x_2) = \frac{\sin\left(\dfrac{x - x_1}{2n}\right)}{\sin\left(\dfrac{x + x_2}{2n}\right)} \tag{1.126b}$$

$$\Gamma(\phi', \theta) = \frac{\sin(\phi'/n) - \sin(\theta/n)}{\sin(\phi'/n) + \sin(\theta/n)} \tag{1.127}$$

Also,

$$D^{\pm}(\beta; K) = -\frac{\exp(-j\pi/4)}{2n\sqrt{2\pi k}} \cot\left(\frac{\pi \pm \beta}{2n}\right) F[Ka^{\pm}(\beta)] \tag{1.128}$$

where the transition function $F(X)$ and $a^{\pm}(\beta)$ are defined in Appendix A.1.

The field on the surface needs to be evaluated in applications like mutual coupling calculations of two or more antennas situated on the surface on an aircraft. For details of calculations of surface fields, the reader is referrred to [22]. The uniform high-frequency solution is expressed as

$$u^d(P_0) \sim -\frac{2\sqrt{2}}{\sqrt{\pi}} \exp(-j\pi/4) M_n(n\pi, \phi'; \theta_0, \theta_n) \cdot \frac{S_n(\theta_0; n\pi + \pi, -\phi')}{S_n(\theta_n; n\pi - \phi', -\pi)} \cdot a(n\pi - \phi')$$

$$\cdot \frac{\sigma(\phi', n\pi - \pi)}{1 + \cos\theta_n} \cdot \{1 + \sin\theta_n[L(\theta_0) + L(\theta_n) + \sigma_n(n\pi - \pi, \phi')]\}$$

$$\cdot \frac{F[k\rho a(n\pi - \phi')] - F[k\rho a(\pi - \theta_n)]}{a(n\pi - \phi') - a(\pi - \theta_n)} \cdot \frac{\exp(-jk\rho)}{\sqrt{k\rho}} \tag{1.129}$$

1.11 DIFFRACTION BY AN IMPEDANCE WEDGE AND IMPEDANCE HALF PLANES

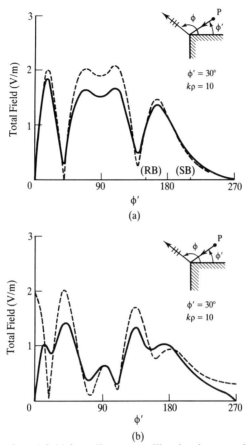

FIGURE 1.26 Plots of total field for a line source illuminating a wedge with 90° interior angle ($n = 3/2$) with equal face impedances: (a) TE case, $\sin \theta = 0.25$ and $\sin \theta = 0$; (b) TM case, $\sin \theta = 4$ and $\sin \theta = \infty$ [22].

The only quantities that remain to be defined are $\sigma_n(\alpha, \alpha_1)$ and $L(\theta_0)$. They are given by

$$\sigma_n(\alpha, \alpha_1) = \frac{1}{n} \frac{\sin(\alpha/n)}{\cos(\alpha_1/n) - \cos(\alpha/n)} \quad (1.130)$$

$$L(\theta_w) = f_n(3\pi/2 - w\pi - \theta_w) + f_n(\pi/2 - w\pi + \theta_w) \quad (1.131)$$

Numerical Computations and Typical Results. Based on the above theory, numerical calculations were carried out in [22]. Figures 1.26 and 1.27 show the plots of the total far field of a line source in the presence of a right-angled wedge and a full plane with an edge discontinuity. The solution is continuous for all observation angles. The differences between the PEC and non-PEC cases are also projected in Figure 1.26(a) and (b). The presence of non-PEC surface

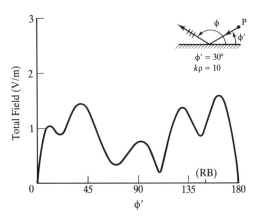

FIGURE 1.27 Total field patterns for line source illumination of a surface impedance junction: $\sin\theta_0 = 0.25$ and $\sin\theta_1 = 4$ on a flat plane [22].

impedance results in a weaker interaction between the incident and reflected fields.

Recently, Greisser and Balanis [136] examined the asymptotic wedge solution for plane wave normal incidence for interior wedge diffraction. Different contributions, namely, a geometrical optics term, a diffracted field term, and surface wave contributions have been estimated. For further details and analytical expressions, the reader is referred to [136].

Impedance Half Plane Solution. The solution presented above and in [22] provides an asymptotic high-frequency form of the Maliuzhinets work [21] for the impedance wedge in terms of a diffraction coefficient similar in form to the UTD diffraction coefficient due to Pathak and Kouyoumjian [7]. The solution is continuous at and around the incident and reflected shadow boundaries, and for a normal incidence is on the edge of the wedge. This solution works fine for a half plane with its interior wedge angle set to zero. Figure 1.28 shows the plots of total fields for a line source illumination of a half plane with equal face impedances for TE and TM cases with $\sin\theta_e = 0.25$ and $\sin\theta_e = 4$. This solution with wedge angle set to zero does not analytically reduce to exactly the same expression as the direct half plane solution reported elsewhere [23].

1.11.3 UTD Solution for the Half Plane Two Face Impedance Problem

Volakis [23] presented a uniform expansion of the Wiener–Hopf solution of the impedance half plane problem at normal and oblique incidence. The diffraction coefficient reduces to the known PEC coefficient with $\sigma = \infty$. Skipping the Wiener–Hopf [23] solution for this problem, the diffracted field is given by

$$E^d \sim E^i \cdot \hat{\beta}'_0 \hat{\beta}_0 D(\phi, \phi'; \pi/2, \eta) + \hat{\phi}'\hat{\phi} D(\phi, \phi'; \pi/2, 1/\eta) A_1(s) \quad (1.132a)$$

1.11 DIFFRACTION BY AN IMPEDANCE WEDGE AND IMPEDANCE HALF PLANES

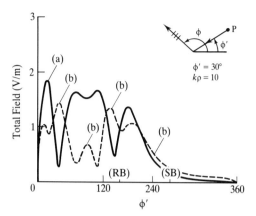

FIGURE 1.28 Total field patterns for line source illumination of a half plane with equal face impedances: curve (a) TE case, $\sin \theta = 0.25$; curve (b) TM case, $\sin \theta = 4$. (From [22] © 1985 IEEE.)

where

$$D(\phi, \phi'; \pi/2, \eta) = \frac{e^{-j\pi/4}}{\sqrt{2\pi k}} \left(\frac{1 - 2\eta \cos \phi/2 \cos \phi'/2}{\cos \phi + \cos \phi'} \right) K_+(\phi; \pi/2, \eta) K_+(\phi'; \pi/2, \eta) \quad (1.132b)$$

$$K_+(\phi, \beta - 0, \eta) = \frac{2[2 \cos \xi (1 - \cos \phi)]^{1/2}}{\left(\sqrt{2} \sin \frac{\phi - \xi}{2} + 1 \right)\left(\sqrt{2} \sin \frac{\phi + \xi}{2} + 1 \right)}$$

$$\cdot \frac{\psi_n(\pi - \phi + \xi)\psi_\pi(\pi - \phi - \xi)}{\psi_\pi(\pi/2)\psi_\pi(\pi/2)} \quad (1.132c)$$

The Pauli–Clemmow method can be employed to get the singularity-free diffraction coefficient. The singularity-free expression is given by

$$D(\phi, \phi'; \pi/2, \eta) = -\frac{e^{-j\pi/4}}{2\sqrt{2\pi k}} C_1(\phi, \phi'; \eta) \sec \beta^-/2 F(2kL \cos^2 \beta^-/2)$$

$$+ C_2(\phi, \phi'; \eta) \sec(\beta^+/2) \quad (1.133)$$

where

$$\beta^\pm = \phi \pm \phi' \quad (1.134)$$

The C functions are given by

$$C_1(\phi, \phi'; \eta) = \frac{2\eta \sin \phi/2 \sin \phi'/2 + 1}{2 \sin \phi/2 \sin \phi'/2} \cdot K_+(\phi; \pi/2, \eta) K_+(\phi'; \pi/2, \eta) K_+(\phi'; \pi/2, \eta)$$

(1.135a)

$$C_2(\phi, \phi'; \eta) = \frac{2\eta \sin \phi/2 \sin \phi'/2 - 1}{2 \sin \phi/2 \sin \phi'/2} \cdot K_+(\phi; \pi/2, \eta) K_+(\phi'; \pi/2, \eta) \quad (1.135b)$$

For the PEC half plane, the expressions simplify.
In this case,

$$K_+(\phi; \pi/2, \eta \to 0) = \sqrt{2} \sin \phi/2 \qquad (1.136a)$$

$$K_+(\phi; \pi/2, \eta \to \infty) = 1/\sqrt{\eta} \qquad (1.136b)$$

and it turns out that

$$D(\phi, \phi'; \pi/2, \eta)|_{\eta \to 0 \text{ and } \eta \to \infty} = D^{s,h}(\phi, \phi'; \beta_0 = \pi/2) \qquad (1.136c)$$

where $D^{s,h}$ are the hard and soft coefficients given in Appendix A.1.

At the shadow boundary, $C_1(\pi + \phi, \phi'; \eta) \to 1$ and at the reflection boundary,

$$C_2(\pi - \phi', \phi'; \eta) \to \Gamma_s = \frac{\eta \sin \phi' - 1}{\eta \sin \phi' + 1}.$$

It may be noted that: (1) the diffracted field at the RB is one half of the reflected field with a sign discontinuity such that the total field remains continuous; (2) the diffracted field at the SB is one half of the incident field again with a sign discontinuity such that the total field remains continuous; and (3) the diffracted and total fields at the observation points away from the ISB and RB are practically independent of the impedance boundary conditions.

The case of oblique incidence has also been treated and is available in [23]. Similar conclusions also apply in the oblique incidence case. Figure 1.29 shows the plot of total field patterns for the impedance half plane with incident angle of 30° degrees and η as a parameter for a particular value of ks' due to electric and magnetic line sources.

1.11.4 UAT Solution for the Half Plane Two Face Impedance Problem

The Uniform asymptotic solution is obtained when the asymptotic evaluation is done by Van der Waerden's method. The UAT treatment of the problem is available in [33]. We present the key expressions and some results here.

Rewriting Maliuzhinets' exact solution [21] for the total electric or magnetic

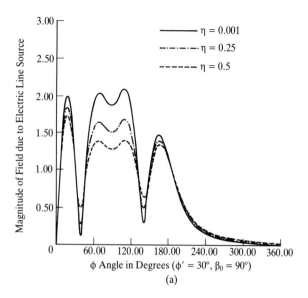

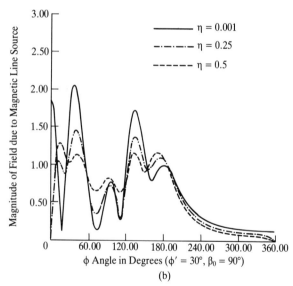

FIGURE 1.29 Total field patterns of a line source in the presence of an impedance half plane at ϕ' and with η as parameter. Values of η are 0.001, 0.25, and 0.5; $ks' = 10.0$: (a) $E^i_{\beta_0}$ due an electric line source; (b) E^i_ϕ due to a magnetic line source [23].

field for a plane wave illuminating a straight wedge of arbitrary wedge angle in the spectral domain, one gets

$$u^t(\rho, \phi) = \frac{-i}{4\pi} \int_\Gamma X(\phi_0, \omega) \exp(-ik\rho \cos(\omega - |\phi|)) \, ds \qquad (1.137)$$

where

$$X(\phi_0, \xi) = \frac{\cos(\phi_0/2)}{\Psi(\phi_0)} \left[\frac{\Psi(\xi - \pi)}{\sin(\xi - \pi/2) - \sin(\phi_0/2)} \mp \frac{\Psi(\xi + \pi)}{\sin(\xi + \pi/2) - \sin(\phi_0/2)} \right]$$

(1.138)

Γ is the Sommerfeld contour and the rest of the symbols have been defined before.

As stated earlier, the contour integration in eqn. (1.137) using Van der Waerden's method [137] constructs the uniform asymptotic theory for scattering by the straight impedance wedge. The impedance half plane solution is available in [33]. The uniform asymptotic expansion for the total field is given by

$$u^t(\rho, \phi) \sim u^i(\rho, \phi)[F(\xi_i) - \hat{F}(\xi_i)] + u^r(\rho, \phi)[F(\xi_r) - \hat{F}(\xi_r)]$$
$$+ u^s(\phi_0, Z^+) + u^s(\phi_0, Z^-) + u^d(\rho, \phi) + O(k^{-3/2}) \qquad (1.139)$$

The first and second terms are the incident and reflected fields, the third and fourth terms are the surface wave contributions from $+$ and $-$ surfaces with surface impedances Z^+ and Z^- respectively and the fourth term is the diffracted field. The $\xi_{i,r}$ are detour parameters, the difference in the path lengths of the straight path from the source to the observation point and the path via the edge. The detour parameters $\xi_{i,r}$ are given by

$$\xi_{i,r} = \pm\sqrt{2k\rho} \cos\left(\frac{\phi \mp \phi_i}{2}\right) \qquad (1.140)$$

where i and r refer to the upper and lower signs. Figure 1.30 shows the total and diffracted field patterns due to a plane wave and anisotropic line incidence whose pattern is given by $P_0(\phi') = 1 + 0.3 \sin \phi'$. The surface waves are not excited at $\phi' = \pi/2$, which is the case of normal incidence. The total field remains finite and continuous at and around the shadow boundaries. With the lit face perfectly absorbing, the total field approximates the incident field for $\phi' > \pi/2$. There is a variation of the total field when the lit face is PEC due to the presence of the reflected field. In Figure 1.30(a) the diffracted field (neglecting the transition function) of curves [136] agrees well with the results of UAT expansion.

The interaction of electromagnetic waves with a curved PEC wedge has also been investigated by Lee and Deschamps [15] using a UAT format to the order

1.11 DIFFRACTION BY AN IMPEDANCE WEDGE AND IMPEDANCE HALF PLANES

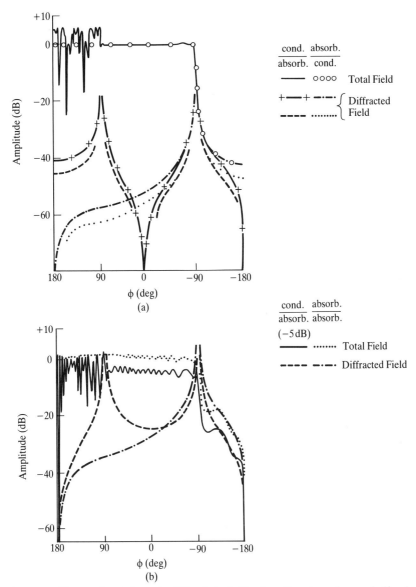

FIGURE 1.30 Plots of total and diffracted fields for the impedance half plane problem [33]: (a) plane wave incidence and H-polarization with $\phi' = 90°$, $\rho = 666.7\lambda$; (b) anisotropic line source incidence and E-polarization with $\phi' = 90°$, $S = 6.667\lambda$, $\rho = 10\lambda$.

of $k^{-1/2}$. The reader is referred to [15] for the mathematical details of the solution. No application of this theory has been presented in [15].

1.12 DIFFRACTION FROM THE EDGES OF IMPEDANCE DISCONTINUITIES ON A HALF PLANE

In this section we discuss the special case of the high-frequency scattering from the edges of impedance discontinuities on a flat plane. This problem can be formulated using the discussion in Section 1.11.2 with wedge angle equal to 2π, i.e. as a full plane. This also has been separately treated and presented [24] before the impedance wedge solution [22] though the solutions are essentially the same. Hence we present some scattering results on the flat plane impedance discontinuity. Figure 1.31(a) shows the geometry of a double impedance discontinuity on a flat plane and Figure 1.31(b) gives the plot of scattered far-field power with $kd = 10$, $\sin \theta_1 = \sin \theta_3 = 0$ and $\sin \theta_2 =$ (a) 0.25 (b) $-j0.25$ (c) $+j0.25$. The results are compared with a moment method solution

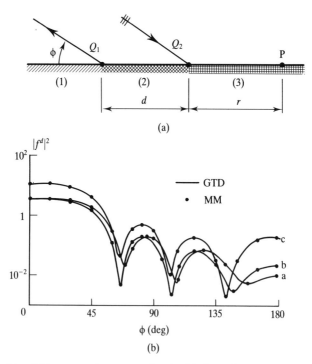

FIGURE 1.31 Far-field scattered pattern of a double impedance discontinuity on a flat plane [24]: (a) the geometry of the double impedance discontinuity on a flat plane; (b) the scattered far-field patterns: $kd = 10$; $\sin \theta_1 = \sin \theta_3 = 0$; $\sin \theta_2 =$ (1) 0.25, (2) $-j0.25$, and (3) $j0.25$.

1.12 DIFFRACTION FROM IMPEDANCE DISCONTINUITIES ON A HALF PLANE

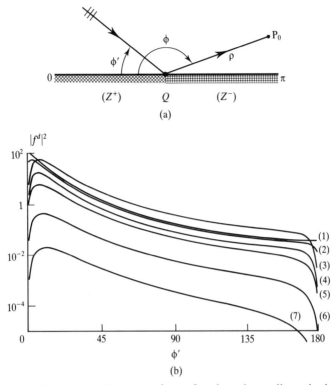

FIGURE 1.32 Far-field scattered pattern of a surface impedance discontinuity on a flat plane [24]: (a) the geometry of a surface impedance discontinuity on a flat plane; (b) the far-field scattered patterns: $k\rho = 20$, $\sin \theta^+ = 0.5$, $\sin \theta_- = $ (1) 0.0, (2) 0.01, (3) 0.10, (4) $-j0.1$, (5) 0.1, (6) 0.25, (7) 1.0.

and good agreement has been obtained. Figure 1.32(a) shows the geometry of a surface impedance discontinuity and Figure 1.32(b) the far-field scattered power pattern for $k\rho = 20$; $\sin \theta^+ = 0.5$; $\sin \theta^- = $ (a) 0 (b) 0.01 (c) 0.1 (d) $-j0.1$ (e) 0.1 (f) 0.25 (g) 1.0. In each case the solution is uniform at and around the shadow boundaries.

A separate treatment for diffraction coefficients related to a discontinuity formed by impedance and resistive half planes was given by Uzgoren, Buyukaksoy, and Serbest [138]. The two-part mixed boundary value problem has been formulated by a Fourier transform technique leading to a Wiener–Hopf (WH) equation. This WH equation is reduced to a scalar form then solved by standard techniques. The diffraction coefficients are obtained by asymptotic evaluation of the integral. The diffraction coefficient for the discontinuity formed by resistive and impedance half planes in the related medium is given by

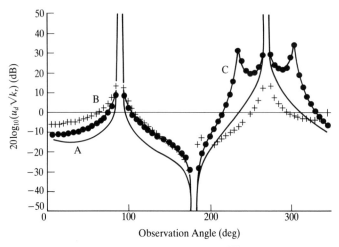

FIGURE 1.33 Scattered field amplitude $20\log_{10}(u_d\sqrt{kr})$ versus observation angle for angle of incidence equal to $\pi/2$; $A_\eta = 0.1$; $B + \eta = 2 + j3$; $C + \eta = j0.2$; $\xi = 0.4, 0.6$.

$$D_{1,2}(\phi', \phi) = -\frac{e^{j\pi/4}}{\sqrt{2\pi}} \frac{\chi^-(k\cos\phi')\chi^-(k\cos\phi)}{(\cos\phi + \cos\phi')\left[\pm 2\eta\cos\dfrac{\phi'}{2}\right]\cos\dfrac{\phi}{2} - (2\zeta - \eta)\beta^+(k\cos\phi'\beta^+(k\cos\phi))} \tag{1.141}$$

The diffraction coefficient in eqn. (1.141) is used to plot the scattered field amplitude $20\log_{10}(u_d\sqrt{kr})$ with observation angles in Figure 1.33 when the angle of incidence is $\pi/2$. The parameter values are:

$$A \,\square\, \eta = 0.1, \quad B + \eta = 2 + j3, \quad C \,\Diamond\, \eta = j0.2.$$

$\xi = 0.4$, 4, and 0.6, respectively. The field expressions are not continuous at the observation angles $\pi/2$ and $3\pi/2$ whereas the diffraction coefficient in [139] does not have a singularity.

1.13 DIFFRACTION BY THICK HALF PLANES

Most of the literature involving the half plane diffraction phenomenon assumes that the half plane is infinitely thin. Much less attention has been given to arrive at a solution of diffraction by a thick half plane. Jones [139] formulated the problem and solved it under the condition $(kl) \ll 1$ where l is the thickness. It was shown by Jones that when the thickness of the plane is less than one-tenth

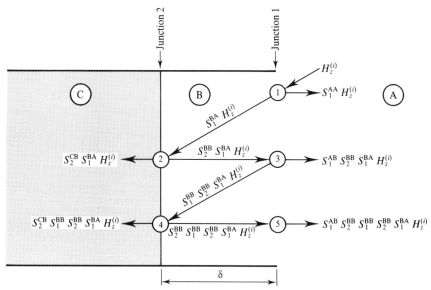

FIGURE 1.34 The development of generalized scattering matrix techniques through multiple scatterings: example of a dielectric-loaded parallel-plate waveguide [141].

of the wavelength, the plane behaves as a combination of a semi-infinite waveguide with a magnetic line source at the open end when the plane of polarization of the indicent wave is perpendicular to the half plane. The solution for the thick half plane in a computationally easy form was suggested by Lee and Mittra [140] using a generalized scattering matrix procedure. The scattered field is expressed as the resultant of the multiple scattering process, as shown in Figure 1.34. Figure 1.35 shows the scattering patterns of a thick half plane computed by taking the number (N) of modes equal to 1, 3, 5, and 7 in the generalized scattering matrix technique. The scattering patterns of the half plane with varying thicknesses with the angle of incidence equal to 30° are shown in Figure 1.36. This method becomes inaccuarate for large thicknesses of the half plane. For details of the scattering matrix approach the reader is referred to [140].

1.13.1 Diffraction by a Thick PEC Half Plane

The diffraction analysis for the thick PEC half plane (Fig. 1.37) was done in [35] for both polarizations using a combination of angular spectrum method (ASM) and generalized scattering matrix formulation (GSMF). The angular spectrum method was introduced by Booker and Clemmow [141, 142] to study the scattering from a PEC half plane, two parallel half planes, and an infinite stack of parallel half planes. The scattered field is expressed in terms of an integral whose integrand is the angular spectrum of the surface current flowing

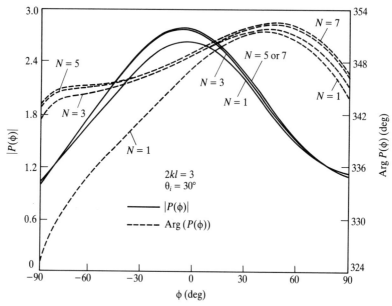

FIGURE 1.35 Scattering pattern of a thick half plane using a generalized matrix technique for number of modes equal to 1, 3, 5, and 7; angle of incidence = 30° [141].

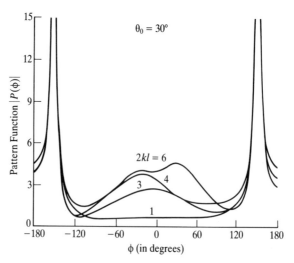

FIGURE 1.36 Scattering patterns of a half plane with thickness; angle of incidence = 30° [141].

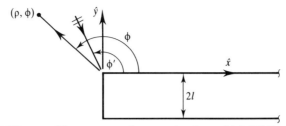

FIGURE 1.37 Electromagnetic scattering by a thick PEC edge.

on the planar structure. Analytical expressions for the current spectra can be found by using boundary and certain continuity conditions. The current thus obtained can be used in the integral to determine the scattered field. We will briefly describe the analytical development of the angular spectrum method. GTD and angular spectrum methods give the same results for sufficiently thick edges. It appears that a combination of ASM and GTD can predict the field scattered by a half plane of arbitrary intermediate thickness.

Angular Spectrum Method. For E_z incedence (Fig. 1.37) the total field scattered by the thick half plane is given by

$$E_z^s = E_1^s + E_2^s \tag{1.142}$$

where 1 and 2 stand for upper and lower half planes and E_1^s and E_2^s are scattered fields from the upper and lower planes of the thick half planes. The two contributions can be expressed in the angular spectrum as [35]

$$E_{z1}^s = \int_C P_1(\cos \alpha) e^{-jk\rho \cos(\phi \mp \alpha)} \, d\alpha \quad \text{for} \quad -2l < y < 0 \tag{1.143}$$

and

$$E_{z2}^s = \int_C P_2(\cos \alpha) e^{\mp j2kl \sin \alpha} e^{-jk\rho \cos(\phi \mp \alpha)} \, d\alpha \quad \text{for} \quad 2l < y < 0 \tag{1.144}$$

where $P_{1,2}(\cos \alpha)$ is the angular spectrum of the currents flowing on the upper and lower half planes.

The angular spectrum of the current is available from Clemmow's work and is given by

$$P_{1,2}(\cos \alpha) = \frac{1}{4\pi j} \frac{2 \sin \alpha/2 \sin \phi_0/2}{\cos \alpha + \cos \phi_0} \cdot \left[\frac{L_1(\cos \phi_0)}{U_1(\cos \alpha)} \mp \frac{L_2(\cos \phi_0)}{U_2(\cos \alpha)} \right] \tag{1.145}$$

The functions $U_{1,2}(\lambda)$ and $L_{1,2}(\lambda)$ are regular and holomorphic in the upper and lower λ-half planes. These are known as split functions.

The total scattered field can be expressed as

$$\begin{Bmatrix} E_z^s \\ H_z^s \end{Bmatrix} = \left\{ S(\phi, \phi_0) \frac{\exp(-jk\rho_0)}{k\rho} \right. \tag{1.146}$$

The function $S(\phi, \phi_0)$ is given by

$$S(\phi, \phi_0) = S_0(\phi, \phi_0) + [L_n(\phi)]^T [\Gamma_{mn}] \cdot \{I - [R_{mn}][\Gamma_{mn}]\}^{-1} [C_m(\phi_0)]\} \tag{1.147}$$

Single and double subscripts denote column vectors and square matrices. I is identity matrix and $S(\phi, \phi_0)$ is the diffraction coefficient.

Numerical Results for a Thick PEC Half Plane. Figure 1.38 compares calculated and measured backscattering cross sections for an edge with thickness $2l = 0.62$ in. for E_z and H_z incidence. It may be noted that the E_z incidence the agreement between the theoretical and experimental results is very good but for H_z incidence the measured values in the main lobe region are about 2 dB lower than the calculated values by either method. As pointed out in [35] the agreement seems to be better in cases of thicker edges. Figure 1.39 shows normalized plots of the echo width versus edge thickness via the angular spectrum method and UTD (a) E_z incidence (b) H_z incidence. The results differ for H_z for relatively thin edges (thickness less than a quarter of a wavelength) and UTD fails for thicknesses below one-tenth of a wavelength. For other results see reference [35]. The solution using ASM along with GSMF is applicable everywhere in the region.

1.13.2 Diffraction by a Thick Half Plane with Two Face Impedances

The problem of diffraction by a thick half plane with two face impedances [36] can be solved in an exactly similar manner as the case of PEC thick plane. The thick half plane is equivalent to two impedance half planes containing a recessed stub.

The scattered field is given by [36]

$$E_{z0,e}^s = \sum_{m'=1,2...} R_{m\acute{o},e\acute{n}\acute{o},c}^e \frac{m'\pi}{2l}(y+l)e^{jkx\lambda_{m'}}$$

$$+ \eta \sum_{m=0,1...} R_{m0,eno,e}^h \frac{m\pi}{2l}(y+l)e^{-jkx\lambda_m} \tag{1.148}$$

1.13 DIFFRACTION BY THICK HALF PLANES

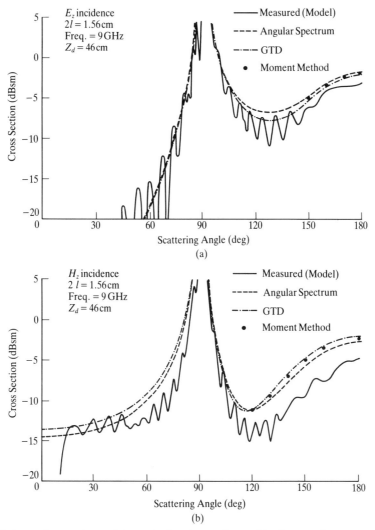

FIGURE 1.38 Comparison of theoretical and experimental backscattering cross sections for an edge with thickness $2l = 0.62$ in.: (a) E_z incidence; (b) H_z incidence [34].

where

$$R^e_{m\acute{0},\,e n\acute{0}} = -\frac{1}{j} A^e_{0,\,e} e^{-j(m'n/2)} \frac{1}{\sqrt{1-\lambda_m^2}(\lambda'_n + \lambda'_m)} \cdot \frac{U_3(\lambda_m;\eta);\, L_{1,\,2}(\lambda_n)}{L_3(\lambda'_n;\eta) U_{1,\,2}(\lambda'_m)} \quad (1.149\text{a})$$

$$R^e_{m\acute{0},\,e n\acute{0},\,e} = R^e_{m\acute{0},\,e n\acute{0},\,e} = 0 \quad (1.149\text{b})$$

$$R^h_{m0,\,e n0,\,e} = A^m_0 e^{-j(m\pi/2)} \frac{\sqrt{1+\lambda_n}}{\sqrt{1-\lambda_m}} (\lambda_n + \lambda_m) \cdot \frac{U_3(\lambda_m;\eta) L_{1,\,2}(\lambda_n)}{L_3(\lambda_n;\eta) U_{1,\,2}(\lambda_m)} \quad m \neq 0 \quad (1.149\text{c})$$

$$R^h_{m0,\,e n0,\,e} = R^h_{m0,\,e n0,\,e} = 0 \quad (1.149\text{d})$$

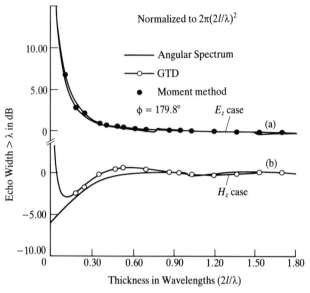

FIGURE 1.39 Normalized plots of the echo-width versus edge thickness for a thick edge via the angular spectrum method and UTD: (a) E_z incidence; (b) H_z incidence [34].

and

$$R_{0n_e} \approx -2 \sin\frac{\theta}{2} \frac{L_{2+}(\lambda_n)}{U_{2+}(1)} \frac{1}{L_3(\lambda_n;\eta)L_3(1;\eta)} \quad |\eta| > 1 \quad (1.149e)$$

$$R_{0n_e} \approx 0 \quad \text{otherwise} \quad (1.149f)$$

$$U_n(\lambda) = \sqrt{2kl}\sqrt{1-\lambda}\,U_{n+}(\lambda)$$

$$U_{n+}(\lambda) = \frac{1}{(\eta\sqrt{1-\lambda^2}+1)L_{n+}(\lambda;\eta)}$$

The expressions in eqns. (1.148) and (1.149) reduce to the known ones for the PEC case when $\eta = 0$ or $\eta \to \infty$ corresponding to the E_z and H_z incidence, respectively.

Figure 1.40 shows the backscattered echo width versus aspect angle for $2l = 0.01, 0.1, 0.2, 0.3$, and 0.4 wavelengths for the normalized surface impedance η equal to (a) $j2$ ohms and (b) $(2+j2)$ ohms. The numerical results do not take into account surface waves on the thick impedance half plane. It is observed that the echo width varies significantly with angle when the aspect angle increases beyond 105°. Figure 1.39 compares the echo width normalized to $2\pi(2l/\lambda)^2$ of the thick PEC half plane with electrical thickness for E_z and H_z cases using the angular spectrum method, UTD, and moment method.

1.13 DIFFRACTION BY THICK HALF PLANES 79

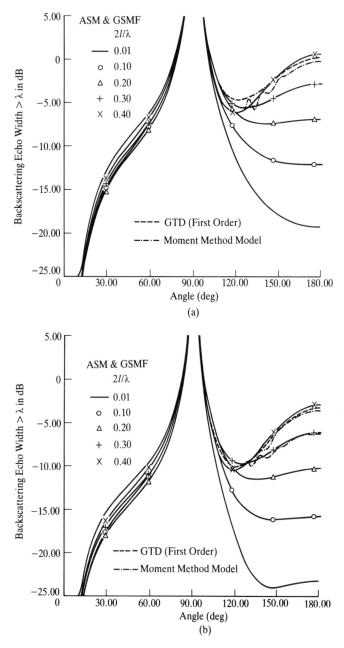

FIGURE 1.40 Backscattered echo width versus aspect angle for different thicknesses $(2l/\lambda)$ equal to 0.01, 0.1, 0.2, 0.3, and 0.4 wavelengths: (a) $\eta = j2$; (b) $\eta = 2 + j2$ [36].

1.14 DIFFRACTION BY A DIELECTRIC HALF PLANE

The problems of electromagnetic scattering from dielectric half planes have been investigated by A. D. Rawlins [143], Anderson [38], A. Chakrabarty [39], J. L. Volakis and T. B. A. Senior [40], sheet simulation of the thin dielectric layer by T. B. A. Senior and J. L. Volakis [41]. The problem of isotropic dielectric/ferrite half plane and related configurations has been discussed by R. G. Rojas and P. H. Pathak [145].

One practical application which motivated the work on the problem of diffraction by dielectric sheets is in connection with acoustic or electromagnetic noise abatement using noise barriers and in surface wave antennas. In these applications, the residual field in the shadow region is solely due to the diffraction by the edge of the half plane consisting of the noise reduction material.

1.14.1 Diffraction of a Plane Wave by a Thin Dielectric Half Plane

In this section we will summarize the solution of the above two-dimensional problem for two orthogonal polarizations as presented by different workers. Kaminetsky and Keller [67] determined approximately the diffracted fields about certain dielectric structures but their results are not applicable for the diffraction analysis of the dielectric half plane.

Anderson's Solution. The scattering of an incident plane electromagnetic wave by a dielectric half plane is shown in Figure 1.41. With a suppressed time variation of $\exp(j\omega t)$ the incident field for an E-polarized case is given by

$$E_z^i = \exp[jkr\cos(\phi - \phi')] \tag{1.150}$$

where the angle of incidence ϕ' lies between 0 and π and $k = 2\pi/\lambda$ is the wave number.

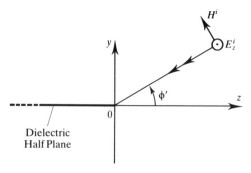

FIGURE 1.41 Electromagnetic scattering of an incident plane wave by a dielectric half plane.

The total field at the observation point (r, ϕ) is a vector sum of the incident field E_z^i and the scattered field E_z^s.

A thin dielectric half plane of thickness t and relative permittivity ε_r is approximately equivalent to an infinitely thin polarization current sheet of surface density $[jB E_z(x, 0)]$ with the susceptance $B = \omega t \varepsilon_0 (\varepsilon_r - 1)$ where ω is the angular frequency and ε_0 is the permittivity of free space. The condition for this polarization current sheet approximation is $kt\sqrt{\varepsilon_r} \ll 1$. The problem is essentially to solve the wave equations subject to the pertinent boundary conditions and the Sommerfeld radiation condition and at a later stage to test the uniqueness of the solution. The resulting diffraction integral is then evaluated asymptotically to obtain an expression for the diffracted field for the dielectric half-plane problem.

For a dielectric–dielectric boundary, the boundary conditions are

$$E_z(x, 0^+) = E_z(x, 0^-), \qquad -\infty \leq x \leq \infty \tag{1.151}$$

$$H_x(x, 0^+) = H_x(x, 0^-), \qquad 0 \leq x \leq \infty \tag{1.152}$$

$$H_x(x, 0^+) - H_x(x, 0^-) = -jBE_z(x, 0), \qquad -\infty \leq x \leq 0, \tag{1.153}$$

0^+ and 0^- are quantities which are just greater and just less than zero.

The scattered field E_z^s satisfies the wave equation

$$\left[\frac{\delta^2}{\delta x^2} + \frac{\delta^2}{\delta y^2} + k^2 \right] E_z^s(x, y) = 0 \tag{1.154}$$

There are several methods of solving the wave equation in (1.154) and they are described in [143, 144]. Two approaches are normally used: (1) the wave equation is Fourier transformed and the pertinent boundary conditions is used to obtain an integral equation of the Wiener–Hopf type and (2) the general solution of the transformed wave equation is obtained, inverted, and the boundary conditions applied to obtain a pair of dual integral equations. This formulation has the advantage that it is amenable to direct solution using complex variable theory. On inverting the general solution of the wave equation and using (1.151), the scattered field is given by

$$E_z^s(x, y) = \int_{-\infty}^{\infty} A(\zeta) \exp[-j(x\zeta \pm y\sqrt{k^2 - \zeta^2})] \, d\zeta \tag{1.155}$$

The asymptotic evaluation of this integral using a contour and the steepest descent path is available in [38]. It turns out that the total field $E_>^t(r, \theta)$ is asymptotically given by

$$E_>^t(r, \theta) = e^{jkr \cos(\phi - \phi')} + c_1 R(\phi_0') e^{jkr \cos(\theta + \phi')} + E_>^d(r, \theta) \tag{1.156}$$

where

$$E^d_>(r, \theta) = \frac{u e^{-j(kr+\pi/4)}}{\sqrt{\pi \sigma_0^-(\phi')\sigma_0^-(\theta)}}$$
$$\times \{F[\sqrt{2kr}\cos\tfrac{1}{2}(\phi-\phi')] \pm F[\pm\sqrt{2kr}\cos\tfrac{1}{2}(\theta+\phi')]\}$$
$$+ c_2 G(\phi') e^{-kuy} e^{jkx\sqrt{1+u^2}} \quad (1.157)$$

with

$$R(\phi'_0) = \frac{u}{j\sigma_0^-(\phi')\sigma_0^-(\pi-\phi')} = \left[j\frac{\sin\phi'}{u} - 1\right]^{-1} \quad (1.158)$$

$$G(\phi') = \frac{2u \sin\frac{\gamma}{2} \cos\frac{\phi'}{2}}{(\cos\phi' - \cosh\gamma)\gamma_0^-(\phi')} \cdot \lim_{\phi \to (\pi+j\gamma)} \frac{\phi - (\pi+j\gamma)}{\sigma_0^-(\phi)} \quad (1.159)$$

and

$$c_1 = \begin{cases} 0, & 0 \le \theta \le \pi - \phi' \\ 1, & \pi - \phi' \le \theta \le \pi \end{cases} \quad c_2 = \begin{cases} 0, & \theta \ne \pi \\ 1, & \theta \approx \pi \end{cases} \quad (1.160)$$

The $\pm$ sign in eqn. (1.157) is for $\theta \le \pi - \phi'$.
It turns out [38] that the expression in eqn. (1.159) reduces to

$$G(\phi') = \frac{2u \sinh\frac{\gamma}{2} \cos\frac{\phi'}{2}}{\cosh\gamma(\cosh\gamma - \cos\phi') \cdot \frac{\sigma_0^-(j\gamma)}{\sigma_0^-(\phi')}} \quad (1.161)$$

The total field $E^t_<(r, \theta)$ is then asymptotically given by

$$E^t_<(r, \phi) = [1 + c_3 R(\phi')] e^{jkr\cos(\phi-\phi')} + E^d_<(r, \phi) \quad (1.162)$$

where

$$E^d_<(r, \theta) = \frac{u e^{-j(kr+\pi/4)}}{\sqrt{\pi \sigma_0^-(\phi')\sigma_0^-(2\pi-\phi)}} \cdot F[-\sqrt{2kr}\cos\tfrac{1}{2}(\phi-\phi')]$$
$$\mp F[\pm\sqrt{2kr}\cos\tfrac{1}{2}(\phi-\phi')] + c_2 G(\phi') e^{kuy} e^{jhx\sqrt{1+u^2}} \quad (1.163)$$

where

$$c_3 = \begin{cases} 0, & \phi > \pi + \phi' \\ 1.0, & \phi < \pi + \phi' \end{cases} \quad (1.164)$$

The upper/lower sign is for $\phi \le \pi + \phi'$.
Now consider the expression for the total field at an observation point at a

1.14 DIFFRACTION BY A DIELECTRIC HALF PLANE

large distance from the edge and away from the shadow boundaries ($\phi = \pi \pm \phi'$). The total field $E^t_>(r, \phi)$ and $E^t_<(r, \phi)$ can be expressed as

$$E^t(r, \phi) = E^g(r, \phi) + E^d(r, \phi) \tag{1.165}$$

where

$$E^{GO}(r, \phi) = \begin{cases} e^{jkr\cos(\phi-\phi')}, & \begin{aligned} 0 &\le \theta < \pi - \phi' \\ \pi + \phi' &\le \theta \le 2\pi \end{aligned} \\ e^{jkr\cos(\phi-\phi')} + R(\phi')e^{jkr\cos(\phi-\phi')}, & \pi - \phi' \le \phi \le \pi \quad (1.166a) \\ T(\phi')e^{jkr\cos(\phi-\phi')}, & \pi \le \phi < \pi + \phi' \quad (1.166b) \\ \text{with} \quad T(\phi') = 1 + R(\phi') & \quad (1.167) \end{cases}$$

and the diffracted field is given by

$$E^d(r, \phi) \sim \mp \sqrt{\frac{2}{\pi}} \frac{ue^{j\pi/4}}{\sigma_0^-(\phi')\sigma_0^-(\zeta)} \frac{\cos\frac{\phi}{2}\cos\frac{\phi'}{2}}{\cos\phi + \cos\phi'} \frac{e^{-jkr}}{\sqrt{kr}} + c_2 G(\phi') e^{-ku|y|} e^{jkx\sqrt{1+u^2}}$$

(1.168)

The amplitude of the cylindrical wave in eqn. (1.165) can be identified as the diffraction coefficient of the dielectric half plane. The second term in eqn. (1.165) is a surface wave term with

$$\zeta = \begin{cases} \phi, & 0 \le \phi \le \pi \\ 2\pi - \phi, & \pi \le \phi \le 2\pi \end{cases} \tag{1.169}$$

The $\pm$ signs above are for $\phi \le \pi$. $T(\phi')$ and $R(\phi')$ are the plane wave Fresnel transmission and reflection coefficients for the dielectric sheet for the angle of incidence.

At and near the shadow boundaries, the Fresnel function is approximated by their low-argument expressions. The general proof of uniqueness for the diffraction problem of dielectric obstacles with edges is difficult. It has been demonstrated [38] that a limiting case of the dielectric half plane is unique. Equations (1.166) and (1.167) can be reduced [38] to unique equations for PEC cases.

Example 1.7 [34] A plane wave of unit amplitude incident normally upon a thin dielectric sheet of thickness $t = \lambda/16$ and relative permittivity $\varepsilon_r = 5$. The transmission coefficient is found to be $|T| = 0.786 \; \underline{/0.667 \text{ radians}}$. Plot the amplitude of the electric field $|E_<(r, \theta)|$ at a distance of $y = -3\lambda$ behind the dielectric sheet. Plot the geometrical optics field.

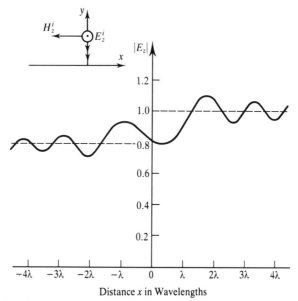

FIGURE 1.42 Electric field amplitude at $y = -3\lambda$ behind a dielectric half plane; $t = \lambda/16$, $\varepsilon_r = 5$ [34].

Solution

$$t = \lambda/16; \quad \varepsilon_r = 5.0$$

$$u = \pi t \lambda (\varepsilon_r - 1) = \pi/4$$

$$R(\phi') = \left[j \frac{\sin \phi'}{u} - 1 \right]^{-1}$$

The GO field is given by the first term of eqn. (1.166a,b).

Figure 1.42 gives the plot of the amplitude of the electric field $|E_<(r, \theta)|$ at a distance $y = -3\lambda$ behind the dielectric half plane with thickness $t = \lambda/16$, $\varepsilon_r = 5$. The geometrical optics field is shown by the dotted line in Figure 1.42. The interference fringes beyond $|x| > \lambda$ are due to the superposition of diffracted field and geometrical optics.

Volakis and Senior [41] presented a uniform diffraction coefficient for the thin dielectric half plane for an H-polarized incident wave

$$D(\phi, \phi') = -\frac{e^{j\pi/4}}{2\sqrt{2\pi k}} \left[\sec \frac{\phi + \phi'}{2} + \sec \frac{\phi - \phi'}{2} \right] \hat{H}(k \cos \phi) \quad (1.170)$$

Figure 1.43 shows the echo width of a thin dielectric half plane with different relative dielectric constant and two different thicknesses t. The GTD results

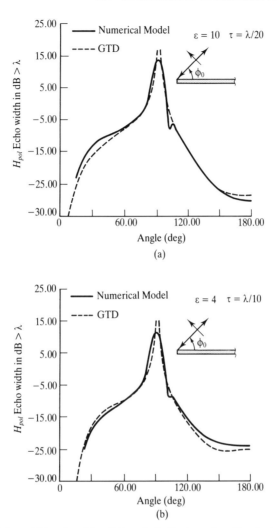

FIGURE 1.43 Echo width of a thin dielectric half plane for H-polarization GTD and numerical results compared: (a) $t = \lambda/20$, $\varepsilon_r = 10.0$; (b) $t = \lambda/10$, $\varepsilon_r = 4$ [41].

with H polarization with a numerical model were compared with numerical results. Thickness t equal to (a) $\lambda/20$ and (b) $\lambda/10$. It is found that they match pretty well. Some comments about this solution and that due to Chakrabarty [39] are in order. But the solution presented in [39] is inconsistent with the physical behavior of the currents. The solution in [39] violates reciprocity except at normal incidence ($\phi' = \pi/2$) while the solution in [38] is reciprocal for all angles of incidence.

It appears to me the dielectric half plane scattering theory has not been used in practical applications.

1.15 DIFFRACTION BY A DIELECTRIC WEDGE

Unlike metal wedges, where exact solution to the canonical problem and a dyadic diffraction coefficient was available quite sometime ago, the canonical problem of a dielectric wedge has lacked an exact solution. One of the major difficulties lies in the two different propagating waves [145], one inside and one outside the wedge, which make it hard to satisfy boundary conditions. Function theoretic methods [146] have been employed in [147–150] to solve this problem but the solutions are either in terms of complicated integrals or in error [152]. Ray-optical methods [151, 152] and perturbation schemes [152, 153] for high-frequency diffraction have been used in this problem. But ray-optical methods have limitations in constructing guided and evanescent modes and solutions from perturbation schemes are restricted to limited ranges of dielectric constant and/or wedge angle. Hybrid techniques, a topic reserved for Section 1.17, have also been used to treat dielectric wedge problems. Using a hybrid technique, a numerical diffraction coefficient has been obtained for the dielectric wedge. These are discussed in Section 1.17.1.2. Very recently, a complete physical optics solution [154] has been obtained by PO approximation of the dual integral equation formulated in the spatial frequency domain. This solution has been constructed using GO optics terms, including multiple reflection inside the wedge and the edge-diffracted field. The diffraction coefficient [154] is expressed as two finite series of cotangent functions weighted by the Fresnel reflection coefficients. The error in this solution for the E-polarized wave has also been discussed in a companion paper [155].

Though there are some previously attempted solutions, we present here the one due to Kim, Ra, and Shin [154, 155]. In this work, the complete PO solution for scattering by a dielectric wedge of arbitrary dielectric constant and wedge angle is obtained by Fourier transforming the Fredholm integral equation for the same problem. Two sets of integral equations are arrived at. One set gives the total field in all the space and the other set helps in checking the accuracy of the solution. The GO contribution is extracted easily from dielectric interfaces taking care of the reflected and refracted rays at the interfaces. The PO solution is obtained by approximating the exact fields by the GO solution along the dielectric interfaces and then substituting one set of the dual integral equations. This PO solution consists of GO and edge-diffracted terms. The diffraction coefficients are obtained in the UTD format as two finite series of cotangent functions weighted by the Fresnel reflection coefficients.

The geometry and the coordinate system of the dielectric wedge scattering is shown in Figure 1.44. The interior wedge angle is θ_d and the relative dielectric constant is ε_r. The details of the mathematical development are given in [154]. We present the key equations necessary for computation of scattered fields.

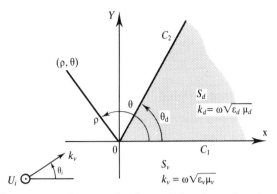

FIGURE 1.44 Electromagnetic scattering by a dielectric wedge; interior wedge angle θ_d, relative dielectric constant ε.

1.15.1 Dual Integral Equations

The two functions $P(\alpha, \beta)$ and $Q(\alpha, \beta)$ satisfy the dual integral equations [155] given below.

$$\text{In } S_d: u_i(\rho, \theta) - F^{-1} \frac{P(\alpha, \beta) + \dfrac{1}{r} Q(\alpha, \beta)}{\alpha^2 + \beta^2 - k_d^2} = 0 \qquad (1.171a)$$

$$F^{-1} \frac{P(\alpha, \beta) + Q(\alpha, \beta)}{\alpha^2 + \beta^2 - k_v^2} = 0 \qquad (1.171b)$$

where k_d is the wave number in the dielectric, S_v is the volume exterior to the dielectric wedge and S_d is the volume occupied by the dielectric wedge. F^{-1} is the inverse Fourier integral.

$P(\alpha, \beta)$ and $Q(\alpha, \beta)$ are given by

$$P(\alpha, \beta) = i\beta \int_0^\infty d\rho\, u(\rho, 0)\, e^{-i\alpha\rho} + i(\alpha \sin\theta_d - \beta \cos\theta_d)$$

$$\times \int_0^\infty d\rho\, u(\rho, \theta_d)\, e^{-i(\alpha\cos\theta_d + \beta\sin\theta_d)\rho} \qquad (1.172a)$$

$$Q(\alpha, \beta) = i\beta \int_0^\infty d\rho\, u(\rho, 0)\, e^{-ik\rho} - \int_0^\infty d\rho\, \frac{1}{\rho} \frac{\delta}{\delta\theta} u(\rho, \theta_d)\, e^{-i(\alpha\cos\theta_d + \beta\sin\theta_d)\rho}$$

$$(1.172b)$$

1.15.2 Total Field at Any Point

The total field $u(\rho, \theta)$ at any observation point may be expressed as

$$u(\rho, \theta) = \begin{cases} u^i(\rho, \theta) + F^{-1} - \dfrac{P(\alpha, \beta) + Q(\alpha, \beta)}{\alpha^2 + \beta^2 - k_d^2} & \text{in } S_d \quad (1.173a) \\[2ex] u^i(\rho, \theta) - F^{-1} - \dfrac{P(\alpha, \beta) + Q(\alpha, \beta)}{\alpha^2 + \beta^2 - k_v^2} & \text{in } S_v \quad (1.173b) \end{cases}$$

1.15.3 Geometrical Optics Field

An additional task with dielectric in contrast with a conducting wedge is that it is needed to know the ray trajectory through the dielectric wedge before the GO ray field expressions can be found.

The GO field at any point of observation (ρ, θ) is expressed as

$$\text{In } S_v: u_g(\rho, \theta) = W(\theta_i, 2\pi)e^{ik_v\rho\cos(\theta-\theta_i)} + W(\theta_r, 2\pi)R_0 e^{ik_v\rho\cos(\theta-\theta_r)}$$

$$+ \sum_{n=1}^{N_e} W(\theta_{2n,t}, 2\pi)K_{2n,t}e^{ik_v\rho\cos(\theta-\theta_{2n,t})}$$

$$+ \sum_{n=1}^{N_0} W(\theta_d, \theta_{2n-1,t})K_{2n-1,t}e^{ik_v\rho\cos(\theta-\theta_{2n-1,t})} \quad (1.174a)$$

$$\text{In } S_d: u_g(\rho, \theta) = \sum_{n=1}^{N} W(0, \theta_d)K_n e^{ik_d\rho\cos(\theta-\theta_n)} + S(\theta_{N+1})K_{N+1}e^{ik_d\rho\cos(\theta-\theta_{N+1})}$$

$$(1.174b)$$

where

$$S(\theta_{N+1}) = W(0, \theta_{N+1}) \quad \text{and} \quad W(\theta_{N+1}, \theta_d) \quad \text{for} \quad N_e = N_0 \quad \text{and} \quad N_E \neq N_0$$

$$K_n = T_0 R_1 R_2 \cdots R_{n-1}$$

$$K_{n,t} = K_n T_n$$

where
N_e = total number of internal reflections at interface C_1
N_0 = total number of internal reflections at interface C_2
$K_n, K_{n,t}$ = the amplitudes of the GO components $u_n, u_{n,t}$

For certain angles of incidence total internal reflections occur and transmitted angles $\theta_{2n,t}$ or $\theta_{2n-1,t}$ become complex. In such situations, the corresponding window function is $W(\theta_d, 2\pi)$.

The nth Fresnel reflection coefficient R_n for E- and H-polarizations are given by

$$R_n = T_n - 1 = \frac{\tau \sin \theta_i - \sqrt{\varepsilon - \cos^2 \theta_i}}{\tau \sin \theta_i + \sqrt{\varepsilon - \cos^2 \theta_i}} \quad \text{for } n = 0 \quad (1.175a)$$

$$= \frac{\sin(\theta_i - n\theta_d) - \tau\sqrt{\dfrac{1}{\varepsilon} - \cos^2(\theta_i - n\theta_d)}}{\sin(\theta_i - n\theta_d) - \tau\sqrt{\dfrac{1}{\varepsilon} - \cos^2(\theta_i - n\theta_d)}} \quad \text{for } n \neq 0 \quad (1.175b)$$

At the tip, the reflections and refractions do not occur. Therefore, a window function $W(\theta_p, \theta_q)$ is defined as

$$W(\theta_p, \theta_q) = \begin{cases} 1, & \text{for } \theta_p \leq \theta \leq \theta_q \\ 0, & \text{elsewhere} \end{cases} \quad (1.176)$$

1.15.4 Physical Optics Field

In order to get an expression for a PO field we substitute the GO field from eqn. (1.166a), $P(\alpha, \beta)$ and $Q(\alpha, \beta)$ from eqn. (1.172a,b) in the expression for total field in (1.173b) and then integrate along the dielectric interfaces. It turns out that the PO approximations to $P(\alpha, \beta)$ and $Q(\alpha, \beta)$ denoted by $P_p(\alpha, \beta)$ and $Q_p(\alpha, \beta)$ are given by

$$P_p(\alpha, \beta) = \frac{\beta}{\alpha - k_v \cos \theta_i} + R_0 \frac{\beta}{\alpha - k_v \cos \theta_r} + \sum_{n=1}^{N_0} K_{2n,t} \frac{\beta}{\alpha - k_v \cos \theta_{2n,t}}$$

$$+ \sum_{n=1}^{N_0} K_{2n-1,t} \cdot \frac{(\alpha \sin \theta_d - \beta \cos \theta_d)}{(\alpha \cos \theta_d + \beta \sin \theta_d) - k_v \cos(\theta_{2n-1,t} - \theta_d)} \quad (1.177a)$$

$$Q_p(\alpha, \beta) = \frac{\tau k_v \sin \theta_i}{\alpha - k_v \cos \theta_i} + R_0 \frac{\tau k_v \sin \theta_r}{\alpha - k_v \cos \theta_r} + \sum_{n=1}^{N_0} K_{2n,t} \frac{\tau k_v \sin 2n,t}{\alpha - k_v \cos \theta_{2n,t}}$$

$$+ \sum_{n=1}^{N_0} K_{2n-1,t} \cdot \frac{-\tau k_v \sin(\theta_2 n - 1.t - \theta_d)}{(\alpha \cos \theta_d + \beta \sin \theta_d) - k_v \cos(\theta_{2n-1,t} - \theta_d)} \quad (1.177b)$$

The PO approximation $u_p(\rho, \beta)$ of $u(\rho, \beta)$ is obtained by replacing $P(\alpha, \beta)$ and $Q(\alpha, \beta)$ by $P_p(\alpha, \beta)$ and $Q_p(\alpha, \beta)$ in eqn. (1.172a,b) and then taking a Fourier transform we get

$$u_p(\rho, \theta) = u_g(\rho, \theta) + v_1(\rho, \theta) \quad \text{in } S_v \quad (1.178a)$$
$$= u_g(\rho, \theta) + v_2(\rho, \theta) \quad \text{in } S_d \quad (1.178b)$$

where

$$v_1(\rho, \theta) = -\frac{i}{4\pi} \int_{SDP} d\omega f_1(\omega) e^{ik_v\rho \cos(\omega - \theta)} \quad (1.179a)$$

$$v_2(\rho, \theta) = -\frac{i}{4\pi} \int_{SDP} d\omega f_2(\omega) e^{ik_d\rho \cos(\omega - \theta)} \quad (1.179b)$$

where $f_1(\omega)$ and $f_2(\omega)$ are the diffraction coefficients whose expressions are given in the next section.

1.15.5 Diffraction Coefficients of the Dielectric Wedge

The complex diffraction coefficients $f_1(\rho, \theta)$ and $f_2(\rho, \theta)$ are given by

$$f_1(\omega) = -\cot\left(\frac{\omega - \theta_i}{2}\right) - R_0 \cot\left(\frac{\omega - \theta_r}{2}\right)$$

$$\sum_{n=1}^{N} (-1)^n K_{n,t} \cot(\omega - \theta_{\theta,t}) \quad (1.180a)$$

$$f_2(\omega) = (-1)^{N+1} K_{N+1} \cot\left(\frac{\omega - \theta_{N+1}}{2}\right) \quad (1.180b)$$

The interesting features of the diffraction coefficients are: (1) they are expressed as cotangent functions in the UAT format; (2) the factor and the argument of each cotangent function are exactly the same as the amplitude and propagating angle of the corresponding GO term; and (3) $f_2(\omega)$ has only one term.

The real and imaginary parts of the diffraction coefficients are plotted in Figure 1.45. The total internal reflections generated inside the wedge give rise to the imaginary part of the diffraction coefficients and the real part blows up at transition angles $\theta_{1,t}$ only ε is equal to 2, see Figure 1.45(a). A uniform asymptotic expansion of $v_1(\rho, \theta)$ around the transition angle $\theta_{n,t}$ is given by [154]

$$v_1(\rho, \theta) = \frac{1}{2} \frac{e^{ik_v\rho + (i\pi/4)}}{\sqrt{2\pi k_v \rho}} \left[f_1(\theta) + (-1)^n K_{n,t} \cot\left(\frac{\theta - \theta_{n,t}}{2}\right) \right.$$

$$\left. + (-1)^n K_{n,t} \tan\left(\frac{\theta - \theta_{n,t}}{4}\right) - (-1)^n \frac{k_{n,t}}{2} \operatorname{sgn}(\theta - \theta_{n,t}) F(\xi) \right] e^{ik_v\rho}$$

$$(1.181a)$$

where $f_1(\theta)$ is obtained by substituting θ for ω in eqn. (1.180a).

$$\operatorname{sgn}(\theta) = -1 \quad \text{for} \quad \theta < 0 \quad \text{and} \quad +1 \quad \text{for} \quad \theta > 0$$

$$\xi = \sqrt{k_v\rho} \left| \sin\frac{\theta_{n,t} - \theta}{2} \right| \quad \text{and} \quad F(\xi) = \frac{2}{\sqrt{\pi}} e^{-i2\xi^2} \int_{(1-i)\xi}^{\infty} e^{-x^2} dx \quad (1.181b)$$

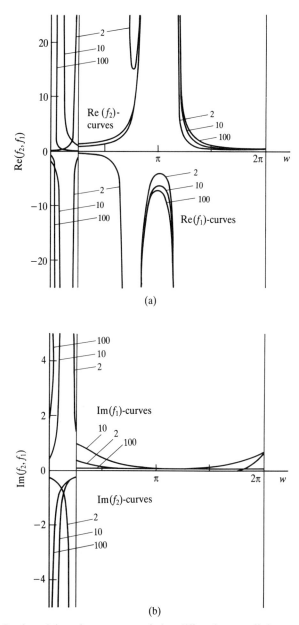

FIGURE 1.45 Real and imaginary parts of the diffraction coefficients: $f_1(\omega)$ in $\theta_d \leq \omega \leq 2\pi$ and $f_2(\omega)$ in $0 \leq \omega \leq \theta_d$ for $\theta_d = 45°$ and $\theta_i = 150°$, $\varepsilon = 2$, 10, and 100: (a) real part; (b) imaginary part [155].

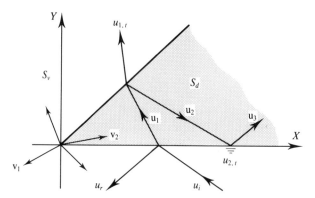

FIGURE 1.46 Diffraction by a dielectric wedge: components of the PO field for $\theta_d = 45°$ and $\theta_i = 150°$ [155].

1.15.6 Some Results and Discussions

The above formulation has been used to do numerical calculations in [155] for some typical cases. Some results are presented here. The wedge angle is $\theta_d = 45°$, are the relative dielectric constants $\varepsilon_r = 2$, 10, and 100, and the angle of incidence $\theta_i = 150°$. Figure 1.46 shows the components of the PO fields for $\theta_d = 45°$ and $\theta^i = 150°$ and Figure 1.47 shows the GO field amplitude patterns at a distance of 5λ from the edge of the dielectric wedge for the above data. The diffracted fields for the same set of data are plotted in Figure 1.48. It shows the wedge region. The PO fields which are the superposition of GO and edge-diffracted fields are plotted in Figure 1.49. Again, the parameter values are the same as before. The field patterns are smooth around the transition regions. It may be noted here that the diffraction coefficients $f_1(\omega)$ and $f_2(\omega)$ are not the same at $\omega = 0$ and $\pi/4$, as shown in Figure 1.45. These are responsible for the abrupt discontinuities in the PO field (Fig. 1.45) at the interfaces of the wedge.

1.16 DIFFRACTION BY DOUBLE AND MULTIPLE WEDGES

So far we have been discussing the cases where only a single edge is involved. Double and multiple edges are very real in many practical applications, e.g. wave propagation loss and trajectory calculations in irregular terrain and clutter modeling from land and sea. The terrain propagation modeling using high-frequency techniques is discussed in Chapter 8 and the clutter modeling in Chapter 9.

It has been shown [156] that conventional UTD is not sufficient for cases where there are multiple wedges and particularly when the edge of one wedge is in the shadow boundary of another and its mechanical application may lead to a lot of error [156]. Various approaches were proposed [157–166] to

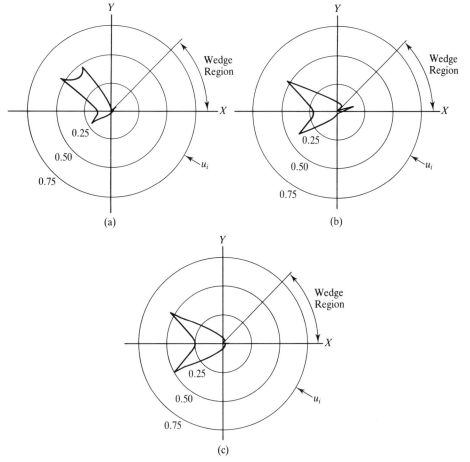

FIGURE 1.47 Geometrical optics field amplitude patterns; observation point 5λ from the edge, $\theta_d = 45°$ and $\theta_i = 150°$; ε_r equal to (a) 2, (b) 10, and (c) 100 [155].

overcome the difficulties. These provide insight into the double wedge diffraction problems but these solutions are valid under restricted conditions. The solution in [158] is restricted to cases where either the source and the two edges, or the two edges and the field point are in the same plane. The treatment and results in [159] are on the assumption of narrow spacing between the two edges. The solution reported in [159] works only for far-field observation points. The treatment using the spectral theory of diffraction in [163] is only capable of handling knife-edges with the source, observation point, and the two edges in the same plane.

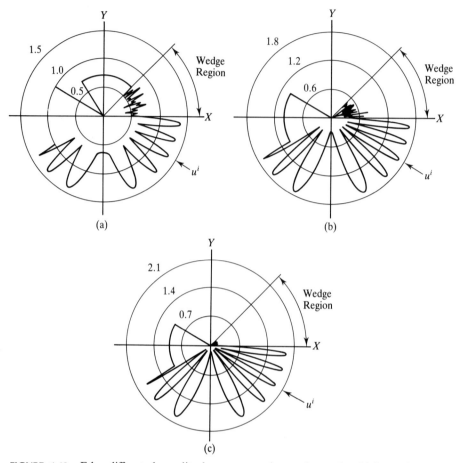

FIGURE 1.48 Edge-diffracted amplitude patterns: observation point 5λ from the edge, $\theta_d = 45°$ and $\theta_i = 150°$; ε_r equal to (a) 2, (b) 10, and (c) 100 [155].

1.16.1 Uniform Double PEC Wedge Diffraction Coefficient

The double diffraction (DD) coefficient for PEC wedges described in [162–164] is simple to apply. The DD coefficient is expressed as a product of two single diffraction coefficients in terms of double Fresnel integrals. It incorporates direct, singly diffracted and doubly diffracted rays. Only the diffraction coefficient for normal incidence to the edge is considered here. For oblique incidence, the coefficient need be extended to a dyadic formulation. The geometry of the double wedge is shown in Figure 1.50. The exterior wedge angles of wedges 1 and 2 are respectively given by $m\pi$ and $n\pi$, where m and n are in general nonintegers.

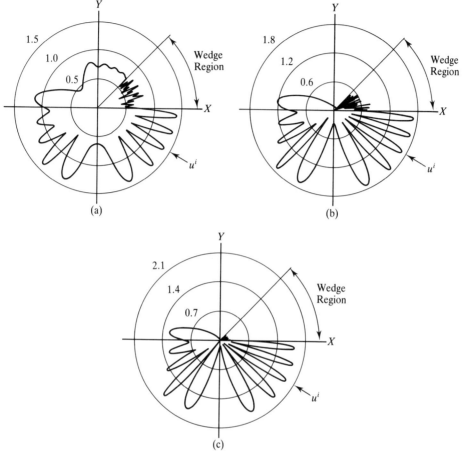

FIGURE 1.49 The PO field amplitude patterns at a distance of 5λ from the edge for $\theta_d = 45°$ and $\theta_i = 150°$; (a) $\varepsilon = 2$, (b) $\varepsilon = 10$, (c) $\varepsilon = 100$ [155].

The total field is given by

$$E_t(P) = E_i(P) + E_{d1}(P) + E_{d2}(P) + E_{dd}(P) \quad (1.182)$$

where E_{d1}, E_{d2}, E_{dd} are respectively the diffracted fields from edges 1 and 2 and the doubly diffracted fields.

The doubly diffracted fields E_{dd} at the observation point are given by

$$E_{dd} = E_i(Q_1) D_1 D_2 \Gamma_{dd}(\rho, \rho', \rho'') e^{-jk\rho} \quad (1.183)$$

$$E_{dd} = E_i(Q_1) \sum_{N=1}^{4} \sum_{M=1}^{4} d_1(N) D(M) \Gamma(\rho, \rho', \rho'') e^{-\rho} \quad (1.184)$$

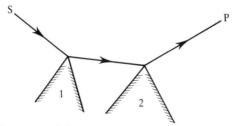

FIGURE 1.50 Geometry of the double wedge consisting of two PEC wedges.

The expressions in eqns. (1.183) and (1.184) are valid when the second wedge is not in the transition region of the first.

The spreading factor A_{dd} is given by

$$A_{dd}(s) = \begin{cases} \dfrac{1}{\sqrt{\rho\rho'}} & \text{for plane and cylindrical wave incidence} \\ \sqrt{\dfrac{\rho''}{\rho\rho'(\rho + \rho' + \rho'')}} & \text{for spherical wave incidence} \end{cases}$$

where ρ, ρ', and ρ'' are respectively the distance of the observation point from the edge of the second wedge, distance between two edges, and the distance from the source to the edge of the wedge (Fig. 1.46).

The solution for double diffraction (DD) is shown [164] to be of the same form as in eqn. (1.159). But some of the 16 products in eqn. (1.182) have to be replaced by more complicated functions. This is decided by the inequalities

$$kL_1 a_1(N) \ll 1 \tag{1.185a}$$

$$kLa_{(M)} \ll 1 \tag{1.185b}$$

The details of the derivation of the dyadic diffraction coefficient based on a Green's function for double diffraction is available in [165]. The double diffracted field is given by

$$E_{dd} = E_i(Q_1) D_1 D_2 A_{dd}(\rho, \rho', \rho'') e^{-jk\rho} \tag{1.186}$$

The double diffraction coefficient and the variables used in double diffraction are shown in Tables 1.4 and 1.5. An efficient method of computing the double Fresnel integral using exponential and trigonometric functions is discussed in the appendix of [164]. An early work to predict the RCS of a double wedge using Keller's GTD is available in [167].

TABLE 1.4 Double Diffraction Coefficient D^2

$$D^2_{s,h} = \frac{-j}{4\pi nm}\sqrt{BC}\sum_{M=1}^{4}\sum_{N=1}^{4}(\mp 1)^{(N+M)}\cdot\exp[-jk\{(B+C\delta^2)a_2(M)+Ca_1(N)\}]$$

$$\cdot[\{I_1+I_2\}\Gamma+\{I_1-I_2\}\tilde{\Gamma}]$$

where the upper sign applies to the soft boundary condition and with

$$\Gamma(N,M) = \cot\left(\frac{\Pi(M)+\beta(M)}{2n}\right)|\sqrt{a_2}|\cdot\cot\left(\frac{\Pi(N)+\gamma_0(N)}{2m}\right)|\sqrt{a_1}i|$$

$$\tilde{\Gamma}(N,M) = -\frac{\chi a_2}{\sqrt{2n}\sin^2\left(\frac{\Pi(M)+\beta(M)}{2n}\right)}\cdot\cot\left(\frac{\Pi(N)+\gamma_0(N)}{2m}\right)|\sqrt{a_1}|$$

$$I_1 = e^{-2jkC\sqrt{a_1}\sqrt{a_2}\delta}\int_{S_{1+}}^{\infty}\int_{S_{2+}}^{\infty}e^{j(s^2+t^2)}\,dt\,ds$$

$$I_2 = e^{+2jkC\sqrt{a_1}\sqrt{a_2}\delta}\int_{S_{1-}}^{\infty}\int_{S_{2-}}^{\infty}e^{j(s^2+t^2)}\,dt\,ds$$

where

$$S_{1\pm} = \psi\sqrt{kC}(\sqrt{a_1}\pm\sqrt{a_2}\delta)$$

$$S_{2\pm} = -\psi\chi\sqrt{\frac{C}{B}}\{\psi\sqrt{kC}(\sqrt{a_1}\pm\sqrt{a_2}\delta)-s\}\delta+|\sqrt{kBa_2}|$$

and $a_1 = a_1(N)$, $a_2 = a_2(M)$.

1.16.2 Diffraction Coefficient of a Double Impedance Wedge

In some practical applications one encounters a double impedance wedge (Fig. 1.51(a)) with surfaces having arbitrary impedances. This problem was treated by Tiberio et al. [169] wherein an extended spectral ray method (ESRM) was employed to derive a high-frequency solution for the field doubly diffracted by two joint edges whose outer faces do not intersect. This solution is uniformly valid for all incidence and observation angles. Herman and Volakis [169] presented high-frequency diffraction coefficients for the double impedance wedge up to third-order terms based on the extended spectral ray method (ESRM). It accounts for all surface wave contributions. The results have been validated in [169] against the already available results [172] for double impedance geometries like the thick impedance half plane and impedance insert in a full plane. These geometries do not intersect each other.

The extended spectral ray method (ESRM) is an extension and generalization of the spectral theory of diffraction (STD) briefly discussed in Section 1.10. The

TABLE 1.5 Variables of Double Diffraction Coefficient D^2

$$\beta = \frac{\rho\rho'}{\rho+\rho'}, \quad C = \frac{(\rho+\rho')\rho''}{\rho+\rho'+\rho''}, \quad \delta = \frac{\rho}{\rho+\rho'},$$

$$\psi = \text{sgn}(\sqrt{a_1}), \quad \chi = \text{sgn}(\sqrt{a_2}).$$

Term M	$\Pi(M)$	$\beta(M)$	$\sqrt{a_2}(M)$
1	$-\pi$	$\phi + \phi'$	$\sqrt{2}\cos\left(\dfrac{2n\pi M_p^- - (\phi+\phi')}{2}\right)$
2	$-\pi$	$\phi - \phi'$	$\sqrt{2}\cos\left(\dfrac{2n\pi M_p^- - (\phi-\phi')}{2}\right)$
3	π	$\phi + \phi'$	$\sqrt{2}\cos\left(\dfrac{2n\pi M_p^+ - (\phi+\phi')}{2}\right)$
4	π	$\phi - \phi'$	$\sqrt{2}\cos\left(\dfrac{2n\pi M_p^+ - (\phi-\phi')}{2}\right)$

Term N	$\Pi(N)$	$\gamma_0(N)$	$\sqrt{a_1}(N)$
1	$-\pi$	$\gamma + \gamma'$	$\sqrt{2}\cos\left(\dfrac{2n\pi N_2^- - (\gamma+\gamma')}{2}\right)$
2	$-\pi$	$\gamma - \gamma'$	$\sqrt{2}\cos\left(\dfrac{2n\pi N_p^- - (\gamma-\gamma')}{2}\right)$
3	π	$\gamma + \gamma'$	$\sqrt{2}\cos\left(\dfrac{2n\pi N_p^+ - (\gamma+\gamma')}{2}\right)$
4	π	$\gamma - \gamma'$	$\sqrt{2}\cos\left(\dfrac{2n\pi N_p^+ - (\gamma-\gamma')}{2}\right)$

where $N_p^\mp$ and $M_p^\mp$ are different from N and M, which are integers varying from 1 to 4.

basis of ESRM is that, as noted by Tiberio [168], the field diffracted by an edge is of the same form and can be interpreted as an infinite sum of inhomogeneous plane waves.

Double Impedance Wedge Solution. The geometry of the double impedance wedge is shown in Figure 1.51. The single diffracted field $u_d^1(\phi, \phi_0)$ from edge Q_1 of the impedance wedge can be written as

$$u_d^1(\phi, \phi_0) = \int_C F(\alpha; \phi, \phi_0) e^{-jk\rho \cos\alpha} \, d\alpha \quad (1.187)$$

1.16 DIFFRACTION BY DOUBLE AND MULTIPLE WEDGES

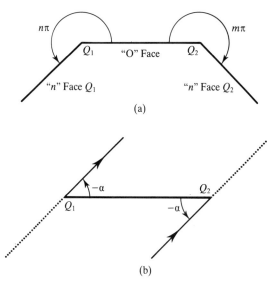

FIGURE 1.51 (a) Geometry of the double impedance wedge. (b) Geometrical interpretation of spectral plane waves.

where
ϕ_0 = angle of incidence
ϕ = angle of observation
$F(\alpha; \phi, \phi_0)$ = a known spectral function
C = the steepest descent path

An asymptotic evaluation of the integral in eqn. (1.179) leads to

$$u_1^d(\phi, \phi_0) \approx D(\phi, \phi_0) \frac{e^{-jk\rho}}{\sqrt{\rho}} \qquad (1.188)$$

Each singly diffracted imhomogeneous wave from Q_1 will be doubly diffracted from Q_2 and the diffraction integral involving the doubly diffracted field is given by

$$u_{21}^d(\phi_2, \phi_0) = \frac{e^{-jk\rho}}{\sqrt{\rho}} e^{-jk\omega \cos \phi_2} \cdot \int_C F(\alpha; 0, \phi_0) D(\phi_2, \alpha) e^{-jk\omega \cos \alpha} \, d\alpha \qquad (1.189)$$

An asymptotic evaluation of eqn. (1.182) will give the expression for the doubly diffracted rays. The reader is referred to the appendix of [169] for details of the uniform asymptotic evaluation. The doubly diffracted field is hence given by

$$u^d_{121}(\phi, \phi_0) = \frac{j2\sqrt{2}e^{-j2k\omega}e^{j3\pi/4}e^{-jk\rho}}{(k\pi)^{3/2}w(nm)^3 n} \cdot \frac{e^{-jk\rho}}{\sqrt{\rho}} \cdot \frac{\Psi_2\left(\dfrac{n\pi}{2} + \pi\right)\Psi\left(\dfrac{m\pi}{2} + \pi\right)}{\Psi\left(\dfrac{n\pi}{2} - \phi_0\right)\Psi\left(\dfrac{n\pi}{2} - \phi\right)\Psi\left(\dfrac{n\pi}{2}\right)}$$

$$\cdot \frac{a_1 a_3^2 a_4}{(a_3 - a_4)}[A\{1 - F_{k\rho}(kwa_1)\} + B\{1 - F_{k\rho}(kwa_2)\}$$

$$+ C\{1 - F_{k\rho}(kwa_3)\}] \cdot [F_{k\rho}(kwa_3) - F_{k\rho}(kwa_4)] \cdot \frac{e^{-jkx}}{4}$$

$$\cdot \left\{\frac{1}{1-\cos\dfrac{\pi-\phi_0}{n}} - \frac{1}{1-\cos\dfrac{\pi+\phi_0}{n}} - \frac{\sin\left(\dfrac{\phi_0}{n}\right)C_{On}(O)}{\cos\dfrac{\pi}{n} - \cos\left(\dfrac{\phi_0}{n}\right)}\right\}$$

$$\cdot \left\{\frac{1}{1-\cos\left(\dfrac{\pi-\phi}{n}\right)} - \frac{1}{1-\cos\left(\dfrac{\pi+\phi}{n}\right)} - \frac{\sin\left(\dfrac{\phi}{n}\right)C_{On}(O)}{\cos\dfrac{\phi}{n} - \cos\dfrac{\phi}{n}}\right\}$$

$$\cdot \left\{\frac{-2\sin\dfrac{\pi}{m}}{\left[1-\cos\dfrac{\pi}{m}\right]^2} - \frac{C_{Om}(O)}{1-\cos\dfrac{\pi}{m}}\right\} \tag{1.190}$$

$$C_{On}(\alpha) = \frac{\sin\left(\dfrac{\theta^+}{n}\right) - \sin\left(\dfrac{\theta^-}{n}\right) - 2\sin\dfrac{\pi}{n}\cos\left(\dfrac{\alpha}{n}\right) + \sin\left(\dfrac{\pi-\theta^+}{n}\right) + \sin\left(\dfrac{\theta^- - \pi}{n}\right)}{4\sin\left(\dfrac{\alpha+\pi-\theta^+}{2n}\right)\sin\left(\dfrac{\alpha+\theta^+}{2n}\right)\cos\left(\dfrac{\alpha+\pi-\theta^-}{2n}\right)\cos\left(\dfrac{\alpha+\theta^-}{2n}\right)} \tag{1.191}$$

It is rather easy to compute the doubly diffracted fields for such a structure. Two special cases of the double wedge whose outer faces do not intersect will be discussed here. These are the thick impedance half plane and the impedance insert in a full plane. A thick half plane is composed of a double wedge structure both wedges having exterior wedge angles with an external angle 1.5π. Figure 1.52 shows the backscattering and bistatic echo width for H-polarization with observation angle θ with thickness as parameter compared with results of a entirely different approach [37]. In Figure 1.53 are shown bistatic scattering patterns for H-polarization with incident angle $\phi' = 60°$ from a perfectly

1.16 DIFFRACTION BY DOUBLE AND MULTIPLE WEDGES 101

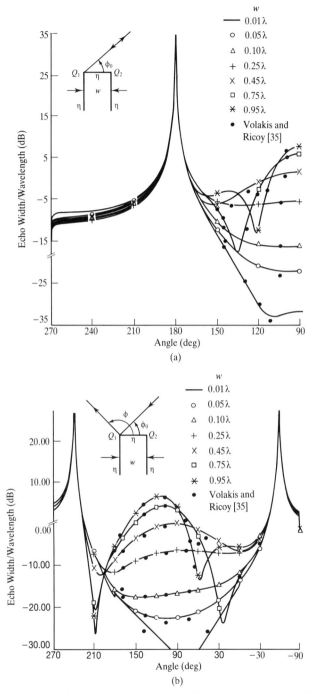

FIGURE 1.52 H-polarization scattering patterns from a perfectly conducting thick half plane of varying width: (a) backscatter and (b) bistatic $\phi_0 = 60°$ [35].

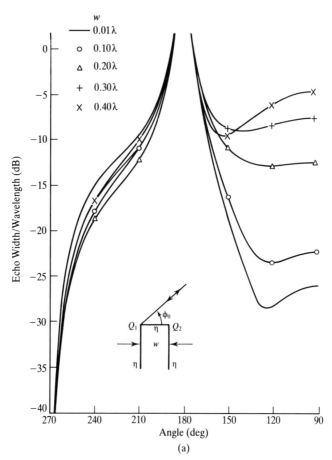

FIGURE 1.53 E-plane backscattering patterns from thick impedance half plane with variable widths: (a) $\eta = 2 - j2$; (b) $\eta = 2 + j2$ [36].

conducting thick half plane with width as a parameter. The results are again compared with an entirely different approach and they agree very well even though for H-polarization the multiply diffracted rays contribute significantly. The thickness of the half plane varies from 0.95λ down to 0.01λ. Figure 1.54 shows the bistatic patterns for an impedance insert: (a) H-polarization, $\phi_0 = 1°$; the values of insert impedances as parameter are $\eta_0 = -j0.25, +j0.25$ and $+0.25$; the width of the strip is 1.6λ; outer impedance $\eta_1 = 0$; and (b) E-polarization; insert impedance is $\eta_0 = 2 - j1$; the strip width $= 0.5\lambda$. The outer impedance is the parameter. Figure 1.55 shows a typical plot of a backscatter pattern of a double wedge with varying wedge angles ($m = n$). The outer faces are PEC and the common face has an impedance of $\eta = 0.5 - j0.25$; width $= 0.5\lambda$; (a) is for E-polarization (b) is for H-polarization. The three curves are for $m = n = 1.5, 1.25,$ and 1.0.

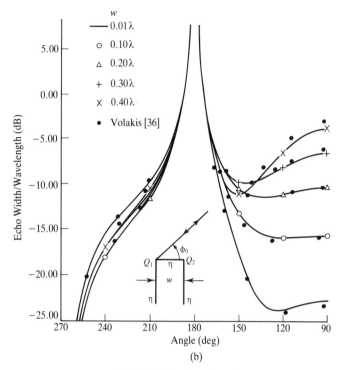

FIGURE 1.53 (*continued*).

1.17 HYBRID TECHNIQUES

Each technique has its own advantages and disadvantages. Hybrid techniques are combinations of two or more techniques to make use of the advantages of each approach in solving complex problems effectively. For example, the advantages of the method of moments and GTD can be combined to solve the scattering problem of a wedge. The MM currents are postulated near the edge and the GTD currents are matched at a distance from the edge on the surface. This reduces the memory requirement as well as computer processing time and provides an efficient and cost-effective solution in some problems. Therefore, at least in principle, using an appropriate combination of techniques complicated problems may be amenable to efficient solutions. A discussion on hybrid techniques has been presented in a recent book [13, section 2.9]. An apparent potential of this method is that one can numerically obtain the diffraction coefficient of configurations for which a canonical solution and hence a diffraction coefficient is not available.

1.17.1 Method of Moments Combined with GTD

Two different approaches of combining the moment method and GTD have been presented in the literature almost at the same time in 1975. In one method

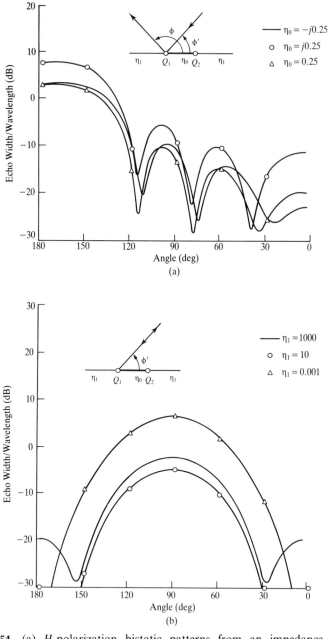

FIGURE 1.54 (a) H-polarization bistatic patterns from an impedance insert with incidence at $\phi_0 = 1°$. The insert impedances are $\eta_0 = -j0.25, j0.25$, and 0.25. The outer impedance is $\eta_1 = 0$ and the width of the strip is 1.6b λ. (b) E-polarization backscatter patterns from an impedance insert with varying outer impedances. The insert impedance is held constant at $\eta_0 = 2 - j1.0$; the outer impedances are $\eta_1 = 1000, 10, 0.001$; the strip width is 0.5λ [36].

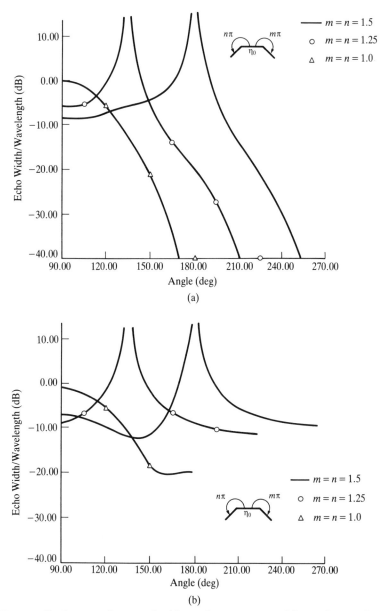

FIGURE 1.55 Backscatter from a double wedge structure with varying wedge angle ($m = n$). The outer faces are perfectly conducting and the common face has impedance $\eta = 0.5 - j0.25$ and width 0.5λ: (a) E-polarization; (b) H-polarization [170].

[44], the impedance matrix of a structure obtained using MOM is modified using GTD to accomodate any change in the shape and size of the existing structure. The change in the structure might be due to an attachment. As an example, the usual impedance matrix which characterizes a wire antenna is modified such that a metallic object or a discontinuity on the object is accounted for. We call this technique hybrid method I. In the other approach [45], as mentioned before, the MM currents are postulated near the edge or edges on the structure and the GTD fields are matched at a distance from the edge or edges. Thus one can go far beyond the capability of a single method to handle complex structures. We designate this hybrid technique II.

(A) Hybrid Technique I: Modified Impedance Matrix of a Body

Using the MM the generalized impedance matrix is given by [44]

$$Z_{mn} = W_m, L(J_n) \qquad (1.192)$$

It can be shown that with a and b as complex scalars [44],

$$\langle \mathbf{J}, a\mathbf{E}_1 + b\mathbf{E}_2 \rangle = \langle a\mathbf{J}, \mathbf{E}_1 + b\mathbf{J}, \mathbf{E}_2 \rangle \qquad (1.193)$$

If we let $a\mathbf{E}_1$ represent $L(J_n)$ in eqn. (1.192) as in a strictly MM solution. Then $b\mathbf{E}_2$ in (1.193) represents an additional field contribution to Z_{mn}. This is also due to J_n but due to some physical process not already accounted for in a strictly MM solution.

The modified impedance matrix may be written as

$$Z'_{mn} = Z_{mn} + Z^g_{mn} \qquad (1.194)$$

where

$$Z^g_{mn} = \langle W_m, bL(J_n) \rangle \qquad (1.195)$$

Hence, the relationship among voltage, current, and impedance is modified to

$$[Z'](I') = (V) \qquad (1.196)$$

where $[Z']$ = generalized impedance matrix modified according to the physical process not accounted for by strictly MM solution.
(I') = Current as modified by the physical process.

Modification of (V) is neglected since it is considered insignificant in the examples we intend to discuss in this book.

This approach has been tested to cases of monopole on a conducting wedge and monopole at the center of a circular disk. These examples will be discussed in Chapter 6 on high-frequency treatment of antennas on structures.

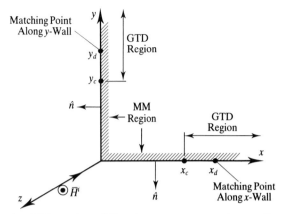

FIGURE 1.56 Basic PEC wedge diffraction geometry and coordinate system, and MM-GTD solution applied to the PEC wedge [45].

(B) Hybrid Technique II: Diffraction by a 90° Dielectric Wedge

This technique based on [45] has already been described [13, 171, section 2.9] in a recent text. Figure 1.56 shows the wedge geometry and the MM currents around the edge and GTD matching points.

The GO incident and reflected fields are expressed as

$$H_z^i = e^{ik\rho \cos(\psi - \psi_0)}, \quad \text{for } |\psi - \psi_0| < 180° \tag{1.197a}$$

$$H_z^r = e^{ik\rho \cos(\psi - \psi_0)}, \quad \text{for } |\psi - \psi_0| < 180° \tag{1.197b}$$

The field diffracted by the wedge can be expressed as

$$H_z^d \sim [D(\psi - \psi_0, n) + D(\psi + \psi_0, n)] \frac{e^{-ik\rho}}{\sqrt{\rho}} \quad \text{for } \rho > \lambda/4 \tag{1.198}$$

where

$$D(\phi, n) = \frac{e^{-i\pi/4}}{\sqrt{2\pi k}} \frac{\frac{1}{n \sin \frac{\pi}{n}}}{\cos \frac{\pi}{n} - \cos \frac{\phi}{n}} \tag{1.199}$$

The total magnetic field can be expressed as

$$H_z^t \sim H_z^i + H_z^r + C \frac{e^{-ik\rho}}{\sqrt{\rho}} \tag{1.200}$$

Using the boundary condition, the total surface current along the wedge walls is given by

$$\mathbf{J}^{\text{GTD}} = (\hat{n} \times \hat{z})[H_z^i + H_z^r + H_z^d] \simeq \mathbf{J}^i + \mathbf{J}^r + (\hat{n} \times \hat{z})C \frac{e^{-ik\rho}}{\sqrt{\rho}}, \quad \text{for } \rho \leq \lambda/4 \quad (1.201)$$

where C is an unknown constant involving a diffraction coefficient of the wedge. If the diffraction coefficient in an example is unknown it can be numerically determined using the method described below.

The integral equation to be solved for this purpose is the well-known magnetic field integral equation (MFIE) given by [44]

$$\mathbf{J} = -H_z^i + \hat{z} \cdot \nabla \times \int_S \mathbf{J} \frac{H_0^{(2)}(k|\rho - \rho'|)}{4i} dl' \quad (2.202)$$

where S is the PEC surface on which the current has to be solved for.

For the 90° wedge shown in Figure 1.56(a), the total surface current density is given by

$$\mathbf{J} = \begin{cases} J_y^{\text{GTD}}, & y_c \leq y \leq \infty, \quad x = 0 \\ J_y^{\text{MM}}, & 0 \leq y \leq y_c, \quad x = 0 \\ J_x^{\text{MM}}, & 0 \leq x \leq x_c, \quad y = 0 \\ J_x^{\text{GTD}}, & x_c \leq x \leq \infty, \quad y = 0 \end{cases} \quad (1.203)$$

The MM current can be expressed as a superposition of orthogonal pulse functions $P_m(l - l_m)$ with weight α_m as

$$J^{\text{MM}} = \sum_{m=1}^{N} \alpha_m P_m(l - l_m) \quad (1.204)$$

By doing point matching at the middle point of all the N pulse currents and using eqn. (1.202) one obtains a set of simultaneous equations given by

$$\sum_{m=1}^{N} l_{mn}\alpha_n + L_n\left(J_x^i + J_x^r + D_x \frac{e^{-ikx}}{\sqrt{x}}\right)\bigg|_{y=0}$$
$$+ L_n\left(J_y^i + J_y^r + D_y \frac{e^{-iky}}{\sqrt{y}}\right)\bigg|_{x=0} = g_n \quad \text{for } 1 \leq n \leq N \quad (1.205)$$

This can be written as

$$\sum_{m=1}^{N} l_{mn}\alpha_m + D_x L_n\left(\frac{e^{-ikx}}{\sqrt{x}}\right) + D_y L_n\left(\frac{e^{-iky}}{\sqrt{y}}\right) = g'_n, \quad 1 \leq n \leq N \quad (1.206)$$

where

$$g'_n = -H^i_z(x_n, y_n) - L_n(J'_x + J^r_x) - L_n(J^i_y + J^r_y) \quad (1.207a)$$

$$L_n\left(\frac{e^{-ikx}}{\sqrt{x}}\right) = -\int_{x_c}^{\infty} \frac{ik}{4} \frac{e^{-ikx'}}{\sqrt{x'}} H^{(2)}_1(k|\rho_n - \rho'|) \, dx' \quad (1.207b)$$

$$L_n\left(\frac{e^{-iky}}{\sqrt{y}}\right) = -\int_{y_c}^{\infty} \frac{ik}{4} \frac{e^{-iky'}}{\sqrt{y'}} H^{(2)}_1(k|\rho_n - \rho'|) \, dy' \quad (1.207c)$$

$$L_n(J^i_x + J^r_x) = \begin{cases} -2\int_{x_c}^{\infty} \frac{ik}{4} e^{jkx'\cos\psi_0} H^{(2)}_1(k|\rho_n - \rho'|) \, dx', & 0 \leq \psi_0 \geq 180° \\ 0 & \text{otherwise} \end{cases}$$

$$L_n(J^i_y + J^r_y) = \begin{cases} -2\int_{\infty}^{y_c} \frac{ik}{4} e^{jky'\sin\psi_0} H^{(2)}_1(k|\rho - \rho'|) \, dy', & \text{for } 0 \leq \psi_0 \geq 180° \\ \text{otherwise } 0 \end{cases}$$

In actual practise the integrals can be evaluated over a finite limit (typically a few wavelengths) since the current decays away from the edge. The techniques have been applied to solve a number of problems, namely, scattering from a 90° PEC wedge, square, and circular cylinders and the radiation pattern of an axial waveguide mounted on a finite cylinder. Some of these applications and the results are described in Chapter 6, where the high-frequency analysis of antennas on structures is presented.

1.17.2 Diffraction by a 90° Dielectric Wedge

The problem of scattering by a 90° dielectric wedge can be solved by hybrid method II on the same lines as described above for the PEC case. The basic formulation is available in [45]. The solution for the PEC case can be obtained by equating the relative dielectric constant to large value. Table 1.6 shows the variation of D_x and D_y with aspect angle.

Figure 1.57 respectively show the amplitude and phase of the surface current density for a typical dielectric wedge scattering.

1.18 SURFACE WAVE DIFFRACTION

1.18.1 Introduction

As we discussed earlier, the GO predicts zero fields in the shadow region. In practice, it is established that the fields in the shadow region are nonzero even if the target is smooth and does not have any edges or discontinuities. The shadow region gets a small fraction of the incident energy intercepted by the

TABLE 1.6 Variation of D_x and D_y with Aspect Angle

($f = 10$ GHz; $N = 10$; $\varepsilon_r = 4.0$; $D = X_c$; $X_c = 2\lambda$; $l = 6\lambda$)

Aspect Angle (deg)	D_x Re	D_x Im	D_y Re	D_y Im
0.0	$0.261E - 03$	$-0.249E - 02$	$-0.414E - 04$	$0.125E - 01$
22.5	$0.901E - 04$	$-0.179E - 02$	$0.117E - 02$	$0.124E - 01$
45.0	$-0.815E - 03$	$-0.841E - 03$	$0.112E - 01$	$0.309E - 02$
67.5	$0.295E - 02$	$0.465E - 03$	$0.991E - 03$	$-0.521E - 02$
90.0	$0.966E - 02$	$0.291E - 02$	$0.157E - 01$	$0.462E - 02$
112.5	$0.295E - 02$	$0.465E - 03$	$0.991E - 03$	$-0.521E - 02$
135.0	$-0.815E - 03$	$-0.841E - 03$	$0.112E - 01$	$0.309E - 02$
157.5	$-0.910E - 04$	$-0.179E - 02$	$0.117E - 02$	$0.124E - 01$
180.0	$-0.261E - 04$	$-0.249E - 02$	$-0.414E - 02$	$0.124E - 01$
202.5	$-0.901E - 04$	$-0.179E - 02$	$0.117E - 02$	$0.124E - 01$
225.0	$-0.815E - 03$	$-0.841E - 03$	$0.112E - 01$	$0.309E - 02$
247.0	$0.294E - 02$	$0.465E - 03$	$0.991E - 03$	$-0.521E - 02$
270.0	$0.996E - 02$	$0.291E - 02$	$-0.158E - 01$	$0.462E - 02$

target, by virtue of the surface waves that creep along the surface of the object, and continuously shades as it creeps round and round till it dies out. The estimation of surface and creeping waves is important in many applications, e.g. in situations where an antenna is located on or near the surface of conducting bodies such as aircraft, ships, and couplings between antennas which are along line-of-sight as well as not along line-of-sight on such bodies. The problem of finding the optimum position of a receiving antenna in the presence of other transmitting antennas on such structures is of great concern. Even an apparently weak field may cause serious interference with presently available sensitive receivers. It is found that at high frequencies, the surface wave coupling paths become extremely important. Doubt has been raised whether these have been adequately taken care of in the currently available interantenna coupling models, such as those used in the codes IEMCAP [173] and AAPG [174]. A striking example is that Gennelo and Pesta [175] have shown a large discrepancy, as large as 35–40 dB at 18 GHz between the experimentally observed and theoretically predicted values of the isolation between two horn antennas on an F-16 aircraft. Their conclusion is that if a valid EMC analysis has to be done on an aircraft with EHF or SHF terminals on board, the shading algorithm would call for a modification.

The major limitations of such isolation predictions are:

1. Inadequate source representations which are compatible with HF asymptotic methods, as an example, accurate near-field expressions are not available for a slot array which is flush mounted on an aircraft and radiating in the presence of the aircraft.

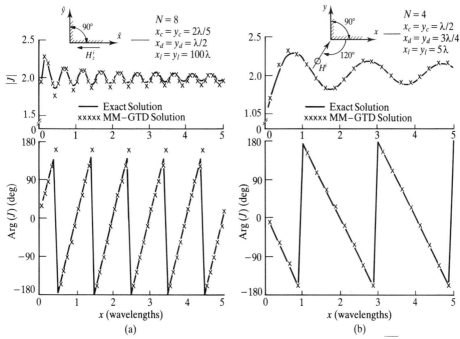

FIGURE 1.57 Variation of the amplitude and phase of the surface current density along X-wall [45]; $N = 4$; $X_c = \lambda/2$; $D = 3\lambda/4$; $l = 5\lambda$; $\psi_0 = 180°$: (a) amplitude; (b) phase.

2. The recent advances in the uniform theory of diffraction developed by Pathak and Wang [178] in predicting the field due to a small dipole moment on a curved surface is limited to a doubly curved surface whose one principal radius of curvature is very much larger than the other. This means the analysis is restricted to only cylindrical or nearly cylindrical surfaces. A theory which is applicable for any relative values of the two principal radii of curvature is needed.

3. Existing models are inadequate to predict how a structure on or above which a given antenna is radiating causes depolarization of the electromagnetic energy.

These limitations in the present state of the art make the available models inaccurate for shading loss prediction and leave the area open to further investigations.

This section aims to survey the available models, to characterize surface waves, to estimate the shading loss between two locations on a curved surface, and to discuss creeping waves on dielectric and dielectric-coated conducting surfaces. The applications of the theory will be presented in Chapter 6.

1.18.2 A Brief Review of Analytical Models

The task of finding the expression for the surface fields due to sources on smooth objects is essentially the asymptotic evaluation of the integrals involved in the field expressions for the canonical problems. The canonical problems are sources on conducting spheres, infinite cylinders, and infinite cones. To describe the phenomenon of a part of the energy trespassing into the shadow region when an incident ray strikes a smooth conducting surface, Levy and Keller [177] postulated surface-diffracted rays and obtained their surface diffraction coefficients and also their attenuation constants. This was possible by the evaluation of the surface diffraction integral asymptotically and comparing the result with the expression for the diffracted field involving the surface wave diffraction coefficient. A similar treatment was also presented by Franz and Klante [178].

Hasserjian and Ishimaru [179] obtained an approximate expression for the magnetic field induced by slots on a conducting circular cylinder of infinite length and a large electrical radius. The approximations are derived from the formal solution of the cylinder problem and are valid for all distances from the slot. The field expressions for an electrically small slot dipole at any orientation are expressed in terms of the field of the slot on a flat conducting sheet multiplied by a curvature term. The curvature term is a function of ray path and radius of curvature and hence renders the analysis applicable to other convex surfaces. The analysis has been extended to half-wavelength transverse and longitudinal slots on a cylinder. They retained terms up to the lowest order in $(1/kt)$ and $1/m$ in their work. Here, t refers to the path length between the source and receiving points and $m = (k\rho_g/2)^{1/3}$ where ρ_g is the surface radius of curvature in the direction of the ray. They compared the amplitude and phase of the magnetic field varying along the axial distance from the axial and transverse $\lambda/2$ slots on the cylinder.

A similar analysis was reported earlier [180] by Wait to obtain the surface field of a large conducting sphere excited by a short thin dipole.

However, these investigations do not consider surface torsional effects on a cylinder when the transmitting and receiving points have different values of z and ϕ and hence are not accurate for field and/or coupling calculations for arbitrary entry and receiving points of the ray paths on the cylinder. Hwang and Kouyoumjian [181] were the first to report consideration of torsion in such an analysis. They predicted a GTD solution for a surface current excited by a slot in a conducting convex cylinder. The GTD expressions for the same problem of a slot on a cylinder are considerably improved by adding higher-order terms. To be accurate, Chang, Felsen, and Hessel [182] and Chan et al. [183] considered terms up to $1/(kt)^2$ and an additional term in $1/m$ in the problem of excitation of a slot on a cone. On the contrary, Lee [184] developed a Green's function for a cone and Lee and Naini [185] developed an approximate asymptotic Green's function for the surface magnetic field due to a magnetic dipole on a general convex cylinder. In this work, an approximate solution for the surface field was constructed via a heuristic modification of the canonical solution for the surface field of a slot in a conducting sphere. This modification is nothing but the addition of a term to the sphere solution such that it provides accurate numerical results for conical and cylindrical geometries as well. The solution has been presented in a form compatible with ray interpretation and is based on classical work of V. A. Fock [186] and the current GTD recipes. This was applied to calculate the mutual admittance between two slots on a cone and good agreement with experimental results was obtained. The problem of calculation of mutual impedance between antennas is discussed in Chapter 7.

It was found from the analysis [186–189] that the effect of torsion is confined to the source location and depends on the direction of the surface ray with respect to the principal directions of the surface ray path away from the source. Initially in [181–188], the tangential magnetic current moments (or dipole) were decomposed into circumferential and axial directions and ray fields were separately determined. It was later discovered [190, 191] that this decomposition is not necessary and a uniform GTD solution was formulated directly in the ray-fixed coordinate system.

But the solution of Mittra and Safavi-Naini [193] for the problem of radiation from tangential magnetic dipoles on a perfectly conducting circular cylinder does reduce to the geometrical optics (GO) field in the deep shadow region and in the nonprincipal planes. But since their solution does not identify the torsion to be present only on the source, it can be extended to almost cylindrical surfaces such as long thin cone. Pathak and Wang [192] proposed a uniform GTD solution for sources on an arbitrary convex surface and presented the expressions for surface fields, the cylinder being a special case. Also, both the formulations due to Pathak and Wang [192] and Hasserjian and Ishimaru [179] yield almost identical results from an engineer's point of view. However, the authors claim that it is more complete and accurate (since the higher-order terms are considered) than the solution presented [181] when the observation point is in the deep shadow and caustic regions.

Hasserjian and Ishimaru Model. In this model [179], appropriate expressions are obtained for the magnetic field induced by slots on conducting circular cylinders of infinite length and of radius large compared to the wavelength. The approximations are derived from the normal solutions of the cylinder problem and are valid for all distances from the slot. The expressions for an arbitrarily oriented short slot dipole on a convex surface is obtained as a product of the field of the slot on a conducting sheet and a curvature term. The curvature term is a function of the *ray path length* and the radius of curvature of the surface.

The potential function Ψ_1 for the slot is given by

$$\Psi_1 = [P\varepsilon \exp(-iR_1)/(2\pi R_1)] \cdot F(y) \quad (1.208)$$

where $P = Vkdl$, the moment of the magnetic dipole
ε = the dielectric constant of the medium
$R_1 = \sqrt{(a\phi)^2 + z^2}$, the path length spanned
 = $a\phi$, for the equatorial plane
$F(y)$ = a correction term, which accounts for the curvature of the surface and is a function of R and the radius of curvature. $F(y)$ is a varying function.

$F(y)$ is shown [181] to be given by

$$F(y) = 1 - \sqrt{i\pi/2}y' + i7y'^3/1024 + \cdots \quad \text{for } \tau \text{ small}$$

$$= \sqrt{\pi/i}(y'/2)^{1/3} \sum_{m=1}^{\infty} \exp - i(y'/\sqrt{2})^{2/3} t_m m \ldots \quad \text{for } \tau \text{ large}$$

where the t_m are the roots of the derivative of the Airy function $W'(t) = 0$; $y = x_1^{3/2}/(ka)$; $x_1 = ka\phi$.

The magnetic field components are given by

$$H_z \simeq \frac{iP\omega\varepsilon}{2\pi} \frac{\delta^2}{\delta z_1^2} + 1 \frac{e^{-iR_1}}{R_1} F(y) \quad (1.209a)$$

$$H_\phi \simeq \frac{iP\omega\varepsilon}{2\pi} \frac{\delta^2}{\delta x_1 \delta z_1} \frac{e^{-iR_1}}{R_1} F(y) \quad (1.209b)$$

Hence, eqn. (209a,b) can be used to calculate the fields of the short dipole on a cylinder.

Keller's Surface Wave Theory. Levy and Keller [177] evaluated asymptotically the solution for electromagnetic scattering by a conducting cylinder and showed that the vector diffracted creeping ray field at P in terms of the incident field at Q is given by (see Fig. 1.58)

$$\mathbf{E}^d(P) = [\hat{n}_2\hat{n}_1\hat{\vartheta}(1, 2) + \hat{b}_2\hat{b}_1\hat{u}(1, 2)]\mathbf{E}^i(1) \cdot \sqrt{\frac{\rho}{s(\rho + s)}} \cdot \exp(-jks) \quad (1.210)$$

in which $\hat{\vartheta}(1, 2)$ and $\hat{u}(1, 2)$ are equal to

$$\sqrt{d\eta_1/d\eta_2} \exp(-jkt) \cdot \sum D_{ph,s}(1) D_{ph,s}(2) \cdot \exp\left[-\int_1^2 \alpha_p(t') dt'\right] \quad (1.211)$$

where $(\hat{n}_1, \hat{t}_1, \hat{b}_1)$ and $(\hat{n}_1, \hat{t}_1, \hat{b}_1)$ are respectively the mutually orthogonal directions at initial and final points on the surface and ρ is the caustic distance.

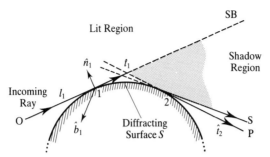

FIGURE 1.58 Diffraction by a curved surface.

$s = Q_2 P$, $\sqrt{d\eta_1/d\eta_2}$ is the divergence factor. $D_{ph,s}(i)$ are the surface diffraction coefficients for hard and soft cases respectively; $\alpha_p(t)$ is the complex attenuation constant given by (if the radius of curvature a at the different points on the path varies slowly). The surface diffraction coefficients and the complex attenuation constants are given by

$$D_p^2 \approx \exp(-i\pi/12)(ka)^{1/3}/\sqrt{kd_p} \tag{1.212a}$$

and

$$\alpha_p \approx [(ka)^{1/3}/2a]3\pi/4(4p + N)^{2/3} \exp(i\pi/6) \tag{1.212b}$$

$n = 1$ and 3 for hard and soft cases. d depends on the mode number; $d_0 = 1.083$ and $d_1 = 0.555$ for the hard case; for the soft case $d_0 = 0.645$ and $d_1 = 0.49$; 0 and 1 stand for zero and first order modes; a is the radius of curvature of a point on the path in the principal plane of interest. If the surface is cylindrical then the radius of curvature is given by $a/\sin^2 \delta$ where δ is the smaller angle between the axis of the cylinder and the direction of entry at a point on the path. The corrections for D and α_p for the cases where the surface has varying curvature have also been obtained and interested readers may refer to Levy and Keller [177] and Franz and Klante [178].

Uniform Theory of Curved Surface Diffraction. Pathak and Wang [192] presented an asymptotic solution for the electromagnetic currents which are induced on an electrically large perfectly conducting smooth convex surface by an infinitesimal magnetic current moment or an electric current moment situated off the surface.

The canonical solution to obtain explicit expressions for the surface fields $d\mathbf{H}_m$ and $d\mathbf{E}_m$ due to a source dp_m on an arbitrarily convex surface are given by (Fig. 1.59)

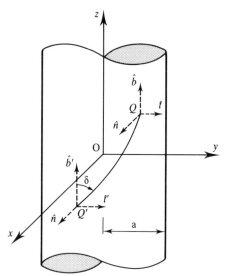

FIGURE 1.59 Diffraction of a surface wave by the curved surface of a cylinder.

$$d\bar{\bar{H}}_m(Q/Q') = \frac{-jk}{4\pi} d\mathbf{p}_m(Q') \Big\{ 2Y_0(\hat{b}'\hat{b}) \Big[\Big(1 - \frac{j}{kt}\Big) \tilde{V}(\xi)$$
$$+ D^2 \Big(\frac{j}{kt}\Big)^2 \Big(\bigwedge_s \tilde{U}(\xi) + \bigwedge_c \tilde{V}(\xi)\Big) + \tilde{T}_0^2 \frac{j}{kt} (\tilde{U}(\xi) - \tilde{V}(\xi)) \Big]$$
$$+ \hat{t}'\hat{t} \Big[D^2 \frac{j}{kt} \tilde{V}(\xi) + \frac{j}{kt} \tilde{U}(\xi) - 2\Big(\frac{j}{kt}\Big)^2 \Big(\bigwedge_s \tilde{U}(\xi) + \bigwedge_c \tilde{V}(\xi)\Big) \Big]$$
$$+ (\hat{t}'\hat{b} + \hat{b}'\hat{t}) \Big[\frac{j}{kt} \tilde{T}_0 (\tilde{U}(\xi) - \tilde{V}(\xi)) \Big] \Big\} DG_0(kt) \qquad (1.213a)$$

and

$$d\tilde{E}_m(Q/Q') = \frac{-jk}{4\pi} d\bar{\mathbf{p}}_m(Q') \Big\{ 2(\hat{b}'\hat{n}) \Big[\Big(1 - \frac{j}{kt}\Big) \tilde{V}(\xi) + \tilde{T}_0^2 \frac{j}{kt} (\tilde{U}(\xi) - \tilde{V}(\xi)) \Big]$$
$$+ \hat{t}'\hat{n} \Big[\tilde{T}_0 \frac{j}{kt} (\tilde{U}(\xi) - \tilde{V}(\xi)) \Big] \Big\} DG_0(kt) \qquad (1.213b)$$

The surface fields $d\mathbf{H}_e$ and $d\mathbf{E}_e$ due to dp_e on an arbitrary convex surface are given by

$$d\tilde{E}_e(Q/Q') = \frac{jk}{4\pi} d\bar{\mathbf{p}}_e(Q') \Big\{ 2Z_0 \hat{n}'\hat{n} \Big[\tilde{V}(\xi) - \frac{j}{kt} \tilde{V}(\xi) + \Big(\frac{j}{kt}\Big)^2 \Big(\bigwedge_s \tilde{V}(\xi) + \bigwedge_c \tilde{U}(\xi)\Big)$$
$$+ \tilde{T}_0^2 \frac{j}{kt} (\tilde{U}(\xi) - \tilde{V}(\xi)) \Big] \Big\} DG_0(kt) \qquad (1.214a)$$

and

$$d\bar{\mathbf{H}}_e(Q/Q') = \frac{-jk}{4\pi} d\bar{\mathbf{p}}_e(Q') \left\{ 2(\hat{n}'\hat{b}) \left[\left(1 - \frac{j}{kt}\right) \tilde{V}(\xi) + \tilde{T}_0^2 \frac{j}{kt} (\tilde{U}(\xi) - \tilde{V}(\xi)) \right] \right.$$
$$\left. + \hat{n}'\hat{t} \left[\tilde{T}_0 \frac{j}{kt} (\tilde{U}(\xi) - \tilde{V}(\xi)) \right] \right\} DG_0(kt) \quad (1.214b)$$

where $d\mathbf{p}_m$ = magnetic dipole moment
$d\mathbf{p}_e$ = electric dipole moment
t = path length = $\sqrt{a(z-z')^2 + (z-z')^2}$
$\xi = \sqrt{a^2(\phi - \phi')^2 + (z-z')^2(k/2\rho_g)^2}$
$\tilde{T}_0 = \cot \delta$
δ = angle of entry ($=\pi/2$ for equatorial excitation)

The coordinates of the points Q and Q' are (a, ϕ, z) and (a, ϕ', z) and

$$\rho_g = 1 - \frac{R_2(Q')}{R_1(Q')} \frac{R_2(Q)}{R_1(Q)}$$

where $R(Q)$ and $R(Q')$ are the principal radii of curvature of the surface along which the wave creeps round and round.

The generalized Fock integrals $U(\xi)$ and $V(\xi)$ for the arbitrary convex surface are given by

$$\tilde{U}(\xi) = \left[\frac{kt}{2m(Q')m(Q)\xi} \right]^{3/2} \tilde{U}(\xi) \quad (1.215a)$$

and

$$\tilde{V}(\xi) = \left[\frac{kt}{2m(Q')m(Q)\xi} \right]^{3/2} \tilde{V}(\xi) \quad (1.215b)$$

in which the Fock functions $U(\xi)$ and $V(\xi)$ are expressed as

$$U(\xi) = \frac{\xi^{3/2} e^{i3\pi/4}}{\sqrt{\pi}} \int_{\infty e^{-j2\pi/3}}^{\infty} d\tau \frac{W_2'(\tau)}{W_2(\tau)} e^{-i\xi\tau} \quad (1.216a)$$

and

$$V(\xi) = \int_{\infty e^{-i2\pi/3}}^{\infty} d\tau \frac{W_2(\tau)}{W_2'(\tau)} e^{-i\xi\tau} \quad (1.216b)$$

Expressions for Field Components for Arbitrary Positions of Transmitter and Receiver. Let the coordinates of the transmitting and receiving points be (a, ϕ, z) and (a, ϕ', z') on the surface of the smooth cylinder. The transmitting and receiving dipole moments are situated at the two points.

Then, the t and z are given by

$$t = \sqrt{a^2(\phi - \phi')^2 + (z - z')^2}$$
$$\xi = \sqrt{a^2(\phi - \phi)^2 + (z - z')^2(k/2\rho_g)^2}$$

The angle of entry with cylinder axis is given by

$$\delta = \tan^{-1}[a(\phi - \phi')/(z - z')]$$

The torsion factor $\tilde{T}_0 = \cot \delta$.

The field components using Keller's approach are already discussed and are already given by eqns. (1.210) and (1.211).

With a little algebra, the different co- and cross-polarized components as obtained from eqn. (1.214a,b) are given by

$$H_{zz'} = -\frac{ik}{2\pi Z_0} d\mathbf{p}_m \{(1 - ikt)V(\xi) + t(i\{kt\} + \tilde{T}_0^2(U(\xi) - V(\xi))\} \frac{\exp(-ikt)}{\sqrt{t}}$$
(1.217a)

$$H_{\phi\phi'} = \frac{ik}{2\pi Z_0} d\mathbf{p}_m \{i/kV(\xi) + i/ktU(\xi) - 2(i/kt)^2 V(\xi)\} \frac{\exp(-ikt)}{\sqrt{t}} \quad (1.217b)$$

$$H^{\phi'z} = \frac{i \exp(-ikt)}{2\pi kt} d\mathbf{p}_m \tilde{T}_0(U(\xi) - V(\xi)) \quad (1.217c)$$

$$E_{z'\rho} = -\frac{ik}{2\pi} d\mathbf{p}_m \left\{ \left(1 - \frac{i}{kt}\right) V(\xi) + \tilde{T}_0^2 i/kt(U(\xi) - V(\xi)) \right\} \exp(-ikt)/\sqrt{t}$$
(1.217d)

For electric and magnetic dipole moments the field components on the surface are given by

(a) Due to Electric Dipole Moment

$$E_{\rho\rho'} = -\frac{ik}{4\pi} d\mathbf{p}_e \left\{ V(\xi) - \frac{i}{kt} V(\xi) + \left(\frac{1}{kt}\right)^2 U(\xi) \right.$$
$$\left. + \tilde{T}_0^2 i/kt(U(\xi) - V(\xi)) \right\} \exp(-ikt)/\sqrt{t} \quad (1.217e)$$

$$H_{\rho'z} = -\frac{ik}{4\pi} d\mathbf{p}_e \left\{ \left(1 - \frac{i}{kt}\right) V(\xi) + \tilde{T}_0^2 i/kt(U(\xi) - V(\xi)) \right\} \exp(-ikt)/\sqrt{t}$$
(1.217f)

and
$$H_{\rho'\phi} = \frac{\tilde{T}_0^2 i}{kt} d\mathbf{p}_e \{U(\xi) - V(\xi)\} \exp(-ikt)/\sqrt{t} \tag{1.217g}$$

By reciprocity, $u_{\rho\rho'} = u_{\rho'\rho}$, $u_{\rho\rho'} = u_{\rho'\phi}$, etc.

When both the transmitting and receiving antennas are in the equatorial plane, the torsion factor ($\tilde{T}$) is equal to zero.

(b) *Due to Magnetic Dipole Moment*

$$H_{\phi\phi'} \text{ remains as it is}$$
$$H_{\phi'z} = 0$$
$$H_{zz'} = -\frac{ik}{2\pi Z_0} d\mathbf{p}_m \{(1 - i/kt)V(\xi) + t(i/kt)\} \exp(-ikt)/\sqrt{t}$$
$$E_{z'\rho} = -\frac{ik}{2\pi} d\mathbf{p}_m \{(1 - i/kt)V(\xi)\} \exp(-ikt)/\sqrt{t}$$
$$E_{\phi'\rho} = 0$$

1.18.3 Source Representation

In problems of antennas on or near a surface, field expressions for the antennas need to be made compatible with GTD [193]. One task is to find the position of the source or sources on the cylinder. The source is often distributed over the length or the aperture and hence some equivalent locations are to be found from which the source appears to radiate. The second point is that in the near field of antennas, the field components in the ray path are very strong. But GTD analysis does not include components in the direction of the ray path. Richmond [193], based on Schelkunoff's work [194], proposed field expressions for both monopole and dipole using piecewise sinusoidal current distribution. These field expressions do not have a field component along the ray path. The GTD compatible field expressions for monopoles and dipoles (Fig. 1.60) are given below.

Monopole Fields. The near-zone fields (see Fig. 1.60(a)) are given by

$$E_\rho = \frac{Z_0}{4\pi\rho \sin h\gamma d} [(I_1 \exp(-\gamma R_1) - I_2 \exp(-\gamma R_2)) \sin h\gamma d (I_1 \cos h\gamma d - I_2)$$
$$\times \exp(-\gamma R_1 \cos \theta_1) + (I_2 \cos h\gamma d - I_1) \exp(-\gamma R_2 \cos \theta_2)] \tag{1.218a}$$

$$E_z = \frac{Z_0}{4\pi \sin h\gamma d} \left[(I_1 - I_2 \cos h\gamma d) \frac{\exp(-\gamma R_2)}{R_2} + (I_2 - I_1 \cos h\gamma d) \frac{\exp(-\gamma R_1)}{R_1} \right]$$
$$\tag{1.218b}$$

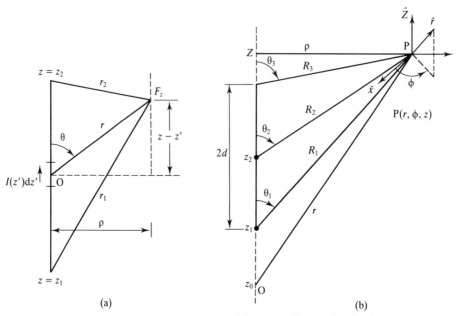

FIGURE 1.60 Linear antennas: (a) monopole and (b) dipole.

Dipole Fields. The GTD-compatible near-zone fields of the dipole (Fig. 1.60(b)) are expressed as

$$E_\rho = \frac{iZ_0 I_0}{4\pi\rho \sin kd} [\exp(-ik\rho)\cos\theta_1 - 2\cos kd \exp(-ikR_2)\cos\theta_2 $$
$$+ \exp(-ikR_3)\cos\theta_3] \qquad (1.219a)$$

$$E_z = -\frac{i\eta I_0}{4\pi \sin kd}\left[\frac{\exp(-ikR_1)}{R_1} - 2\cos kd \exp(-ikR_2)/R_2 + \frac{\exp(-ikR_3)}{R_3}\right]$$
$$(1.219b)$$

$$E_\phi = \frac{i\eta I_0}{4\pi r \sin kd}\left[(z_1 - z_0)\frac{e^{-ikR_1}}{R_1} - 2(z_2 - z_0)\cos kd \frac{\exp(-ikR_2)}{R_2}\right.$$
$$\left. + (z_3 - z_0)\frac{\exp(-jkR_3)}{R_3}\right] \qquad (1.219c)$$

with $I(z) = \hat{z}I_0 \sin[k(d - |z - z_0|)]/\sin kd$ as the assumed current distribution.

Discussion and application of the surface wave theory presented above can be found in a technical report [195].

1.18.4 Surface Waves on Coated Bodies

The properties of electromagnetic creeping waves on a surface having impedance boundary conditions and on dielectric-coated surfaces are of interest to radar engineers. Hence, the creeping wave constants, the number of propagating modes on such surfaces, and their dependence on thickness and complex dielectric constant are of importance.

The creeping waves on impedance boundary cylinders and dielectric-coated cylinders were investigated sometime ago. Approximate results for a purely reactive boundary were presented by Wait [196] and Logan and Yee [197]. The cylinder with finite conductivity has been discussed in [198]. Helstrom [199] provided pole trajectories associated with the creeping waves on a coated cylinder. The lowest-mode creeping wave on a reactive impedance cylinder as well as the dielectric coated cylinder was analyzed by Elliot [200]. For electrically thin coatings, the lowest mode is enough to account for the creeping wave propagation. For electrically thick coatings, higher-order modes have to be taken into account. The lowest-order creeping modes in the case of scattering from a dielectric-coated cylinder have been discussed by Wang [201, 202].

Creeping Wave Constants and Modal Impedance of a Dielectric Coated Cylinder.
R. Paknys and N. Wang [203] took a close and fresh look at the electromagnetic characteristics of the creeping waves associated with a dielectric-coated cylinder (Fig. 1.61). It is found that an infinite number of modes are supported by a coated cylinder. The propagation constants and creeping waves have been determined numerically. Higher-order modes with thick coatings have been investigated. The modes have no cutoff frequencies. A few Elliot type modes can exist on a dielectric coated cylinder and each one of them has a critical radius below which this type of mode cannot exist.

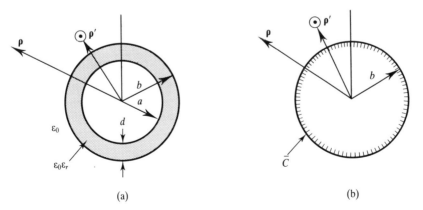

FIGURE 1.61 Geometry and coordinate system of (a) dielectric-coated and (b) impedance cylinders.

The Green's function for an impedance cylinder with a line source excitation is

$$G(\rho, \rho') = \frac{i}{4} \int_{-\infty}^{\infty} \left\{ J_\nu(k_0\rho_<) - \frac{b(\nu)}{d(\nu)} H_\nu^{(2)}(k_0\rho_<) \right\} \cdot H_\nu^{(2)}(k_0\rho_>) e^{i\nu|\phi-\phi'|} d\nu$$

$$+ \frac{\pi}{2} \sum_{n=-\infty}^{\infty} \sum_{\nu_p}' \frac{b(\nu_p)}{\frac{\delta}{\delta\nu} d(\nu)|_{\nu_p}} H_{\nu_p}^{(2)}(k_0\rho') H_{\nu-p}^{(2)}(k_0\rho') H_{\nu_p}^{(2)}(k_0\rho) e^{i\nu_p|\phi-\phi'-2n\pi|}$$

(1.220)

where

$$b(\nu) = J_\nu'(k_0 b) + i\bar{C} J_\nu(k_0 b) \tag{1.221a}$$

$$d(\nu) = H_\nu'^{(2)}(k_0 b) + i\bar{C} H_\nu^{(2)}(k_0 b) \tag{1.221b}$$

where

$$\bar{C} = -\frac{E_\phi}{\eta_0} H_z|_{\rho=b} \tag{1.122c}$$

I_0, M_0, and $d(\nu_p) = 0$ defines the roots ν_p. The impedance boundary condition at $\rho = b$ is

$$\frac{\delta G}{\delta \rho} + ik_0 \bar{C} G = 0 \tag{1.222}$$

The following relations are valid:

For TE case

$$H = H_z = i\omega\varepsilon_0 M_0 G \tag{1.223a}$$

$$\bar{C} = -E\phi/\eta_0 H_z|_{\rho=b} \tag{1.223b}$$

For TM case

$$E = E_z = i\omega\mu_0 I_0 G \tag{1.223c}$$

$$\bar{C} = \eta_0 H\phi/E_z|_{\rho=b} \tag{1.223d}$$

where $\eta_0 = \sqrt{\mu_0/\varepsilon_0}$ M_0 and I_0 is the strength of a magnetic or electric line source. The $n = 0$ term gives the dominant contribution from the integral. The coefficients $\bar{C}_\nu^h$ for TE and $\bar{C}_\nu^s$ for TM cases, which can be interpreted as the modal impedance and admittance for the creeping waves, are given by

$$\bar{C}_\nu^h = i \frac{k_1 \varepsilon_0}{k_0 \varepsilon_1} \frac{J_\nu'(k_1 b) N_\nu'(k_1 a) - J_z'(k_1 a) N_\nu'(k_1 b)}{J_\nu(k_1 b) N_\nu'(k_1 a) - J_\nu'(k_1 a) N_\nu(k_1 b)} \tag{1.224a}$$

$$\bar{C}_\nu^s = j \frac{k_1 \mu_0}{k_0 \mu_1} \frac{J_\nu'(k_1 b) N_\nu(k_1 a) - J_\nu(k_1 a) N_\nu'(k_1 b)}{J_\nu(k_1 b) N_\nu(k_1 a) - J_\nu(k_1 a) N_\nu(k_1 b)} \tag{1.224b}$$

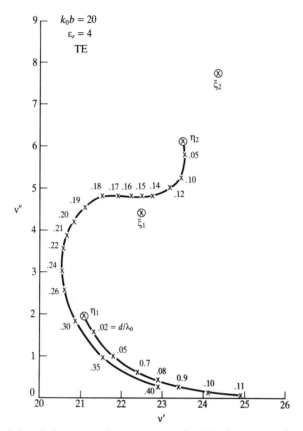

FIGURE 1.62 Azimuthal propagation constants for TE for a coated cylinder [203].

The superscripts s and h denote soft (Dirichlet) and hard (Neumann) boundary condition at $\rho = a$. The quantities $\bar{C}_v^h$ and $\bar{C}_v^s$ are interpreted as the modal impedances and admittances. Figure 1.62 depicts the TE azimuthal propagation constants for the coated cylinder. Figure 1.63 shows the impedance functions associated with the first two TE modes.

1.18.5 Selected Results and Discussions on Surface Wave Propagation

Some results of the decay of field components are available [173, 174, 195]. Figures 1.64 and 1.65 respectively show the decay of field components using Keller's approach and Uniform Theory of Slope Diffraction (UTSD). In Keller's approach it is assumed that the incident fields for both polarizations are of unit amplitude, and for UTSD the sources are of unit electric and magnetic dipole moments. From Figure 1.64 it is found that the rate of decay is dependent on whether it is a hard or a soft case. It is also dependent on the mode (only zeroth-order and first-order modes are considered). The hard case seems to be

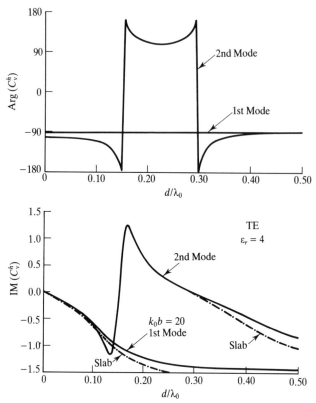

FIGURE 1.63 Impedance functions associated with the first two TE modes [203].

troublesome since the field decays at a relatively slower rate than the soft case. The hard polarization, therefore, can be a potential source of interference.

As mentioned earlier, it is expected that the cross-polarized components should disappear when both transmitting and receiving antennas are in the equatorial plane. But UTSD predicts nonexistence of only some of the cross-polarized components in the equatorial plane. For the equatorial plane, the torsion factor $\tilde{T}_0 = 0$. For magnetic dipole moments as the source, $E_{z'\rho}$ (see eqn. (1.217f)) is nonvanishing. For electric dipole moments as the source, $H_{\rho'z}$ is nonzero.

To be more precise, with a magnetic dipole moment transmitting, $H_{\phi'z}$, $H_{z'\rho}$ and $E_{\phi'\rho}$ vanish when both transmitting and receiving points are in the same equatorial plane; $E_{z'\rho}$ gets modified but does not vanish. (Since $E_{z'\rho}$ is a cross-polarized component it should vanish for the equatorial case. But apparently not according to theory, something that requires further investigations.) Hence, with torsion there are six components and without torsion there are three components with a magnetic dipole moment as the transmitting

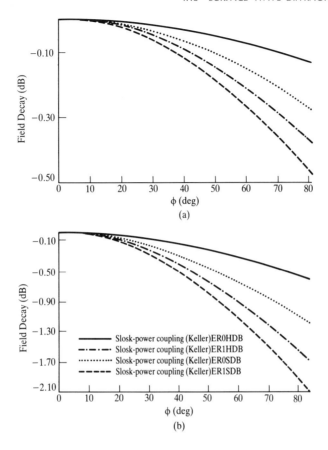

FIGURE 1.64 Theoretical coupling calculations between two elementary dipole moments on a cylinder compared with experimental results with frequency: $a = 0.3885$ m; separation = 20 cm. The analysis uses Keller's approach and the measurements take only the direct contribution [193]. (a) $f = 118$ MHz; $z_1 = 0$; $z_2 = -4.876$ m; $\phi_2 = 80°$; $a = 1.3462$ m. (b) $f = 236$ MHz; $z_1 = z_2 = -2.32$ m; $\phi_2 = 80°$; $a = 1.3462$ m. (c) $f = 2317$ MHz; $4z_1 = z_2 = 0$; $\phi_2 = 88°$; $a = 1.3462$ m. (d) $f = 1000$ MHz; $4z_1 = z_2 = 0$; $\phi_2 = 180°$; $a = 10\lambda$. (*continued on next page*)

source. With the electric dipole moment source the numbers are three and two respectively.

Figure 1.65 shows the decay of different co- and cross-polarized field components using UTSD. It was found that the cross-polarized components $E_{\rho\phi'}$ (and $E_{\rho'\phi}$), $H_{\phi'z}$ (and $H_{\phi z'}$), $H_{\phi'z}$ (and $H_{\phi z'}$) are weaker than the copolar components $H_{zz'}$ (and $H_{z'z}$) and $H_{\phi\phi'}$ (and $H_{\phi'\phi}$) while $H_{z'\rho}$ (and $H_{z\rho'}$) is not. In the equatorial plane the cross-polarized components $H_{\phi'z}$, $E_{\phi'\rho}$ and $H_{\rho'\rho}$, and $H_{\rho'\phi}$ do not get excited but $E_{z'\rho}$ and $H_{\rho'z}$ are not predicted to assume zero values.

Table 1.7 gives the relative values of shading loss of co- and cross-polarized

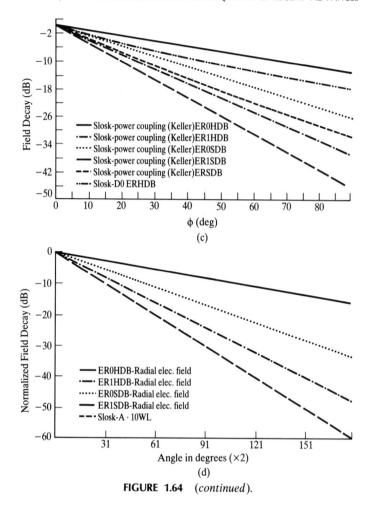

FIGURE 1.64 (*continued*).

components. They differ in intensity considerably for the same path length in equatorial cases, particularly for smaller radii of curvature of the cylinder. This is because even though the path length is the same for the nonequatorial case the effective radius of curvature depends on the angle of entry. Also, the above analytical model is accurate for large radii of curvature of the surface.

Keller's model and AAPG/IEMCAP cannot predict cross-polarized fields on a structure when torsion is present and hence can only partly take care of the coupling phenomenon.

As a result of the investigations reported in a technical report [197] the following conclusions could be arrived at:

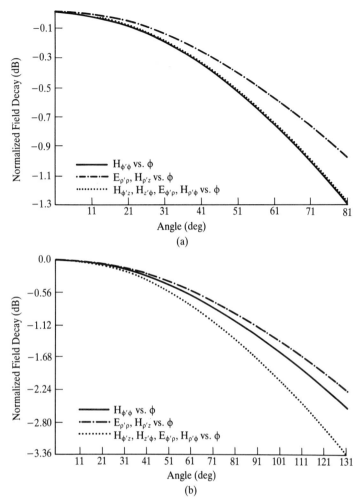

FIGURE 1.65 Normalized Field Decay in dB using UTSD versus angle in degrees [193]. (a) $f = 118$ MHz; $4z_1 = 0$; $z_2 = -4.87$ m; $\phi_2 = 132°$; $a = 1.3462$ m. (b) $f = 229$ MHz; $z_1 = 0$; $z_2 = -4.8768$ m; $\phi_2 = 83°$; $a = 1.3462$ m. (c) $f = 1000$ MHz; $4z_1 = 2.324$ m; $\phi_2 = 83°$; $a = 1.3462$ m. (d) $f = 1000$ MHz; $z_1 = 0$; $z_2 = 7.28$ m; $\phi_2 = 132°$; $a = 1.3462$ m. *(continued on next page)*

1. The existing theories are valid for large kR.
2. In Keller's approach of associating an attenuation constant and a complex surface diffraction coefficient, the first-order mode decays much faster than the zeroth-order mode for large radius of curvature. It is recommended that for intermediate values of the radius of curvature, only the zeroth-order mode be considered.
3. Keller's formulation cannot predict cross-polarized components. The

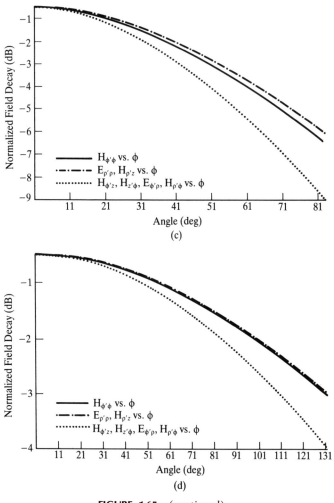

FIGURE 1.65 (*continued*).

surface ray induced by the tangential E^i (soft case) seems to attenuate much faster than the normal E^i (hard case).

4. The UTSD formulation can predict co- as well as cross-polarized components. But inconsistencies remain in the sense that it predicts some nonvanishing components when both the source and the observation points are in the equatorial plane.
5. It is possible experimentally using an HP8510 network analyzer to estimate the direct path contribution from transmitting to receiving antennas on a structure. Use can be made of time-domain gating followed by conversion to the frequency domain. Further details will be discussed in Chapter 6.

TABLE 1.7 Comparison of Shading Loss (dB) for Equatorial and Nonequatorial Cases with the Same Path Length on a PEC Cylinder

(a) $a = 45$ m $= 150\lambda$, path length $= 916.3\lambda$

$H_{zz'}$		$H_{\phi\phi'}$		$E_{\rho\rho'}$	
1	2	1	2	1	2
420.23	394.2	469.96	444.84	367.71	315.71

$E_{z'\rho}$		$H_{\phi'z}$		$E_{\phi'\rho}$		$H_{\rho'\phi}$	
1	2	1	2	1	2	1	2
419.2	394.2	NE	462.98	NE	510.42	NE	459.91

(b) $a = 3$ m $= 10\lambda$, path length $= 61.09\lambda$

$H_{zz'}$		$H_{\phi\phi'}$		$E_{\rho\rho'}$	
1	2	1	2	1	2
187.96	147.46	227.13	185.52	136.44	96.02

$E_{z'\rho}$		$H_{\phi'z}$		$E_{\phi'\rho}$		$H_{\rho'\phi}$	
1	2	1	2	1	2	1	2
187.96	147.53	NE	190.743	NE	213.68	NE	162.16

(c) $a = 0.9$ m $= 3\lambda$, path length $= 18.33\lambda$

$H_{zz'}$		$H_{\phi\phi'}$		$E_{\rho\rho'}$	
1	2	1	2	1	2
134.17	38.516	168.10	86.25	82.65	4.19

$E_{z'\rho}$		$E_{\phi'\rho}$		$H_{\phi'\rho}$		$E_{\phi'\rho}$		$H_{\rho'\phi}$	
1	2	1	2	1	2	1	2	1	2
134.17	55.73	146.49	83.13	NE	76.7	NE	84.36	NE	32.84

1: $z' = -1$ m; $z = 1$ m (for nonequatorial cases)
2: $z' = z = 1$ m (for equatorial cases)
NE: these components do not get excited

1.19 FIELD ESTIMATION AT A CAUSTIC

Though the theories of diffraction are called uniform theories, the ray techniques gives rise to infinite fields at the caustics (focal points and focal planes). So caustics require special treatment in electromagnetic field problems. Normally two methods are currently used: physical optics (see Section 1.8) and the method of equivalent current (MEC). In the early 1950s, Keller [204] also attempted a solution for fields at a caustic.

1.19.1 Method of Equivalent Current (MEC)

The focal planes exist in many practical situations. When a circular aperture is illuminated the fields along its axis or along the axis of a ring–slope discontinuity (junction of the cylinder and cone in a cone–cylinder structure) on a complex body are caustic fields. Conventional ray methods predict singular fields at the caustic. Postulating hypothetical currents is sometimes helpful in predicting electromagnetic fields at the caustic. One type of equivalent current developed to treat caustics of the diffracted fields has been described in [126]. Another type of equivalent currents, developed to supplement the usual GO expressions, was presented in [211] and, like the method in [126], can offer a systematic approach for solving complex problems along with GTD. Here are some examples.

Method of Equivalent Current (MEC) for Diffracted Fields. Since this method has already been described in several textbooks [205, 206] we will only present here the key expressions since it will be necessary to use them in later chapters. The equivalent electric and magnetic edge currents are given by [205]

$$I^e = i\frac{2}{Z_0 k}[G(n, \psi^-) - G(n, \psi^+)](-\sin\phi' E_z^i + \cos\phi' E_y^i)\cdot e^{ika\sin\theta\cos\phi} \quad (1.225a)$$

$$I^m = i\frac{2}{Y_0 k}[G(n, \psi^-) + G(n, \psi^+)](\cos\phi' H_y^i - \sin\phi' H_z^i)\cdot e^{ika\sin\theta\cos\phi} \quad (1.225b)$$

where

$$G(n, \psi) = \frac{1}{n}\sin\frac{\pi}{n}\frac{1}{\cos\frac{\pi}{n} - \cos\frac{\psi}{n}} \quad (1.226)$$

where n is a numeric related to the interior wedge angle and $n = 2\pi - \alpha/\pi$ where α is the half interior angle of the wedge.

The expressions for equivalent currents in (1.225a,b) are strictly for diffracted fields and do not include GO terms. These currents can be employed using the standard procedure for calculation of fields at any point, including the caustic. The example of scattering from a cone–cylinder is discussed in [127]. The GO line currents are treated in the next section.

GO Equivalent Line Currents. To formulate the GO equivalent line currents, consider the reflection of a plane wave from a cylinder as shown in Figure 1.66(a). The equivalent line current is shown in Figure 1.66(b). The GO reflected field can in general be written in the form

$$(E^{GO}) = E^i(Q_R)[R]\sqrt{\rho_1^r \rho_2^r}\frac{e^{-iks}}{s} \quad (1.227)$$

1.19 FIELD ESTIMATION AT A CAUSTIC

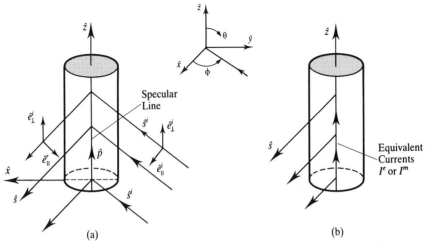

FIGURE 1.66 The reflection from a PEC cylinder: (a) incident wave is a plane wave; (b) due to the equivalent line currents [211].

where $E^i(Q_R)$ = GO incident field column vector on the surface
E^{GO} = GO reflected field column vector
$[R]$ = reflection coefficient matrix (2 × 2).
  The refraction index is given by

$$[R] = \begin{bmatrix} R_{\parallel\parallel} & 0 \\ 0 & R_{\perp\perp} \end{bmatrix}$$

where ρ_1^r, ρ_2^r are the two principal radii of curvature of the reflected wavefront. These are related to the angle of incidence θ_i, the principal radii of curvature of the surface $R_{1,2}$ and the principal radii of curvature of the incident wavefront $\rho_{1,2}^i$ by the relations

$$\rho_1^r = \frac{R_1 \rho_1^i \cos\theta_i}{R_2 \cos\theta_i + 2\rho_1^i} \qquad (1.228a)$$

$$\rho_2^r = \frac{R_2 \rho_2^i}{R_2 + 2\rho_2^i \cos\theta_i} \qquad (1.228b)$$

For reflection from the cylinder, the reflected field can be simplified to

$$E_{\parallel,\perp}^R = (\pm 1) E_{\parallel,\perp}^i \sqrt{\rho_1^r} \frac{e^{-iks}}{\sqrt{s}} \qquad (1.229)$$

Let the fictitious electric and magnetic currents be designated I^e and I^m. The fields due to these currents are given by

$$E_{\parallel} = \frac{ikZ_0}{4\pi} I^e \sqrt{\frac{2\pi}{k}} e^{-i\pi/4} \frac{e^{-jks}}{\sqrt{s}} \tag{1.230a}$$

$$E_{\perp} = \frac{ik}{4\pi} I^m \sqrt{\frac{2\pi}{k}} e^{-i\pi/4} \frac{e^{-iks}}{\sqrt{s}} \tag{1.230b}$$

where Z_0 = intrinsic impedance of the free space
k = the wavenumber
s = the distance of the observation point from the edge

Comparing eqn. (1.229) with eqn. (1.230a,b) one gets,

$$I^{e,m} = \begin{cases} E_{\parallel}^i \dfrac{\sqrt{\rho_1^r}}{Z_0} \sqrt{\dfrac{8\pi}{k}} e^{-j\pi/4} & (1.231a) \\[2ex] -E_{\perp}^i \sqrt{\rho_1^r} \sqrt{\dfrac{8\pi}{k}} e^{-i\pi/4} & (1.231b) \end{cases}$$

The above expressions are for normal incidence to the edge. The expressions for currents for oblique incidence to the edge are

$$I^e = \hat{p} \frac{\hat{p} \cdot \mathbf{E}^i}{Z_0 \sin \beta_0^i} \sqrt{\rho_\tau^r} \sqrt{\frac{8\pi}{k}} e^{-i\pi/4} \tag{1.232a}$$

$$I^m = -\hat{p} \frac{(\hat{p} \times \hat{s}^i) \cdot \mathbf{E}^i}{\sin \beta_0^i} \sqrt{\sigma_\tau^r} \sqrt{\frac{8\pi}{k}} e^{-i\pi/4} \tag{1.232b}$$

where $\beta_0^i = \cos^{-1}(-\hat{s}^i \cdot \hat{p})$
$\hat{p}$ = a unit vector tangent to the specular line
$\hat{s}^i$ = a unit vector along the direction of incidence

The application of this technique to the reflected field computation from complex geometries will be discussed in a later section.

Applications of the Method of Equivalent Current (MEC). Equations (1.231a,b) and (1.232a,b) can be used to treat a variety of problems, e.g. monopole above a ground plane, RCS of a cone and a cone cylinder. The problem of a monopole on a circular or rectangular ground plane is discussed in detail in Chapter 6.

1.20 INCREMENTAL LENGTH DIFFRACTION COEFFICIENT (ILDC)

The conventional two-dimensional plane wave diffraction coefficients described in Appendix A introduce errors in many situations, such as regions of sudden change in the radius of curvature and regions near a caustic. To overcome this limitation of the conventional diffraction coefficient, an incremental length diffraction coefficient (ILDC) was introduced by Mintzner [207]. The ILDC is defined for all angles of incidence and scattering. ILDC identifies with the two-dimensional diffraction coefficient wherever the later is available. A fringe wave ILEDC (also known as Ufimtsev's ILEDC) has also been defined as a generalization of Ufimtsev's work on the physical theory of diffraction. In fact, it can be shown that the two-dimensional diffraction coefficients, such as those used by Keller and Ufimsev, can be considered as special cases of ILDCs.

Before the work on ILDCs by Mitzner [207], the same concept was implicitly used by some workers, e.g. Ufimtsev [209, 210], Millar [212–214] and Ryan and Peters [126]. Mitzner [207] matched the required nonuniform current integrals to similar integrals and determined the PRD diffraction coefficient of the wedge. Michaeli [215] integrated the total wedge current in closed form and determined the GTD incremental diffraction coefficient of the wedge. Knott and Senior [216] extended the equivalent edge current method [126] and obtained approximate GTD incremental diffraction coefficients for arbitrary bistatic scattering.

A very good account of the ILDC is given in [207, 217]. Consider the far field in a three-dimensional scattering problem. This is given by

$$\mathbf{E}^s = \frac{\exp(ikR_0)}{R_0} \mathbf{F} \tag{1.233a}$$

$$\mathbf{H}^s = \eta_0 \hat{e}_r^s \times \mathbf{E}^s \tag{1.223b}$$

where $\mathbf{E}$ is the radiation vector, $\mathbf{F}$ is independent of R_0, and the direction of scattering is normal to the direction $\hat{e}_r^s$.

The radiation vector can be expressed in terms of the surface currents $\mathbf{J}$ and $\mathbf{M}$ on the target as [216]

$$\mathbf{F} = -ik\pi \mathbf{e}_r^s \int_S ds' \exp\{-ik\hat{e}_r^s \cdot \mathbf{r}'\}(\mathbf{J} + \eta_0 \mathbf{e}_r^s \times \mathbf{M}) \tag{1.234}$$

The incident field is a plane wave which can be written as

$$\mathbf{E}_0 = E_0 \mathbf{p} \exp\{-ik\hat{e}_r^s \cdot \mathbf{r}\} \tag{1.235a}$$

$$\mathbf{H}_0 = \frac{-\mathbf{e}_r^i \times \mathbf{E}_0}{Z_0} \tag{1.235b}$$

where the vector **p** determines the polarization of the wave and Z_0 is the impedance of free space. The polarization vector is expressed as

$$\mathbf{p} = p_\perp \mathbf{e}_\perp^i + p_\parallel \mathbf{e}_\parallel^i \tag{1.236}$$

The currents **J** and **M** are linear functions of $E_0 \mathbf{p}$ and are given by

$$\mathbf{K}_q = E_0 \bar{\bar{\mathbf{K}}}_q \cdot \mathbf{p} \quad \text{for} \quad q = e, m \tag{1.237}$$

F can be written in the following form

$$\mathbf{F} = E_0 \frac{1}{k} \bar{\bar{D}} \cdot \mathbf{p} \tag{1.238}$$

The dyadic $\bar{\bar{D}}$ is called the three-dimensional diffraction coefficient associated with the problem; it is dimensionless, it is independent of E_0 and **p**, and it is a function of the frequency, the angle of incidence, and the scattering

$$\bar{\bar{D}} = \bar{\bar{D}}(k; E_r^i; e_r^s) \tag{1.239}$$

Let us now consider scattering from the surfaces currents on an incremental length element of the cylinder which lies between the planes $t = \pm \frac{1}{2} dt$ instead of infinite or full length.

The incremental far-field radiation vector dF_∞ is given by

$$dF_\infty = dF_{\infty \perp} e_\perp^s + dF_{\infty \parallel} e_\parallel^s \tag{1.240}$$

It is related to the three-dimensional diffraction coefficient $d\bar{\bar{D}}_\infty$ by

$$d\bar{F}_\infty = E_0 \frac{1}{k} d\bar{\bar{D}}_\infty \cdot \mathbf{p} \tag{1.241}$$

Since $d\bar{\bar{D}}_\infty$ is a linear function of the length dt of the incremental element. Therefore,

$$d\bar{\bar{D}}_\infty = \frac{e^{-j\pi/4}}{(2\pi)^{1/2}} k \bar{\bar{d}} \, dt \tag{1.242}$$

$\bar{\bar{d}}$ is the dyadic incremental diffraction coefficient for the cylinder. Two-dimensional diffraction coefficients are recovered when the incremental diffraction coefficient is integrated over the bounding curves of the target. Hence, the total cross section of a target can be obtained by integrating the product of incident field and incremental diffraction coefficient with due consideration of phase over the bounding curves of the target.

Michaeli [215] independently extracted the incremental diffraction coefficient. Very recently, an incremental length diffraction coefficient has been presented for the canonical problem of a locally tangent wedge with surface impedance boundary conditions on its faces [218]. The theory has been applied to the case of the scattering of a plane wave from an impedance half plane.

1.20.1 Applications of ILDC

Incremental diffraction coefficients have been used by different authors to estimate scattering from various types of targets such as an infinite wedge [218] a canonical problem, and polygonal plates [208]. The results of ILDC applied to polygonal plates are discussed in Chapter 5.

1.21 RELATIONSHIP BETWEEN GTD, PTD, AND MEC

A discussion of the relationship between GTD, PTD, and PEC has been presented in [215]. The diffracted field at any point due to an edge discontinuity C in the Fresnel and far fields is given by the Fresnel integral

$$E^d ik \int_C [Z(\mathbf{r}')\hat{s} \times (\hat{s} \times \hat{t}) + M(\mathbf{r}')\hat{s} \times \hat{t}] G(\mathbf{r}', \mathbf{r}) \, dl \qquad (1.243)$$

where k = the wave number of the incident wave
Z = the intrinsic impedance of the medium
$\mathbf{r}, \mathbf{r}'$ = the position vectors of the observation point and a point on the curve C
$dl = |d\mathbf{r}'|$ = the incremental length along the curve C
$\hat{t} = d\mathbf{r}'/dt$ = the tangent vector
$\hat{s} = \mathbf{s}/s = (\mathbf{r} - \mathbf{r}')/|\mathbf{r} - \mathbf{r}'|$ = the direction of observation for the elementary edge element at $\mathbf{r}'$
$G(\mathbf{r}', \mathbf{r}) = \exp(-iks)/4\pi s$ = the three-dimensional Green's function

The induced electric and magnetic currents are linear functions of incident magnetic and electric fields. Under the high-frequency limit, the integral in eqn. (1.243) can be evaluated asymptotically and it reduces to a sum of ray contributions where the law of diffraction is satisfied (the smaller angle between the incident direction and the tangent to the edge, and the angle between the direction of observation and the tangent to the edge are equal; this means that the incident and scattered directions lie on the periphery of the same cone). The electric and magnetic edge currents in terms of GTD soft and hard coefficients are given by

$$I = -\frac{2(\mathbf{E}^i \cdot \hat{t})D_s}{ikZ_0 \sin \phi'} \quad (1.244a)$$

$$M = -\frac{2(\mathbf{H}' \cdot \hat{t})D_h}{ikY_0 \sin \phi'} \quad (1.244b)$$

S_s, D_h = soft and hard diffraction coefficients of the edge, see eqns. (A.2) to (A.7).

Therefore it is established that the equivalent currents $\mathbf{I}$ and $\mathbf{M}$ can be expressed in terms of the GTD coefficients. But there is a restriction; the currents cannot be used at or near the caustic, whereas the equivalent currents in eqn. (1.232a,b) can be used at the caustic without encountering a singularity. In order to establish a relationship between MEC and PTD in a mathematical form [215], use is made of the fact that the edge diffraction is mainly due to the surface currents on two narrow strips of surface area S_1 and S_2 on the two sides of the edge. The edge-diffracted field with such currents is given by

$$\mathbf{E}^d \sim ikZ_0 \int_C \sum_{i=1}^{2} \int_0^\infty \hat{s} \times [\hat{s} \times \mathbf{J}_i(l, x_i)] \cdot G(\mathbf{r}'(l, x_i), \mathbf{r}) J_i(l, x_i)\, dx_i\, dl \quad (1.245)$$

$J_i(l, x_i)$ = the Jacobian of the transformation from local Cartesian coordinates on the surface elements dS_i to the variables l, x_i, $\int_0$ means that only the asymptotic contribution at the end point $x_i = 0$ is considered. It turns out that the equivalent edge currents in terms of the PTD surface currents are given by

$$M(l) = \sum_{i=1}^{2} M_i(l), \qquad I(l) = \sum_{i=1}^{2} I_i(l) \quad (1.246)$$

where

$$M_i(l) = \frac{Z_0}{\sin^2 \beta} \hat{t} \cdot [\hat{s} \times \mathbf{K}_i(l)], \quad (1.247a)$$

$$I_i(l) = \frac{1}{\sin^2 \beta} \hat{s} \cdot [(\hat{t} \times \hat{s}) \times \mathbf{K}_i(l)] \quad (1.247b)$$

where

$$\mathbf{K}_i(l) = \int_0^\infty \mathbf{J}_i(l, x_i) \exp(ikx_i \cdot \hat{s})\, dx_i \quad (1.247c)$$

The expressions for $\mathbf{I}$ and $\mathbf{M}$ can be obtained by asymptotic endpoint evaluation of the integral in eqn. (1.129c) for a particular case, for example, a wedge. This is available in section III of [217] so it is not given here.

1.22 DIFFRACTION BY FERRITE HALF PLANES

Uniform asymptotic solutions have been developed for the total field at any observation point from a ferrite half plane [219] and a planar junction of two dielectric/ferrite half planes [220]. The different contributions shown in Figure 1.67 are as follows: (1) the direct contribution from the source; (2) the reflected field from the half plane; (3) the field transmitted through the half plane; (4) the fields diffracted from the edge of the half plane; and (5) the contribution due to edge-diffracted surface waves. Figure 1.68 shows the original configuration and the decomposition of the incident field into even and odd

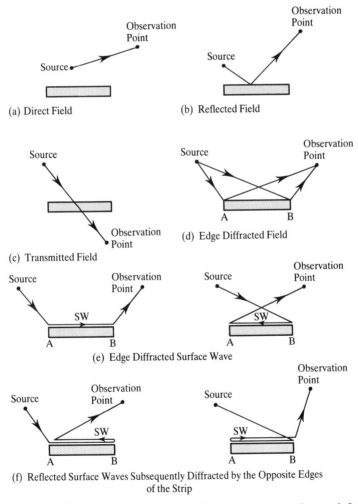

FIGURE 1.67 The different field components that contribute to the total field from a ferrite half plane at an observation point [220].

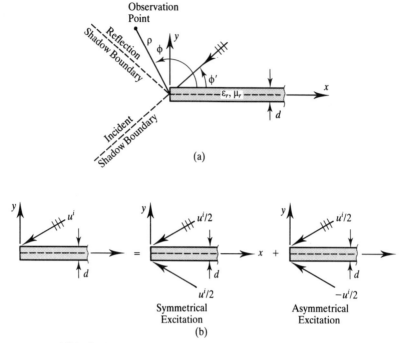

FIGURE 1.68 d/f half plane: (a) original configuration; (b) decomposition of incident field into even symmetric and odd symmetric excitation [220].

symmetric excitations. The solution of this dielectric/ferrite (d/f) half plane problem is obtained by superposing PEC and perfectly magnetically conducting (PMC) solutions (Fig. 1.69). The total field U for the d/f half plane of Figure 1.67.

$$\mathbf{U} = u_z^i + u_z^r + u_z^t + u_{z1}^{sw} + u_{z2}^{sw} + u_z^d \quad 0 \leq \phi \leq 2\pi \quad (1.248)$$

where u_z^i = z-component of incident field
u_z^r = z-component of reflected field
u_z^t = z-component of transmitted field
u_{z1}^{sw} = z-component of surface waves on side 1
u_{z2}^{sw} = z-component of surface waves on side 2
u_z^d = z-component of diffracted field

The details of the derivation are given in [221].
 The incident field u_z^i is given by

$$u_z^i = \mathbf{F}_{0z} e^{-ik\rho \cos(\phi - \phi')} \{ F(\phi - \phi' + \pi) - F(\phi - \phi' - \pi) \} \quad (1.249)$$

where $\mathbf{F}(\cdot)$ is the unit step function.

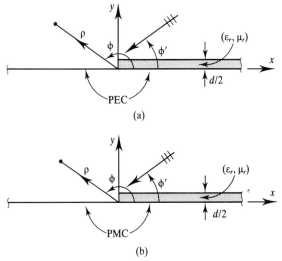

FIGURE 1.69 d/f half plane bisected by a PEC or PMC wall: (a) PEC bisection, which corresponds to even/odd symmetric excitation with the incident field is TE$_z$(TM$_z$) polarized; (b) PMC bisection, which corresponds to even (odd) symmetric excitation of Figure 1.68(b) if the incident field is TM$_z$ (TE$_z$) polarized.

The reflected field f_z^r is given by

$$f_z^r = \{\Lambda_p(\phi')F(\pi - \phi' - \phi) + \Lambda_p(2\pi - \phi')F(\phi - 3\pi + \phi')\} \cdot F_{0z} \cdot e^{-ik\rho\cos(\phi - \phi')} \tag{1.250}$$

where Λ_p is the Fresnel reflection coefficient given by

$$\Lambda_p(\phi') = (\Lambda_b^c(\phi') + \Lambda_p(2\pi - \phi')F(\phi - 3\pi + \phi'))F_{0z} \cdot e^{-ik\rho\cos(\phi - \phi')} \tag{1.251}$$

The transmitted field has the form

$$\mathbf{f}_z^T = \mathbf{T}(\phi')F(\phi - \phi' - \pi) + \mathbf{T}(2\pi - \phi')F(\phi' - \pi - \phi) \tag{1.252}$$

The Fresnel transmission coefficient can be expressed in matrix form as

$$\mathbf{T}(\phi') = \begin{bmatrix} (R_e^m(\phi') - R_e^c(\phi'))/2 & 0 \\ 0 & (R_h^c(\phi') - R_h^m(\phi'))/2 \end{bmatrix} \tag{1.253}$$

The TM- and TE-polarized surface wave field (respectively denoted by f_{z1}^{sw} and f_{z2}^{sw})

$$\mathbf{f} = -\tfrac{1}{2}\bar{r}_s^{m,c}(w_{ms,cs}, \phi')\mathbf{f}_z^i(\rho = 0) \cdot \exp(ik\rho\cos(w_{ms,cs} - \phi))F(\phi_z^{m,c} - \phi)$$
$$\cdot \exp(ik\rho\cos(w_{ms,cs} + \phi))F(\phi - 2\pi + \phi_s^{m,c}) \tag{1.254}$$

The edge-diffracted field from the d/f half plane can be expressed as

$$f_z^d = \frac{-e^{-i\pi/4}}{\sqrt{2\pi k\rho}} e^{ik\rho} \frac{1}{2}\bar{\bar{D}}_z^c(\phi,\phi') + \frac{1}{2}\bar{\bar{D}}_z^m(\phi,\phi') + \sum_{n=1}^{2} \frac{\bar{r}_n[1-F(ikLS_n^2)]}{\sqrt{2S_2}} e^{i\pi/4}$$
$$+ \sum_{n=3}^{4} \frac{\bar{r}[1-F(ikLa_n^2)]}{\sqrt{2a_n}} e^{i\pi/4} f_z^i(\rho=0) \qquad (1.255)$$

where

$$\bar{r}_1 = -\bar{\bar{\Lambda}}_p(\tilde{\phi}'); \qquad \bar{r}_2 = [\bar{\bar{T}}(\tilde{\phi}') - \bar{\bar{I}}] \qquad (1.256a)$$
$$\tilde{\phi}' = \phi' \quad \text{if} \quad 0 < \phi' \le \pi \qquad (1.256b)$$
$$2\pi - \phi' \quad \text{if} \quad \phi' > \pi \qquad (1.256c)$$
$$\bar{r}_3 = \tfrac{1}{2}\bar{r}_s^c(w_{cs},\phi'); \qquad \bar{r}_4 = \tfrac{1}{2}\bar{r}^m(w_{ms},\phi') \qquad (1.256d)$$

where

$$w_1 = \pi - \phi', \quad w_2 = +\phi', \quad w_3 = w_{cs}, \quad w_4 = w_{ms} \quad \text{in}$$

$$S_n = \sqrt{2}e^{i\pi/4}\sin\left(\frac{w_n-\phi}{2}\right); \quad w_n = \pi - \phi' \text{ if } n=1; \quad w_n = w_{cs}, \text{ if } n=2$$

$$a_n = \sqrt{2}e^{i\pi/4}\sin\left(\frac{w_n-\tilde{\phi}}{2}\right); \quad \tilde{\phi} = \phi \text{ if } 0 < \phi \le \pi \text{ and } 2\pi - \phi, \text{ if } \phi > \pi$$

$$(1.256f)$$

The expression for surface wave incidence is also available. This is obtained as the superposition of PEC and PMC bisection solutions.

For a TE_z-polarized surface wave, the total field can be expressed as

$$\mathbf{f} = \mathbf{f}_{zsw}^i + \mathbf{f}_{zsw}^r + \mathbf{f}_{zsw}^d \quad \text{for} \quad 0 \le \phi \le 2\pi \qquad (1.257)$$

where

$$\mathbf{F}_{0z} = [\eta_0 \mathbf{H}_{0z}]$$
$$\mathbf{f}_{zsw}^i = \mathbf{F}_{0z} e^{-ik\rho\cos(w_{cs}+\phi)} F(\phi - 2\pi + \phi_{is}^c) \qquad (1.258a)$$
$$\mathbf{f}_{zsw}^r = -\bar{r}_s^c(w_{cs},-w_{cs}) \mathbf{f}_{zsw}^i(\rho=\phi=0) \cdot \{e^{ik\rho\cos(w_{cs}-\phi)} F(\phi_s^c - \phi)$$
$$+ e^{ik\rho\cos(w_{cs}+\phi)} F(\phi - 2\pi + \phi_s^c)\} \qquad (1.258b)$$

For the case of a TM_z-polarized surface wave, the expressions are of similar to the TE_z case with different details.

The incident field is same as in eqn. (1.249) with w_{cs} replaced by w_{ms}. The diffracted field is still of the form eqn. (1.255) with $S_n(n=1,2)$ replaced by $a_n(n=1,2)$ and $0 \le \phi \le 2\pi$. The reflected surface wave field is given by eqn. (1.258b) with ϕ_s^c and $\bar{r}_s^c(w_{cs},-w_{cs})$ replaced by ϕ_s^m and $\bar{r}_s^m(w_{ms},-w_{ms})$.

1.23 DIFFRACTION BY ANISOTROPIC HALF PLANES

A generalization of the problem of scattering from a half plane with two face impedances is the anisotropic half plane (Fig. 1.70) where, in general, two impedance matrices for two anisotropic faces are involved. The problem of two antidiagonal matrices was attempted by Senior [221] leading to the conclusion that the problem was beyond the known techniques. He also concluded that even the simpler problem of two equal antidiagonal matrices could not be solved. But using the new techniques discussed by Daniele [222] and an earlier work by Khrapkov [223], the current problem can be solved. Hurd and Luneberg [224] solved the canonical problem of a plane wave obliquely incident on a half plane with two faces anisotropic and imperfectly conducting. They factored a 2×2 Wiener–Hopf matrix and obtained an exact closed-form solution. Serbest and Yazici [225] presented explicit expressions for diffracted, reflected, and transmitted fields using the Wiener–Hopf solution that had been obtained. The expressions for diffracted, reflected, and transmitted fields respectively are

$$u_d(r, \phi) = -\frac{e^{-i\pi/4}}{ikZ_0(2\pi)^{1/2}(\cos\phi + \cos\phi')} \frac{\pm kC \sin\phi'}{G^+(-k\cos\phi)G^-(k\cos\phi')}$$
$$+ (AZ_0 - ikB\cos\phi\cos\phi')L^+(-k\cos\phi)L^-(k\cos\phi'))\frac{e^{-ikr}}{kr^2} \quad y > 0$$

(1.258c)

$$u_r = -\frac{C\sin\phi'}{Z_0 - C\sin\phi'} + \frac{AZ_0 + ikB\cos^2\phi'}{AZ_0 + \sin\phi' + ikB\cos^2\phi'}\exp^{-jkr\cos(\phi+\phi')}$$

for region $0 < \phi < 2\pi$ with $\phi_r = \pi - \phi'$ (1.259)

and for region $\phi_t < \phi < 2\pi$ with $\phi_t = \pi - \phi'$ (shadow boundary), the transmitted field is given by

$$u_t = \sin\phi' \frac{1}{AZ_0 + \sin\phi' + ikB\cos\phi'} + \frac{C}{Z_0 - C\sin\phi'} \cdot e^{-ikr\cos(\phi-\phi')} \quad (1.260)$$

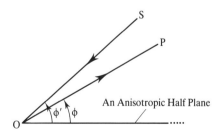

FIGURE 1.70 Diffraction by an anisotropic half plane.

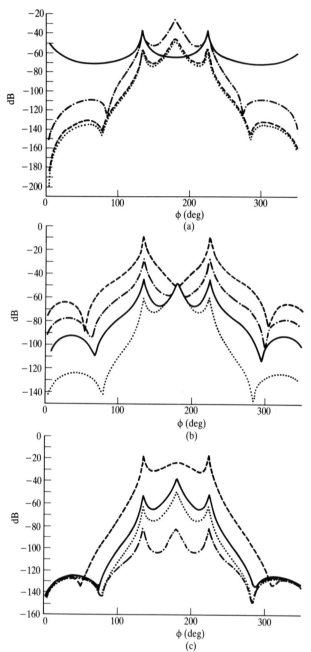

FIGURE 1.71 Scattered field for different values of parameters for the anisotropic half plane. (a) $\mu_r = 1$; $\mu_{nr} = 1.1$; $2kh = 0.25$; incidence angle $= 45°$; $-\varepsilon_r = 4$; $+\varepsilon_r = 4 + j$; $\varepsilon_r = 4 + j2$; $\Diamond \varepsilon_r = 4 + j20$. (b) $\mu_r = 1$; $\mu_{nr} = 1.1$; $+\varepsilon_r = 4 + j$; $2kh = 0.25$; incidence angle $= 45°$; $\Box \mu_r = 3$; $+\mu_r = 0.8$; $-\mu_r = 0.5$; $\Delta\mu_r = 0.01$. (c) $\mu_r = 1.1$; $\varepsilon_r = 4 + j$; $2kh = 0.25$; incidence angle $= 45°$; $\mu_{nr} = 3$; $+\mu_{nr} = 0.8$; $-\mu_{nr} = 0.5$; $\Delta\mu_{nr} = 0.01$ [225].

1.23 DIFFRACTION BY ANISOTROPIC HALF PLANES

The scattered field characteristics are shown in Figure 1.71. It was observed that [225]: (1) for small conductivity both GO and diffracted fields are attenuated in comparison to the zero conductivity case, the layer behaves like a PEC layer when conductivity is increased; (2) in parallel and normal directions of the plane the effects of the magnetic characteristics are nearly the same; and (3) the attenuation increases if the magnetic permeability is decreased.

1.23.1 Discontinuity at the Junction of Two Anisotropic Impedance Half Planes

One important problem in diffraction theory is the diffraction at a junction formed by two anisotropic impedance half planes. It has applications in the scattering of electromagnetic waves in the presence of such discontinuities, in the properties of terrain, and in composite structures like aircraft and missiles.

The diffraction problem of the discontinuity formed by two anisotropic impedance half planes was solved for the first time in [226] using a Fourier transform technique leading to Wiener–Hopf equations. The half planes are conducting in one direction and have different impedances Z_1 and Z_2 in the other direction. The geometry of the problem is shown in Figure 1.72. For interested readers, the details of the solution are given in [226]. We present here the nonuniform expression for the diffracted field; it is given by

$$E_x^d = e^{i\pi/4}\sqrt{2/\pi}(\sigma_1 - \sigma_2)\frac{[A_x^i \sin\theta_0 - A_z^i \cos\theta_0 \sin\phi']}{(\sigma_2 + \sin\phi')(\sigma_2 + \sin\phi)(\cos\phi' + \cos\phi)}$$
$$\cdot \frac{\sin^2\theta}{G^-(X\cos\phi')G^-(X\cos\phi)} \cdot \frac{e^{iX_r}}{\sqrt{X_r}} \quad (1.261)$$

The case of $Z_1 = Z_2$ corresponds to the full plane, there is no diffraction, and only the reflection mechanism comes into picture.

Though the above could not be validated fully because of a lack of experimental results or a solution of the same problem using another method, it was shown that the z-component of the diffracted magnetic field reduces to

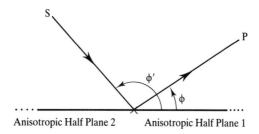

FIGURE 1.72 Geometry of the problem of diffraction at a discontinuity formed by two anisotropic impedance half planes.

the known results for normal incidence. This partially validates their theory. No results have been reported regarding diffraction by a discontinuity.

1.24 DIFFRACTION BY SPECIAL WEDGES

1.24.1 Dielectric-Coated Metal Wedge

The usual formulation used in the literature on scattering from such edges assumes a constant impedance boundary condition. This does not permit one to express both the geometrical optics field and the surface wave that is excited. This is because the impedance associated with a reflection and the surface wave propagation are in general different. The idea of constant impedance has been replaced in [227] by a differential boundary condition in order to solve the scattering problem of a metallic wedge with dielectric coating when the incident magnetic field is parallel to the edge.

The geometry and the coordinate system of the dielectric-covered wedge is shown in Figure 1.73. The PEC wedge is covered on both + and − sides with a dielectric layer of equal thickness. Typical results of this analysis are shown in Figures 1.74 and 1.75.

1.24.2 Infinite Wedges with Tensor Impedance Boundary Conditions

The hybrid method in [44] was extended in [228] to treat the problem of an infinite two-dimensional wedge with tensor impedance boundary conditions using a moment method and a physical optics solution for the currents. The material can also be anisotropic. Unlike in [44], it does not require the impedance matrix to be computed for each angle of incidence at a given

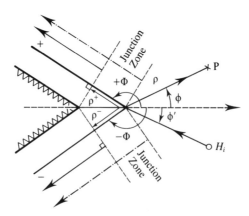

FIGURE 1.73 The geometry and coordinate system of the dielectric wedge covered with a dielectric layer.

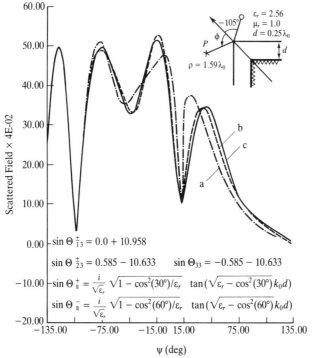

FIGURE 1.74 Actual reflected field H + respectively (a) $(\mathbf{H} = \mathbf{H}_r)$ one mode $\{\sin \theta_{13}^=\}$; (b) $(\mathbf{H} - \mathbf{H}_r)$ three modes $\{\sin \theta_{i,3}^=, i = 1, 2, 3\}$; (c) $(\mathbf{H} - \mathbf{H}_r) \sin \theta_0^=$ [227].

frequency. This approach may have application in treating coated edges for RCS work.

1.24.3 A Heuristic Uniform Slope Diffraction Coefficient for a Lossy Rough Wedge

Diffraction by a rough lossy wedge becomes important in estimating the propagation path loss in radio wave propagation. A heuristic approach was used in [229] to determine the diffraction coefficients for dielectric wedges. This approach was heuristically extended in [230] to present a UTD slope diffraction coefficient for a rough and lossy wedge. One assumption of this approach is that the transmitted portion of the incident and diffracted energy was neglected but reflection coefficients were incorporated. The slope diffraction term was generated in the same way as in UTD slope diffraction [10].

Finite Conductivity Two-Dimensional Uniform Diffraction Coefficient. The finite conductivity of the wedge is incoporated in the PEC wedge diffraction coefficient simply by using the reflection coefficients of the two faces of the wedge in the last two of the four terms of the PEC diffraction coefficient [10]. The finite conductivity diffraction coefficient is given by [229, 231].

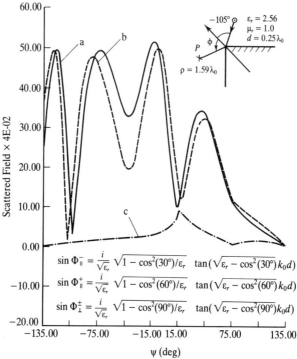

FIGURE 1.75 Total field with relative constant face impedances: (a) $\sin \theta_0^=$; (b) $\sin \theta^=$; (c) difference of diffraction terms (H_{SDP}) corresponding to (a) and (b) [225].

1.25 ACCURACY TESTING OF HIGH-FREQUENCY SOLUTIONS

A reader and a user of high-frequency techniques would obviously ask questions like: How accurate are these asymptotic solutions? Is there any systematic method of testing accuracy of such solutions? To the best of the knowledge of the author, it appears that in the literature researchers have assumed the accuracy to be satisfactory if, perhaps, the solution had a fair agreement with an experiment and/or another method, if relevant, an exact formulation. Apart from the work done by Mittra and Tew [232] and Tew and Mittra [233] there does not seem to be much in the literature on possible accuracy testing schemes of HF solutions and this topic is open to further investigation. They concluded that it is difficult to assess the accuracy of these asymptotic solutions and even to determine their relative accuracy. It was found that the proposed test was most sensitive to the source-free behaviour of the solution and relatively insensitive to the large-path-length behaviour.

The E-field test used in accuracy testing of the high frequency was to examine how well the asymptotic solutions satisfied the boundary condition that the tangential conponent of the E-field should be continuous on the surface of a

1.25 ACCURACY TESTING OF HIGH-FREQUENCY SOLUTIONS

PEC body. The test was performed on the problem of radiation from a slot on a PEC cylinder. The test procedure described in [229] can be summarized as follows

1. A cylindrical transform is defined as

$$H_{z,\phi}(n, k_s) = 1/2\pi \int_0^{2\pi} d\phi \int_{-\infty}^{\infty} dz H_{z,\phi}(\phi, z) e^{-jn\phi} e^{-jk_z z} \quad (1.262a)$$

where, $\gamma = \sqrt{k^2 - k_z^2}$ and $H_{z,\phi}(\phi, z)$ is the tangential magnetic field on the cylinder surface.

2. The electric and magnetic vector potentials A_z and F_z are expanded in terms of unknown coefficients

$$A_\phi, Z_z = \frac{1}{2\pi} \sum_{n=-\infty}^{\infty} e^{jn\phi} \int_{-\infty}^{\infty} f_n(k_z), g_n(k_z) H_n^{(2)}(\rho\sqrt{k^2 - k_z^2}) e^{jk_z z} \, dz \quad (1.262b)$$

where, n and k_z are the transform variables.

3. The unknown coefficients are determined using

$$\mathbf{H} = \nabla \times \mathbf{A} - j\omega\varepsilon_0 \mathbf{F} + 1/j\omega\mu_0 \nabla(\nabla \cdot \mathbf{F}) \quad (1.262c)$$

and they are given by

$$f_n(k_z) = \frac{-1}{\gamma H_n'^{(2)}(\gamma R)} \tilde{H}_\phi(n, k_z) + \frac{nk_z}{\gamma^2 R} \tilde{H}_z(n, k_z) \quad (1.262d)$$

$$g_n(k_z) = \frac{j\omega\mu_0}{\gamma^2 H_n^{(2)}(\gamma R) \tilde{H}_z(n, k_z)} \quad (1.262e)$$

4. The surface field can be obtained from

$$\mathbf{E} = -\nabla \times \mathbf{F} - j\omega\mu_0 \mathbf{A} + 1/j\omega\varepsilon_0 \nabla(\nabla \cdot \mathbf{A}) \quad (1.262f)$$

Several solutions have been tested and compared in [229, 230]. AS-1 is a modal solution and works well for large path lengths. AS-2 is a modification of Fock's solution involving the radiation on a sphere.

Figure 1.76 shows a plot of the resultant E_z-field along the $\phi = 0$ cut. The conclusions are: (1) the expected continuity in E_z is predicted by the analytical solution; (2) the AS-2 and AS-3 solutions are indistiguishable from one another; (3) the AS-1 solution with a direct test does attain the desired shape but has a much greater E_z content in regions where E_z should be zero; (4) the direct computation applied to the AS-2 solution yields a sharp peak at the edge of

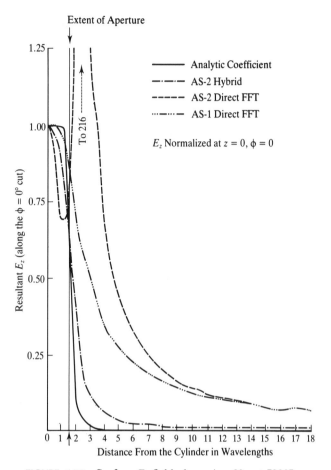

FIGURE 1.76 Surface E_z field along $\phi = 0°$ cut [229].

the slot. The large values extend outside the slot region where the field should go to zero.

It appears the E-field test may provide an accuracy test of the HF asymptotic solutions.

Some comments on the accuracy of the UTD near the 90° edge [232–235] are in order. J. A. Aas [234] compared the results of scattering from a PEC 90° wedge with the exact series expansion using cylindrical wave functions. The exact expression for the total field ($n = 1.5$) in terms of cylindrical wave functions is given by

$$H_z = \frac{4}{n}\frac{1}{2} J_0(k\rho) + \sum_{m=1}^{\infty} j^{m/n} J_{m/n}(k\rho) \cos\frac{m}{n}\phi' \cos\frac{m}{n}\phi \qquad (1.263)$$

1.25 ACCURACY TESTING OF HIGH-FREQUENCY SOLUTIONS

The high-frequency expression for the total field is

$$H_z^{GTD} = H_z^{GO} + D_h \frac{e^{-jk\rho}}{\sqrt{\rho}} \tag{1.264}$$

The magnitudes of the errors $|\Delta E_z| = |E_z - E_z^{GTD}|$ for parallel and $|\Delta H_z| = |H_z - H_z^{GTD}|$ for perpendicular polarizations are presented in Figures 1.77 and 1.78. It is concluded [229] that the error is of the order 35 dB up to a close distance of $\lambda/4$. This is for uniform illumination of the edge without the necessity

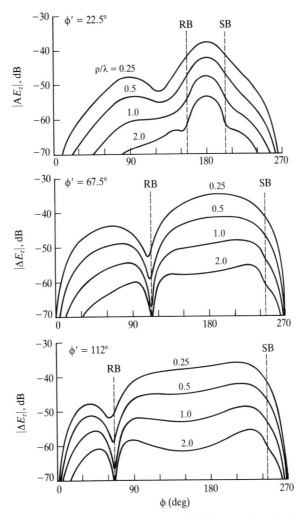

FIGURE 1.77 Relative magnitude of electric error field for parallel polarization and three different angles of incidence [229].

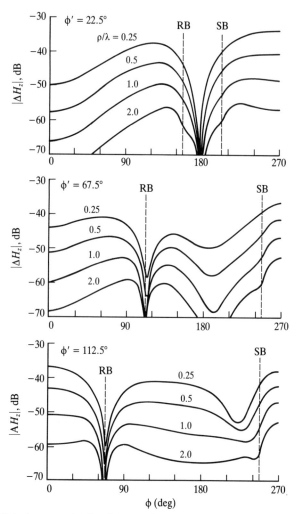

FIGURE 1.78 Relative magnitude of magnetic error for perpendicular polarization and three different angles of incidence [229].

of slope diffraction correction. It is also observed that large errors will creep in if the same calculations are done for the field components transverse to the edge. This means that for field components parallel to the edge of the wedge the UTD can be safely used up to a quarter wavelength distance. In [233], the UTD solution for scattering for a wedge is compared with the exact solution using an alternative form of the diffraction coefficients using a Fresnel argument. The relative magnitudes of magnetic and electric error field for parallel and perpendicular polarizations for three different angles are shown in Figures 1.79 and 1.80.

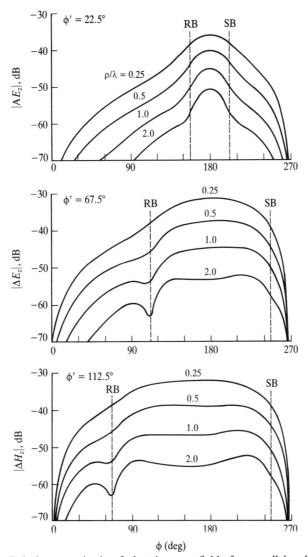

FIGURE 1.79 Relative magnitude of electric error fields for parallel polarization and three different angles of incidence [235].

1.26 BRIDGING THE GAP BETWEEN HIGH- AND LOW-FREQUENCY REGIMES

It is natural to ask how far down the frequency domain we can go with high-frequency techniques. Although an interesting topic of discussion, its answer very much depends on the type of problem being solved. If high-

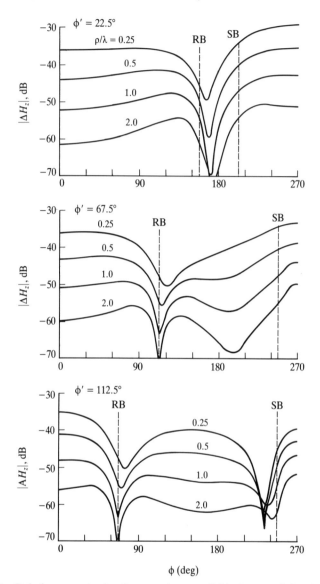

FIGURE 1.80 Relative magnitude of magnetic error fields for parallel polarization and three different angles of incidence [233].

frequency techniques are pushed down to low frequencies, we must take care of the following [237, 238]: (1) interactions of many orders may need to be taken into account; and (2) the principle of locality breaks down and scattering centers cannot be clearly indentified. It is believed that it can be pushed down to the first resonance of the target.

1.27 TIME-DOMAIN HIGH-FREQUENCY TECHNIQUES

Time-domain electromagnetic scattering from various structures has been of interest to researchers for quite some time. There has recently been a renewed interest in light of extensive use of high-resolution radar operating at a very wide band of frequencies. Time-domain techniques can be classified into two broad categories: (1) numerical and (2) analytical. It appears that numerical techniques [239, 242] which can better handle complex objects have been pursued more than the analytical formulations. The early work in time-domain analytical investigations is available in [243–249]. A key contribution in this area was reported by Verrutipong [248]. In this contribution, the time-domain uniform diffraction coefficient was found for the straight wedge with both sides PEC. This is the time-domain version of the uniform theory of diffraction in the frequency domain. Several analytical investigations [248–252] have been reported recently on the time-domain scattering by a disk [249], a transient plane wave reflected by a finitely conducting half space [250], and the PO and UTD scattering from a PEC cylinder [251] with one end open and one end closed. We summarize in this section the recent advances and applications of high-frequency methods in the time-domain.

1.27.1 Time-Domain Physical Optics (TDPO)

Time-domain physical optics (TDPO) is a straightforward extension of frequency-domain physical optics (FDPO). The TD electric and magnetic surface currents are given by

$$\mathbf{J}(t) = 2(\hat{\mathbf{n}} \times \mathbf{H}(t)) \qquad (1.265a)$$

$$\mathbf{K}(t) = 2(\mathbf{E}(t) \times \hat{\mathbf{n}}) \qquad (1.265b)$$

The TD magnetic vector potential $\mathbf{A}(\mathbf{R}, t)$ is obtained by transforming the FD magnetic vector potential $\mathbf{A}(\mathbf{R})$ and is given by

$$\mathbf{A}(\mathbf{R}, t) = \frac{\mu_0}{4\pi} \int_S \mathbf{J}_s(\mathbf{r}', t) e^{-jk_0(r-r')} \, dS' \qquad (1.266)$$

where

$$\mathbf{J}_s(\mathbf{r}', t) = 2(\mathbf{n} \times \mathbf{H}^i) \qquad (1.266a)$$

and

$$\mathbf{H} = \frac{1}{\mu_0} [\Delta \times \mathbf{A}] \qquad (1.266b)$$

1.27.2 Time-Domain Uniform Theory of Diffraction (TDUTD)

We discuss here the TD diffraction coefficient for scattering from a PEC straight wedge with plane wave illumination, as discussed in Verrutipong's work [248].

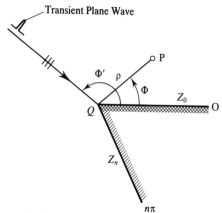

FIGURE 1.81 Time-domain plane wave scattering from a PEC straight wedge: the TDUTD coefficient.

The TD diffraction coefficient for a straight wedge with arbitrary two-face impedance and its applications are still under investigation [252] and are not presented here.

The TD singly diffracted field from the scattering center of a PEC straight wedge (Figure 1.81) is given by

$$\mathbf{E}(\mathbf{R}, t) = \int_{t_0}^{t - R/C} \mathbf{E}^i(\mathbf{R}, \tau) D_{s,h}(t - R/C - \tau) \cdot \Gamma \, PH(t) \quad (1.267)$$

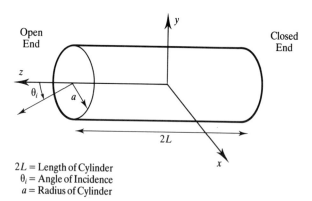

$2L$ = Length of Cylinder
θ_i = Angle of Incidence
a = Radius of Cylinder

FIGURE 1.82 Geometry and coordinate system of the time-domain EM scattering from a PEC cylinder with one end open and one end closed. Diameter = $2a$ meters; length = $2L$ meters [251].

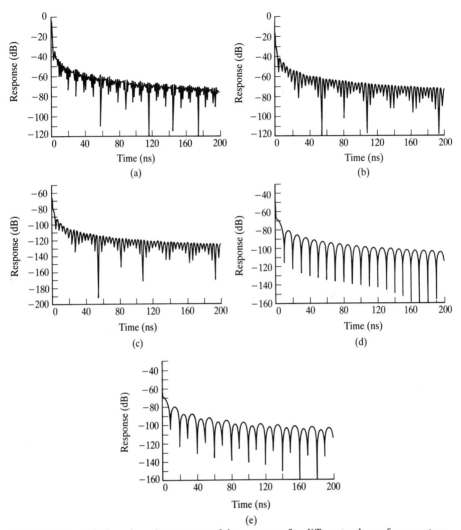

FIGURE 1.83 PO time-domain response of the structure for different values of parameters: $2L = 60$ in.; $a = 4$ in.: (a) $\theta_i = 90°$ (broadside incidence) TM polarization; (b) $\theta_i = 80°$, TM polarization; (c) $\theta_i = 100°$, TE polarization; (d) $\theta_i = 180°$ (closed end–on incidence), TM polarization; (e) $\theta_i = 190°$, TM polarization [251].

The TDUTD coefficient is given by

$$D_{s,h}(t) = \frac{-1}{2n\sqrt{2\pi}} \sum_{m=1}^{4} d_m^{s,h}(\hat{r}', \hat{r}) F(X_m, t), \quad (1.268)$$

where

$$d_1^{s,h}(\hat{r}', \hat{r}) = \frac{1}{\sin \beta_0} \cot \frac{\pi + (\phi - \phi')}{2n} \quad (1.269a)$$

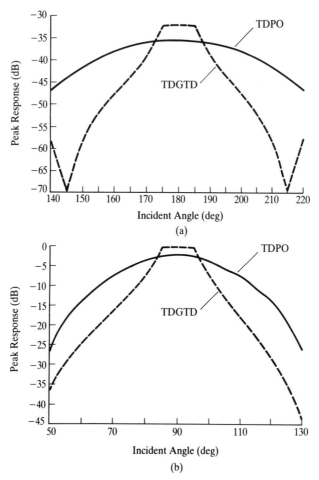

FIGURE 1.84 Comparison of time-domain PO and GTD with angle θ; length = 60 in.; diameter = 4 in.: (a) around closed end-on and (b) around broadside incidence [246].

$$d_2^{s,h}(\hat{r}', \hat{r}) = \frac{1}{\sin \beta_0} \cot \frac{\pi - (\phi - \phi')}{2n} \quad (1.269b)$$

$$d_3^{s,h}(\hat{r}', \hat{r}) = \mp \frac{1}{\sin \beta_0} \cot \frac{\pi + (\phi + \phi')}{2n} \quad (1.269c)$$

$$d_4^{s,h}(\hat{r}', \hat{r}) = \mp \frac{1}{\sin \beta_0} \cot \frac{\pi - (\pi + \phi')}{2n} \quad (1.269d)$$

where the TD Fresnel function in the time domain is given by

$$F(X_m, t) = \frac{X_m}{\sqrt{\pi c t (t + X_m/C)}} \quad (1.269e)$$

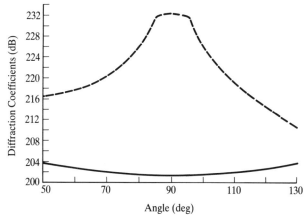

FIGURE 1.85 Diffraction coefficients of the open and closed ends with angle θ; length = 60 in.; diameter = 4 in.; TM polarization: crosses for the closed end (wedge angle 90°); squares for the open end (interior wedge angle 0°).

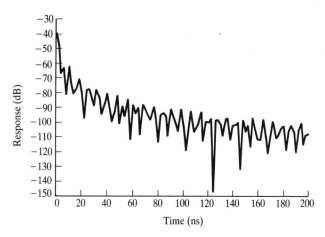

FIGURE 1.86 Comparison of a typical result from (squares) [251] and Dominek's investigations (Crosses) [249]; TM polarization; disk scattering; diameter = 4 in.; $\theta = 60°$.

1.27.3 Results and Discussion

Figure 1.82 shows the geometry and coordinate system of time-domain scattering from a cylinder with one end open and one end closed. Some analytical results for this geometry are shown in Figures 1.83 through 1.86. Figure 1.83 shows the PO time-domain response of the structure for different aspect angles and polarizations. Figure 1.84 shows the comparison of time-domain PO and GTD with aspect angle for (a) around closed end-on incidence and (b) around broadside incidence. The magnitudes of the diffraction coefficients versus θ for the open and closed ends are shown in Figure 1.85. Figure 1.86 shows a comparison of the results from [251] and that due to Dominek [249] showing close agreement.

1.28 SUMMARY

This chapter surveys recent advances in high-frequency techniques so it would be easy for the reader to understand numerous applications covered in Chapters 2 through 9. It discusses the key features of GO, PO, GTD, UTD, UAT, STD, and hybrid methods and it indicates some of the boundary conditions widely used in EM problems. This chapter also deals with thin and thick PEC half planes, impedance half planes, thin dielectric half planes and dielectric wedges. It gives an account of surface and creeping wave analysis, and the incremental diffraction coefficient. Ferrite half planes, dielectric-coated metal wedges, and rough lossy half planes are briefly discussed, along with some typical results from the literature. It looks at accuracy testing of high-frequency solutions and how far down the frequency domain high-frequency techniques can be pushed. The chapter ends with high-frequency PO and GTD solutions in the time domain. These theoretical discussions are used in subsequent chapters.

1.29 PROBLEMS

1.1 Find the geodesic in Figure 1.87 on a surface of revolution which is represented by $x = \rho \cos \phi$, $y = \rho \sin \phi$, $z = f(\rho)$; Also consider the special case of the surface to be that of a circular cone with a half angle θ_0, i.e. $f(\rho) = \rho \cot \theta_0$.

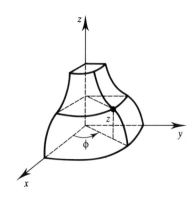

FIGURE 1.87 The surface of revolution.

1.2 The fields of a magnetic line source are given by

$$H_z(\rho) = I^e\sqrt{k/8\pi}\,\exp(i\pi/4)\,\exp(-ik\rho)/\sqrt{\rho}$$

and

$$E_\phi(\rho) = -I^m\sqrt{k/8\pi}\,\exp(i\pi/4)\,\exp(-ik\rho)/\sqrt{\rho}$$

Write down the terms of the LK expansion in eqn. (1.1) for these fields.

1.3 Show that the GO radar cross section of a conducting sphere of radius a is πa^2 square meters. *Hint*: Find the GO reflected field and apply the definition of RCS given by

$$\sigma = \lim_{R\to\infty}(1/4\pi r^2)\left(\frac{|\mathbf{E}^s|^2}{|\mathbf{E}^i|^2}\right)$$

1.4 Consider the problem of determination of a specular point in Section 1.5.1 and Figure 1.5. Apply Snell's law of reflection to obtain the two nonlinear equations (1.22a) and (1.22b).

1.5 Consider an axially symmetric reflector (Fig. 1.6) where the source and observation point are not necessarily restricted in the same azimuthal plane. The equation of the configuration is

$$f(z) = \rho^2$$

Plan a scheme to determine the possible specular points.

1.6 Following the discussion in Section 1.6.2 and the Example 1.5, plot the far-field pattern for the concave hyperboloidal interface (assume $n = 2$).

1.7 Show that the impedance boundary condition $\hat{n}\times(\hat{n}\times\mathbf{E}) = -\eta(n\times\mathbf{H})$ reduces to the conditions $E_x = -\eta H_y$ and $E_y = \eta H_x$ when a plane wave is incident on a material half space at $z = 0$.

1.8 In Figure 1.88 is shown a PEC finite circular ground plane of radius a meters illuminated by a uniform plane wave in the yz-plane obliquely incident at an angle θ_i. Assume a TE_x polarization for the incident field. Determine (a) the PO current density induced on the plate; (b) the far-zone bistatic scattered field based on the PO current density of part (a); and (c) the bistatic and monostatic RDCs of the plate. Plot the normalized monostatic RCS (σ_{3D}/λ^2) in dB for plates with radii $a = \lambda$ and $a = 5\lambda$.

1.9 Repeat Problem 1.8 for TE_z polarization.

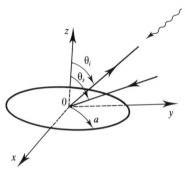

FIGURE 1.88 Electromagnetic scattering by a uniform plane wave by a PEC finite ground plane of radius $a = \lambda$ and 5λ with angle θ_i.

1.10 A plane wave of amplitude 1 V/m is incident normally ($\beta_0 = \pi/2$) on a PEC half plane as shown in Figure 1.28 at an angle of 60°. The polarization is hard. Evaluate the following for an observation point at $\phi = 180°$ and $\rho = 10\lambda$: (a) incident GO field; (b) reflected GO Field; (c) total GO field; (d) diffracted field; and (e) total GO plus diffracted fields.

1.11 An infinitely long wire carrying constant current is held symmetrically at a height h above a PEC disk of radius 5 wavelengths. Determine (a) incident field; (b) reflected field; (c) diffracted field; and (d) total field at any point. Determine what radius the disk will need in order to make the diffracted field -60 dB with respect to the incident field at a distance of 10 wavelengths below the center of the disk and with $h = 5$ wavelengths; frequency = 10 GHz.

1.30 REFERENCES

1. M. Kline and I. Kay *Electromagnetic Theory and Geometrical Optics*, Wiley Interscience, New York, 1965.
2. J. D. Kraus *Electromagnetics*, 3d ed., McGraw-Hill, NY, 1984.
3. A. K. Dominek, L. Peters Jr., and W. D. Burnside, "An Additional Physical Interpretation in the Luneberg–Kline Expansion," *IEEE Trans. Antennas and Propagat.*, **AP-35**(4), 1987, 406–411.
4. A. Taflove, private communication, 1991.
5. P. H. Pathak, "Techniques for High Frequency Problems," Ch. 4 in Y. T. Lo and S. W. Lee (ed), *Antenna Handbook—Theory, Applications and Design*, Van Nostrand Reinhold, New York, 1988.
6. R. G. Kouyoumjian, "The Geometrical Theory of Diffraction and Its Applications," in *Numerical and Asymptotic Techniques In Electromagnetics*, edited by R. Mitra Springer-Verlag, New York 1975.

7. R. G. Kouyoumjian, P. H. Pathak, and W. D. Burnside, "A Uniform GTD for the Diffraction by Edges, Vertices and Convex Surfaces", in *Theoretical Methods for Determining the Interaction Structures*, edited by J. K. Skwirzynski, Sijthaff and Noordhoff, Amsterdam, 1981.
8. J. B. Keller, "Geometrical Theory of Diffraction," *J. Opt. Soc. Amer.*, 52, 1962, 116–130.
9. J. B. Keller, "Diffraction by an Aperture," *J. Appld Phys.*, Vol. 28, April 1957, 426–444.
10. R. G. Kouyoumjian and P. H. Pathak, "A Uniform Geometrical Theory of Diffraction for an Edge in a Perfectly Conducting Surface," *Proc. IEEE*, **62**(11), 1974, 1448–1461. (Comments by J. D. Cashman and authors' replay in *IEEE Trans. Antennas and Propagat.*, **AP-25**(3), 1977, 447–449).
11. D. S. Ahluwalia, R. M. Lewis, and J. Boersma, "Uniform Asymptotic Theory of Diffraction by a Plane Screen," *SIAM J. Appld Math.*, 16, 1968, 783–807.
12. D. S. Ahluwalia, "Uniform Asymptotic Theory of Diffraction by the Edge of a Three-Dimensional Body," *SIAM J. Appld Math.*, 18, 1970, 287–301.
13. A. K. Bhattacharyya and D. L. Sengupta, "Radar Cross Section Analysis and Control," Artech House, Norwood, MA, 1991, pp. 36–38.
14. A. K. Bhattacharyya and D. L. Sengupta, "Radar Cross Section Analysis and Control," Artech House, Norwood, MA, 1991, pp. 51–53.
15. S. W. Lee and G. A. Deschamps, "A Uniform Asymptotic Theory of Electromagnatic Diffraction by a Curved Wedge," *IEEE Trans. Antenna and Propagat.*, **AP-24**(1), 1976, 25–34.
16. P. Y. Ufimtsev, "Approximate Computation of the Diffraction of Plane Electromagnetic Waves at Certain Metal Bodies," *Sov. Phys., Tech. Phys.*, 1957, 1708–1718.
17. P. Y. Ufimtsev, "Method of Edge Waves in Physical Theory of Diffraction," translated by U.S. Airforce Foreign Tech. Divn., Wright-Patterson AFB, OH, Sept. 1971.
18. R. Mittra, Y. Rahmat Samii, and W. L. Ko, "The Spectral Theory of Diffraction," *Appld Phys.*, 10, 1976, 1–13.
19. P. C. Clemmow, "Some Extensions to the Method of Integration by Steepest Descents," *Q. J. Mech. Appld Math.*, 3, 1950, 241–256.
20. B. L. Van der Waerden, "On the Method of Saddle Points," *Appld Sci. Res.*, B, 2, 1952, 33–45.
21. G. D. Maliuzhinets, "Excitation, Reflection and Emission of Surface Waves from a Wedge with Given Face Impedance," *Sov. Phys. Dokl.*, 3, 1959, 752–755.
22. R. Tiberio, G. Pelosi, and G. Manara, "A Uniform GTD Formulation for the Diffraction by a Wedge With Impedance Faces," *IEEE Trans. Antennas and Propagat.*, **AP-33**(8), 1985, 867–873.
23. J. L. Volakis, "A Uniform Geometrical Theory of Diffraction for an Imperfectly Conducting Half-Plane," *IEEE Trans. Antennas and Propagat.*, **AP-34**(2), 1986, 172–180. (Corrections to figures: ibid., **AP-35**(6), 1987, 743–745.)
24. R. Tiberio and G. Pelosi, "High-Frequency Scattering from the Edges of Impedance Discontinuities on a Flat Plane," *IEEE Trans. Antennas and Propagat.*, 590–596.

25. T. B. A. Senior, "The Diffraction Matrix for a Discontinuity in Curvature," *IEEE Trans. Antennas and Propagat.* **AP-20**(3), 1972, 326–333.
26. T. B. A. Senior, "Diffraction Tensors for Imperfectly Conducting Edges," *Radio Science*, **10**(10), 1975, 911–919.
27. T. B. A. Senior, "The Current Induced in a Resistive Hahlf Plane," *Radio Science*, **16**(6), 1981, 1249–1254.
28. T. B. A. Senior, "Skew Incidence on a Right-Angled Impedance Wedge," *Radio Science* **13**(4), 1978, 639–647.
29. T. B. A. Senior, "Solution of a Class of Imperfect Wedge Problems for Skew Incidence," *Radio Science*, **21**(2), 1986, 185–191.
30. T. B. A. Senior, "Backscattering from Tapered Resistive Strips," *IEEE Trans.*, **AP-32**(7), 1984, 747–751.
31. R. L. Haupt and V. V. Liepa, "Synthesis of Tapered Resistive Strips," *IEEE Trans.* **AP-35**(11), 1987, 1217–1225.
32. S-In Young, J.-W. Ra, and T. B. A. Senior, "E-Polarized by a Resistive Half Plane with Linearly Varying Resistivity," *Radio Science*, **23**(3), 1988, 463–469.
33. S. Sanyal and A. K. Bhattacharyya, "Diffraction by a Half-Plane with Two Face Impedances—Uniform Asymptotic Expansion for Plane Wave and Arbitrary Line Source Incidence," *IEEE Trans.*, **AP-34**(5), 1986, 718–723.
34. T. B. A. Senior and J. L. Volakis, "Scattering by an Imperfect Right-Angled Wedge," *IEEE Trans.*, **AP-34**(5), 1986, 681–689.
35. J. L. Volakis and M. A. Ricoy, "Diffraction by a Thick Perfectly Conducting Half Plane," *IEEE Trans.*, **AP-35**(1), 1987, 62–72.
36. J. L. Volakis, "Scattering by a Thick Impedance Half Plane," *Radio Science*, **22**, 1987, 13–25.
37. J. L. Volakis and T. B. A. Senior, "Application of a Class of Generalized Boundary Conditions to Scattering by a Metal-Backed Dielectric Half-Plane," *Proc. IEEE*, **77**(5), 1989, 796–805.
38. I. Anderson, "Plane Wave Diffraction by a Thin Dielectric Half-Plane," *IEEE Trans.*, **AP-27**(5), 1979, 584–589.
39. A. Chakrabarty, "Diffraction by a Dielectric Half Plane," *IEEE Trans.*, **AP-34**(6), 1986, 830–833.
40. T. B. A. Senior and J. L. Volakis, "Sheet Simulation of a Thin Dielectric Layer," *Radio Science*, **22**(7), 1987, 1261–1272.
41. J. L. Volakis and T. B. A. Senior, "Diffraction by a Thin Dielectric Half-Plane," *IEEE Trans.*, **AP-35**(12), 1987, 1483–1487.
42. S. Y. Kim, J. W. Ra, and S.-Y. Shin, "Diffraction by an Arbitrary Angled Dielectric Wedge: Part I—Physical Optics Approximation," *IEEE Trans.*, **AP-39**(9), 1991, 1272–1281.
43. S.-Y. Kim, J.-W. Ra, and S.-Y. Shin, "Diffraction by an Arbitrary-Angled Dielectric Wedge: Part II—Correction to Physical Optics Solution," *IEEE Trans.*, **AP-39**(9), 1991, 1281–1292.
44. G. A. Thiele and T. H. Newhouse, "A Hybrid Technique for Combining Moment Methods with the GTD," *IEEE Trans.*, **AP-23**(1), 1975, 62–69.

45. W. D. Burnside, C. L. Yu, and R. J. Marhefka, "A Technique to Combine the Geometrical Theory of Diffraction and the Moment Methods," *IEEE Trans.*, **AP-23**(4), 1975, 551–558.

46. J. N. Sahalos and G. A. Thiele, "On the Application of the GTD-MM Techniques and Its Limitations," *IEEE Trans.*, **AP-29**(5), 1981, 780–786.

47. W. D. Burnside and P. H. Pathak, "A Summary of Hybrid Solutions Involving Moment Methods and GTD," in *Applications of the Method of Moments to Electromagnetic Fields*, edited by B. J. Strait, SCEEE Press, St. Cloud, Florida, 1980.

48. L. N. Medgyesi-Mitschang, "Hybrid Methods for Analysis of Complex Scatterers," *Proc. IEEE*, **77**(5), 1989, 770–779.

49. L. N. Medguesi-Mitschang and D. S. Wang, "Hybrid Solutions for Large Impedance Coated Bodies of Revolution," *IEEE Trans.*, **AP-34**(11), 1986, 1319–1329.

50. L. N. Medguesi-Mitschang and D. S. Wang, "Hybrid Solutions for Scattering From Large Bodies of Revolution with Material Discontinuities and Coating," *IEEE Trans.*, **AP-32**(7), 1984, 717–723.

51. M. Ideman, "Modal Transfer Coefficients for the Cylindrically and Spherically Curved Concave Sheets," in *Hybrid Formulation of Wave Propagation and Scattering*, edited by L. B. Felsen, NATO ASI Series Applied Sciences, No. 86, Martinus Nijhoff, Boston, MA, 1984, pp. 131–138.

52. H. Serbest, "Diffraction Coefficients for a Curved Edge for Soft and Hard Boundary Conditions," *Proc. IEE, Part H*, **131**(6), 1984, 383–389.

53. A. H. Serbest, A. Buyukaksoy, and G. Uzgoren, "Diffraction of HF EM Waves by Curved Strips," *IEEE Trans.*, **AP-37**(5), 1989, 592–600.

54. M. Ideman and L. B. Felsen, "Diffraction of HF Waves by the Edge of a Perfectly Conducting Sheet Located on a Cylindrical Dielectric Interface," *Wave Motion*, **7**, 1985, 529–556.

55. M. Ideman and L. B. Felsen, "Diffraction of a Whispering Gallery Mode by the Edge of a Thin Concave Cylindrically Curved Surface," *IEEE Trans. Antennas and Propagat.*, **AP-29**(4), 1981, 571–579.

56. E. Topuz, E. Niver and L. B. Felsen, "Electromagnetic Fields near a Concave Perfectly Conducting Cylindrical Surface," *IEEE Trans. Antennas and Propagat.*, **AP-30**(2), 1982, 280–292.

57. T. Ishimaru, L. B. Felsen, and A. Green, "High Frequency Fields Excited by a Line Source Located on a Perfectly Conducting Concave Cylindrical Surface," *IEEE Trans. Antennas and Propagat.*, **AP-26**(6), 1978, 757–767.

58. L. B. Felsen and T. Ishimaru, "High Frequency Surface Fields Excited by a Point Source on a Cylindrical Boundary," *Radio Science*, **14**(2), 1979, 205–216.

59. T. Ishimaru and L. B. Felsen, "High Frequency Fields Excited by a Line Source Located on a Concave Cylindrical Impedance Surface," *IEEE Trans. Antennas and Propagat.*, **AP-27**(2), 1979, 172–179.

60. G. Hessaerjian and A. Ishimaru, "Excitation of a Conducting Cylindrical Surface of Large Radius of Curvature," *IRE Trans.*, **AP-10**, 1962, pp. 264–273.

61. B. R. Levy and J. B. Keller, "Diffraction by a Smooth Object," *Commn. Pure and Appld Math.*, **12**, 1959, 159–209.

62. J. B. Keller and B. R. Levy, "Decay Exponents and Diffraction Coefficients for Surface Waves on Surfaces of Nonconstant Curvature," *IRE AP-IRE (Suppl.)*, **AP-7**, 1959, S52–S61.
63. G. Hesserjian and A. Ishimaru, "Excitation of a Conducting Cylindrical Surface of Large Radius of Curvature," *IRE Trans.*, **AP-10**, 1962, 264–273.
64. J. R. Wait, "Currents Excited on a Conducting Surface of Large Radius of Curvature," *IRE Trans. Microwave Theory and Tech.*, **MTT-4**(3), 1956, 143–145.
65. Y. Hwang and R. G. Kouyoumjian, *The Mutual Coupling Between Slots on an Arbitrary Convex Cylinder*, Ohio State Univ. Eectroscience Lab., Rep. 2902-21, Mar. 1975.
66. P. H. Pathak and N. Wang, "Ray Analysis of the Mutual Coupling between Antennas on a Smooth Convex Surface," *IEEE Trans. Antennas and Propagat.*, **AP-29**(6), 1981, 911–922.
67. K. Kaminetsky and J. B. Keller, "Diffraction Coefficients for Higher Order Edges and Vertices," *SIAM J. Appld Math.*, **10**, 1970, 109.
68. J. Deygout, "Multiple Knife Edge Diffraction of Microwaves," *IEEE Trans.*, **AP-14**(4), 1966, 480–489.
69. L. E. Vogler, "An Attenuation Function for Multiple Knife Edge Diffraction," *Radio Science*, **17**(6), 1982, 1541–1546.
70. D. S. Jones, "Double Knife Edge Diffraction and Ray Theory," *Q. J. Mech. Appld Math.* **26**, 1973, 26.
71. M. I. Herman and J. L. Volakis, "High Frequency Scattering by a Double Impedance Wedge," *IEEE Trans.* **AP-36**(5), 1988, 664–678.
72. M. Schneider and R. J. Luebbers, "A General Uniform Double Wedge Diffraction Coefficient," *IEEE Trans. Antennas and Propagat.*, **AP-39**(1), 1991, 8–14.
73. D. A. McNamara, C. W. I. Pistorius, and J. A. G. Malherbe, *Introduction to the Uniform Geometric Theory of Diffraction*, Artech House, Norwood, MA, 1990 Section 6.5.
74. F. A. Sikta, W. D. Burnside, T. T. Chu, and L. Peters Jr., "First Order Equivalent Current and Corner Diffraction from Flat Plate Structures," *IEEE Trans. Antennas and Propagat.*, **AP-31**(4), 1983, 584–589.
75. T. B. Hansen, "Corner Diffraction Coefficients for the Quarter Plane," *IEEE Trans. Antennas and Propagat.*, **AP-39**(7), 1991, 976–984.
76. P. H. Pathak, "A New Corner Diffraction Coefficient," private communication.
77. M. M. Lipschutz, *Diffential Geometry, Schaum's Outline Series*, McGraw-Hill, New York, 1969.
78. C. E. Weatherburn, *Differential Geometry of Three Dimensions*, Cambridge University Press, Cambridge, 1961.
79. S. W. Lee, *Differential Geometry for GTD Applications*, Electromagnetics Lab, Univ. Illinois, Urbana, Rep. UILU-ENG-77-2264, Oct. 1977.
80. W. V. T. Rusch and O. Sorensen, "On Determining if a Specular Point Exists," *IEEE Trans. Antennas and Propagat.*, **AP-27**(1), 1979, 99–101.
81. R. Mittra and A. Rushdi, "An Efficient Approach for Computing the Geometrical Optics Field Reflected from a Numerically Specified Surface, *IEEE Trans. Antennas and Propagat.*, **AP-27**(6), 1979, 871–877.

82. S. W. Lee and P. W. Cramer, "Determination of Specular Points on a Surface of Revolution," *IEEE Trans. Antennas and Propagat.*, **AP-29**(4), 1981, 662–664.
83. J. B. Keller, R. M. Lewis, and B. D. Seckler," Asymptotic Solution of Some Diffraction Problems," *Commun. Pure Appld Math.*, **IX**, 1956, 207–265.
84. C. E. Schensted, "Electromagnetic and Acoustical Scattering by a Semi-Infinite Body of Revolution," *J. Appld Phys.*, **26**, 1955, 306–308.
85. S. W. Lee, "Electromagnetic Reflection From a Conducting Surface: Geometrical Optics Solution," *IEEE Trans. Antennas and Propagat.*, **AP-23**(2), 1975, 184–191.
86. S. W. Lee and P. W. Cramer "Determination of Specular Points on a Surface of Revolution," *IEEE Trans. Antennas and Propagat.*, **AP-29**(4), 1981, 662–664.
87. C. A. Balanis, *Advanced Engineering Electromagnetics*, John Wiley, New York, 1989.
88. S. W. Lee, M. S. Sheshadri, V. Jamnejad, and R. Mittra, "Refraction at a Curved Dielectric Interface: Geometrical Optics Solution," *IEEE Trans. Microwave Theory and Tech.*, **MTT-30**(1), 1982, 12–19.
89. A. W. Snyder and J. D. Love, "Reflection at a Curved Dielectric Interface—Electromagnetic Tunneling," *IEEE Trans. Microwave Theory and Tech.*, **AP-23**(1), 1975, 134–141.
90. L. B. Felsen and N. Marcuvitz, *Radiation and Scattering of Waves*, Prentice Hall, Englewood Cliffs, NJ, 1973.
91. E. Heyman and L. B. Felsen, "High Frequency Fields in the Presence of a Curved Dielectric Interface," *IEEE Trans.*, **AP-32**(9), 1984, 969–978.
92. [72] Chs. 2 and 3.
93. H. Ansorge, "Electromagnetic Reflection from a Curved Dielectric Interface," *IEEE Trans.*, Vol. **AP-34**(6), 1986, 842–845.
94. A. K. Bhattacharyya and D. L. Sengupta, *Radar Cross Section Analysis and Control*, Artech House, Norwood, MA, 1991, Section 1.2.4.
95. T. B. A. Senior, "Approximate Boundary Conditions," *IEEE Trans. Antennas and Propagat.*, **AP-29**(5), 1981, 826–829.
96. D. S. Wang, "Limits and Validity of the Impedance Boundary Condition on Penetrable Surfaces," *IEEE Trans. Antennas and Propagat.*, **AP-35**(4), 1987, 453–457.
97. [13] Section 1.2.3, pp. 9–10.
98. S. M. Rytov, *J. Exp. Theor. Phys. (USSR)*, **10**(2), 1940, 180 (in Russian).
99. T. B. A. Senior, "Impedance Boundary Conditions for Imperfectly Conducting Surface," *Appld Sci. Res.*, B, **8**, 1960, 418–436.
100. S. N. Karp and F. C. Karal Jr., "Generalized Impedance Boundary Conditions with Applications to Surface Wave Structures," in *Electromagnetic Wave Theory (Part I)*, edited by J. Brown Pergamon, London, 1977.
101. L. N. Medgyesi-Mitschang and J. M. Putnam, "Integral Equation Formulations for Imperfectly Conducting Scatterers," *IEEE Trans. Antennas and Propagat.*, **AP-33**(2), 1985, 206–221.
102. V. N. Kanellopoulos and J. P. Webb, "A Numerical Study of Vector Boundary Conditions for the Finite-Element Solution of Maxwell's Equations," *IEEE Microwave and Guided Wave Letters*, **1**(11), 1991, 325–327.

103. A. Khebir and R. Mittra, "Absorbing Boundary Conditions for Arbitrary Outer Boundary," *IEEE APS Symp.*, San Jose, CA, 1989, Paper No. 2–6, pp. 46–49.

104. P. S. Kildal, "Artificially Soft and Hard Surfaces in Electromagnetics, *IEEE APS Symp.*, **AP-38**(10), 1990, 1537–1544.

105. G. T. Buck et al., *Radar Cross Section Handbook*, Vol. 1, Section 3.3.3, pp. 51–63.

106. E. F. Knott, J. F. Shaeffer, and M. T. Tuley, *Radar Cross Section Handbook*, Section 3.3.3, pp. 57–62.

107. [13] Section 2.3, pp. 36–38.

108. R. F. Harrington, *Time-Harmonic Electromagnetic Fields*, McGraw-Hill, New York, 1961.

109. S. W. Lee, J. Balauf, H. Ling, and R. Chou, "Twelve Versions of Physical Optics: How Do They Compare?" *IEEE AP-S Symp. Dig.*, Syracuse, New York, 1988, pp. 408–411.

110. D. Klement, J. Pressiner, and V. Stein, "Special Problems in Applying the Physical Optics Method for Backscatter Computations of Complicated Objects" *IEEE Trans. Antennas and Propagat.*, **AP-36**(2), 1988, 228–237.

111. R. F. Harrington, *Field Computation by Moment Method*, McMillan, New York, 1968.

112. C. L. Bennett, "Time-Domain Inverse Scatering," *IEEE Trans. Antennas and Propagat.*, **AP-29**(2), 1981, 213–219.

113. J. B. Keller, "Backscattering from a Finite Cone—Theory and Experiment," *IEEE Trans. Antennas and Propagat.*, **AP-8**(2), 1960, 411–412.

114. M. E. Bechtel and R. A. Ross, *Radar Scattering Analysis*, CAL Report No. ER/RIS-10, Cornell Aeronautical Lab., Buffalo, New York, Aug. 1966.

115. S. W. Lee and G. A. Deschamps, "A Uniform Asymptotic Theory of Electromagnatic Diffration by a Curved Wedge," *IEEE Trans. Antennas and Propagat.*, **AP-24**(1), 1976, 25–34.

116. G. A. Deschamps, J. Boersma, and S. W. Lee, "Three-Dimensional Half Plane Diffraction Exact Solution and Testing of Uniform Theories," *IEEE Trans. Antennas and Propagat.*, **AP-32**(2), 1984, 264–271.

117. K. M. Siegel, "Far Field Scattering from Bodies of Revolution," *Appld Sci. Res.*, B, 7, 1958, 293–328.

118. L. B. Felsen, "Plane-Wave Scattering by Small-Angle Cones," *IRE Trans. Antennas and Propagat.*, **AP-5**, 1957, 121–129.

119. L. B. Felsen, "Asymptotic Expansion of the Diffracted Wave for a Semi-Infinite Cone," *IRE Trans. Antennas and Propagat.*, **AP-5**, 1957, 402–404.

120. A. S. Goryainov, "The Diffraction of a Plane Electromagnetic Wave Propagating along the Axis of a Cone," *Radiotekhnika i Elektronika*, **6**, 1961, 47.

121. [104] Section 2.3.1.4.

122. [73] Section 6.5.

123. R. F. Miller, "An Approximate Theory of the Diffraction of an Electromagnetic Wave by an Aperture in a Plane Screen," *IEEE Monograph 152R*, Oct. 1955.

124. R. F. Miller, "The Diffraction of an Electromagnetic Wave by a Circular Aperture," *IEE Monograph 196R*, Sept. 1956.

125. R. F. Miller, "The Diffraction of an Electromagnetic Wave by a Large Aperture," *IEE Monograph 213R*, Dec. 1956.
126. C. E. Ryan and L. Peters Jr., "Evaluation of Edge Diffracted Fields Including Equivalent Currents for the Caustic Regions," *IEEE Trans. Antennas and Propagat.*, **AP-17**(3), 1969, 292–299.
127. F. A. Sikta, W. D. Burnside, T. T. Chu, and L. Peters Jr., "First-Order Equivalent Current and Corner Diffraction Scattering from Flat Plate Structures," *IEEE Trans. Antennas and Propagat.*, **AP-31**(4), 1983, 584–589.
128. T. B. Hansen, "Corner Diffraction Coefficients for the Quarter Plate," *IEEE Trans. Antennas and Propagat.*, **AP-39**(7), 1991, 976–984.
129. K. C. Hill and P. H. Pathak, "On the Contribution of Transition Function Occuring in a New Approximate UTD Diffraction Coefficient," private communication.
130. V. H. Weston, "The effect of a discontinuity in curvature in high frequency scattering," *IEEE Trans. Antennas and Propagat.*, **AP-10**(6), 1962, 775–780.
131. R. Mittra, Y. Rahmat-Samii, and W. L. Ko, "A Spectral Domain Analysis of High Frequency Diffration Problems," in *Electromagnetic Scattering*, edited by P. L. Uslenghi, Academic, New York, 1978, 121–183.
132. R. Mittra, W. L. Ko, and Y. Rahmat-Samii, "Transform Approach to Electromagnetic Scattering," *Proc. IEEE*, **67**, 1979, 1486–1503.
133. B. Noble, *Methods Based on the Weiner-Hopf Technique*, Pergamon Press, London, 1958.
134. R. Mitra, Y. Rahmat-Samii, and W. L. Ko, "Transform Approach to Electromagnetic Scattering," in Part II, No. 9, pp. 649–696 of *Theoretical Methods for Determining the Interaction of Electromagnetic Waves With Structures*, edited by J. K. Skwirzyinski, Sijthoff and Noordhoff, Aphen aan den Riju, Netherlands, 1981.
135. O. M. Bucci and G. Fransceschetti, "Electromagnetic Scattering by a Half Plane with Two Face Impedances," *Radio Science*, **11**, 1976, 49–59. (Replace $\Psi(-\phi_0 + 2\pi)/\Psi(\phi_0)$ by $-\Psi(-\phi_0 + 2\pi)/\Psi(\phi_0)$ wherever it appears on p. 51.)
136. T. Griesser and C. A. Balanis, "Reflections, Diffractions and Surface Waves for an Interior Impedance Wedge of Arbitrary Angle," *IEEE Trans. Antennas and Propagat.*, **AP-37**(7), 1989, 927–935.
137. B. L. Van der Waerden, "On the Method of Saddle Points," *Appld Sci. Res.*, B, **2**, 1952, 33–45.
138. G. Uzgoren, A. Buyukaksoy, and A. H. Serbest, "Diffraction Coefficient Related to a Discontinuity Formed by Impedance and Resistive Half-Planes," *Proc. IEE, Part H*, **136**, 1989, pp. 11–23.
139. D. S. Jones, "Diffraction by a Thick Semi-Infinite Plate," *Proc. Roy. Soc. A*, **217**, 1953, pp. 153–175.
140. S. W. Lee and R. Mittra, "Diffraction by Thick Conducting Half-Plane and a Dielectric-Loaded Waveguide," *IEEE Trans. Antennas and Propagat.*, **AP-16**, 1968, 454–461.
141. H. G. Booker and P. C. Clemmow, "The Concept of an Angular Spectrum of Plane Waves, and its Relation to that of Polar Diagrams and Aperture Distributions," *Proc. Inst. Elec. Eng.*, **97**, 1950, 11–17.

142. P. C. Clemmow, "A Method for the Exact Solution of a Class of Two-Dimensional Diffraction Problems," *Proc. Roy. Soc. A*, **205**, 1951, 286–308.
143. A. D. Rawlins, "Diffraction by an Acoustically Penetrable or an Electromagnetically Dielectric Half Plane," *Int. J. Eng. Sci.*, **15**, 1977, 569–578.
144. R. G. Rojas and P. H. Pathak, "Diffraction of EM Waves by a Dielectric/Ferrite Half Plane and Related Configurations," *IEEE Trans. Antennas and Propagat.*, **37**(6), 1989, 751–763.
145. J. Radlow, "Diffraction by a Right-Angled Dielectric Wedge," *Int. J. Eng. Sci.*, **2**, 1964, 275–290.
146. N. H. Kuo and M. A. Plonus, "A Systematic Technique in the Solution of Diffraction by a Right-Angled Wedge," *J. Math. Phys.*, **46**, 1967, 394–407.
147. E. A. Kraut and G. W. Lehman, "Diffraction of Electromagnetic Waves by a Dielectric Wedge," *J. Math. Phys.*, **10**, 1969, 1340–1348.
148. A. A. Aleksandrova and N. A. Khizhnyak, "Diffraction of Electromagnetic Waves by a Dielectric Wedge," *Solv. Phys. Tech. Phys.*, **19**, 1975, 1385–1389.
149. L. Lewin and L. Sreenivasiah, *Diffraction by a Dielectric Wedge*, Sci. Rep. 47, Dept. Elec Eng., Univ. of Colorado, 1979.
150. S. J. Maurer and L. B. Felsen, "Ray-Optical Techniques for Guided Waves," *Proc IEEE*, **55**, 1967, 1718–1729.
151. P. Balling, "Surface Fields on the Source-Excited Dielectric Wedge," *IEEE Trans. Antennas and Propagat.*, **AP-21**, 1973, 113–115.
152. L. Kaminetsky and J. B. Keller, "Diffraction by Edges and Vertices of Interfaces," *SIAM J. Appl. Math.*, **28**, 1975, 839–856.
153. A. D. Rawlins, "Diffraction by a Dielectric Wedge," *J. Inst. Math. Appl.*, **19**, 1977, 231–279.
154. S.-Y. Kim, J. W. Ra, and S.-Y. Shin, "Diffraction by an Arbitrary-Angled Dielectric Wedge: Part I—Physical Optics Approximations," *IEEE Trans. Antennas and Propagat.*, **39**(9), 1991, 1272–1280.
155. S.-Y. Kim, H. W. Ra, and S.-Y. Shin, "Diffraction by an Arbitrary-Angled Dielectric Wedge: Part II—Correction to Physical Optics Solution," *IEEE Trans. Antennas and Propagat.*, **39**(9), 1991, 1282–1292.
156. S. W. Lee, Y. Rahmat-Samii and R. C. Menendez, "GTD, Ray field and Comments on Two Papers," *IEEE Trans. Antennas and Propagat.*, **AP-26**, 1978, 352–354.
157. R. J. Leubbers, "Propagation Prediction for Hilly Terrain using GTD Wedge Diffraction," *IEEE Trans. Antennas and Propagat.*, **AP-32**, 1984, 951–955.
158. R. Tiberio and R. G. Kouyoumjian, "An Analysis of Diffraction at Edges Illuminated by Transition Region Fields," *Radio Science*, **17**(2), 1982, 323–336.
159. R. Tiberio and R. G. Kouyoumjian, "Calculation of High Frequency Diffraction by Two Nearby Edges Illuminated At Grazing Incidence," *IEEE Trans. Antennas and Propagat.*, **AP-34**, 1984, 1186–1196.
160. A. Michaeli, "A New Asymptotic High Frequency Analysis of Electromagnetic Scattering by a Pair of Parallel Wedges: Closed Form Results," *Radio Science*, **20**, 1985, 1537–1548.
161. A. Michaeli, "A Uniform GTD Solution for the Far Field Scattering by Polygonal Cylinders and Strips," *IEEE Trans. Antennas and Propagat.*, **AP-35**, 1987, 983–986.

162. R. Tiberio, G. Manara, G. Pelosi, and R. G. Koumjoujian, "High Frequency Electromagnetic Scattering of Plane Waves From Double Wedges, *IEEE Trans. Antennas and Propagat.*, **AP-37**, 1989, 1172–1180.

163. Y. Rahmat-Samii and R. Mittra, "A Spectral Domain Interpretation of High Frequency Diffraction Phenomenon," *IEEE Trans. Antennas and Propagat.*, **AP-25**, 1977, 676–687.

164. M. Schneider and R. J. Leubbers, "A General, Uniform Double Wedge Diffraction Coefficient," *IEEE Trans. Antennas and Propagat.*, **AP-39**(1), 1991, 8–14.

165. M. Schneider, "A Uniform Solution of Double Knife-Edge Diffraction," MS theis, Penn. State Univ., Dec. 1987.

166. M. Schneider, "A Uniform Solution of Double Wedge-Edge Diffraction," PhD dissertation, Penn. State Univ., Dec. 1988.

167. R. B. Buchanon and R. L. Moore, "Radar Cross Section of a Double Wedge," *IEEE Trans.* **AP-16**(3), 1968, 375–376.

168. R. Tiberio, G. Manara, G. Pelosi, and R. G. Kouyoumjain, "High-Frequency Diffraction by a Double Wedge," presented at *IEEE AP-S Symp. and URSI Meeting, Vancouver, Canada*, 1985.

169. M. I. Herman and J. L. Volakis, "High Frequency Scattering by a Double Impedance Wedge," *IEEE Trans. Antenna and Propagat.*, **AP-36**(3), 1988, 664–678.

170. M. I. Herman and J. L. Volakis, "High Frequency Scattering by a Resistive Strip and Extensions to Conductive and Impedance Strips," *Radio Science*, **22**, 1987, 335–349.

171. [13] Section 2.9.

172. G. A. Thiele and T. H. Newhouse, "A Hybrid Technique for Combining Moment Methods with the Geometrical Theory of Diffraction," *IEEE Trans. Antennas and Propagat.*, **AP-23**(1), 1975, 62–69.

173. J. L. Bogdanor, R. A. Pearlman, and M. D. Siegel, *Intrasystem Electromagnetic Compatibility Analysis Program*, User's Manual Usuage Section, AD-A-008-527, Dec. 1974; User's Manual Eng. Section AD-A-009-526, Dec. 1974; User's Manual Eng. Section AD-A-009-526, Dec. 1974; NTIS, Springfield, VA.

174. Hans P. Widmer, *A Technical Description of the AAPG Program*, Rep. ECAC-CR-83-048, Nov. 1983, Annapolis, MD.

175. G. Gennelo and A. Pesta, "Aircraft Coupling Model Evaluations at SHF/EMF," *IEEE EMC Symp., Wakefield, MA*, Aug. 20–22, 1985, pp. 72–74.

176. P. H. Pathak and N. Wang, "Ray Analysis of Mutual Coupling between Antennas on Convex Surface," *IEEE Trans. Antennas and Propagat.*, **AP-29**(6), 1981, 911–922.

177. B. R.. Levy and J. B. Keller, "Diffraction by a Smooth Object," *Commn. Pure Appld Math.*, **12**, 1959, 159–209.

178. W. Franz and K. Klante, "Diffraction by Surfaces of Variable Curvature," *IRE Trans. Antennas and Propagat.*, **AP-7**, 1959, S68–S70.

179. G. Hessarjian and A. Ishimaru, "Excitation of a Conducting Surface of a Large Radius of Curvature," *IRE Trans.*, **AP-10**, 1962, 264–273.

180. J. R. Wait, "Currents Excited on a Conducting Surface of Large Radius of Curvature," *IRE Trans. Microwave Theory and Tech.*, **MTT-4**(3), 1956. 143–145.

181. Y. Hwang and R. G. Koumyoumjian, *The Mutual Coupling between Slots on an Arbitrary Convex Cylinder*, Ohio State Univ. Electroscience Lab., Dept. of Elec. Eng., Rep. 2902-21, Mar. 1975. Prepared under grant NG-36-0008-138 for NASA.
182. Z. W. Chang, L. B. Felsen, and A. Hessel, *Surface Ray Methods for Mutual Coupling in Conformal Arrays on Cylinder and Conical Surfaces*, Polytechnic Inst. of New York, Final Rep. Sept. 1975–Feb 1976. Prepared under contract N00123-76-C-0236.
183. K. K. Chan, L. B. Felsen, A. Hessel, and J. Shmoys, "Creeping Waves on a Perfectly Conducting Cone," *IEEE Trans. Antenas and Propagat.*, **AP-25**(5), 1977, 661–670.
184. S. W. Lee, "Mutual Impedance of Slots on a Cone: Solution by Ray Techniques," *IEEE Trans. Antennas and Propagat.*, **AP-26**(6), 1978, 768–773.
185. S. W. Lee and S. Naini, "Approximate Asymptotic Solution of Surface Field due to a Magnetic Dipole on a Cylinder," *IEEE Trans. Antennas and Propagat.*, **AP-26**, 1978, 593–597.
186. V. A. Fock, *Electromagnetic Diffraction and Propagation Studies*, Pergamon, New York, 1965.
187. P. H. Pathak and R. G. Kouyoumjian, "Effects of Torsional Surface Waves on the Radiation from Aperture in Convex Cylindrical Surface," *URSI Annual General Meeting, Boulder, CO*, 1974.
188. R. G. Kouyoumjian and P. H. Pathak, "The Radiation from Torsional Surface Waves on Convex Cylinders," presented at the *URSI General Assembly, Lima, Peru*, 1975.
189. N. Wang, *Near-Field Solution for Antennas On Elliptic Cylinders*, Ohio State Univ. Electroscience Lab., Dept. Elec. Eng., Rep. 784685-1, July 1977.
190. P. H. Pathak and N. Wang, *An Analysis of the Mutual Coupling Between Antennas on a Smooth Convex Surface*, Final Rep., 784583-7, Ohio State Univ. Electroscience Lab., Dept. Elec. Eng., Oct. 1978.
191. R. Mittra and S. Safavi-Naini, "Source Radiation in the Presence of Smooth Convex Bodies," *Radio Science*, **14**(2), 1979, 217–237.
192. P. H. Pathak and N. Wang, "Ray Analysis of Mutual Coupling between Antennas on a Convex Surface," *IEEE Trans.*, **AP-29**(6), 1981, 911–922.
193. J. H. Richmond, *Radiation and Scattering by Thin-Wire Structures in the Complex Frequency Domain*, Ohio State Univ. Electroscience Lab., Dept. of Elec. Eng., Report 2902-10, July 1973. (Presented under grant NGL 36-008-138 for NASA.)
194. S. A. Schelkunoff and T. A. Friis, *Antennas, Theory and Practice,*, John Wiley, 1952, pp. 368–370, 401, 407.
195. A. K. Bhattacharyya, *Shading Loss Calculations*, Techn. Note TN-EMC-87-01, Jan. 1985. Prepared for Defense Research Establishment, Ottawa, Dept. Elec. Eng., Concordia Univ., Montreal, Canada.
196. J. R. Wait, "On the Excitation of Electromagnetic Surface Waves on a Curved Surface," *IRE Trans. Antennas and Propagat.*, **AP-8**, 1960, 445–448.
197. N. A. Logan and K. S. Lee, "A Simple Expression for Propagation Constants Associated with a Reactive Convex Surface," *IEEE Trans. Antennas and Propagat.*, **AP-10**, 1962, 103.
198. W. Streifer, "Creeping Wave Propagation Constants for Impedance Boundary Conditions," *IEEE Trans. Antennas and Propagat.* **AP-12**, 1964, 764–766.

199. C. W. Helstrom, "Scattering from a Cylinder Coated with a Dielectric Material," in *Electromagnetic Theory and Antennas*, edited by E. C. Jordan, MacMillan, New York, 1963.
200. R. S. Elliot, "Azimuthal Surface Waves on Circular Cylinder," *J. Appld Phys.*, **26**(4), 1955, 368–376.
201. N. Wang, "Regge, Poles, Natural Frequencies and Surface Wave Resonances of a Circular Cylinder with a Constant Surface Impedance," *IEEE Trans. Antennas and Propagat.*, **AP-30**, 1982, 1244–1247.
202. N. Wang, "Electromagnetic Scattering from a Cylinder Coated with a Dielectric Layer," *National Radio Science Meeting, Albequerque, NM*, May 1982.
203. R. Paknys and N. Wang, "Creeping Wave Propagation Constants and Modal Impedance for a Dielectric Coated Cylinder," *IEEE Trans. Antennas and Propagat.*, **AP-34**(5), 1986, 674–680.
204. J. B. Keller, "Diffraction by an Aperture," *J. Appld Phys.*, **28**, 1957, 426–444.
205. [13] Section 2.8.
206. [72] Ch. 7.
207. K. M. Mintzer, *Incremental Length Diffraction Coefficients*, Tech. Rep. AFAL-TR-73-296, Northrop Corp., Aircraft Division, April 1974, DTIC AD 918861.
208. R. A. Shore and A. D. Yaghjian, "Incremental Diffraction Coefficients for Planar Surfaces," *IEEE Trans. Antennas and Propagat.*, **AP-36**(1), 1988, 55–70. (Correction: ibid., **AP-37**(10), 1989, 1342).
209. P. Ya. Ufimtsev, "Approximate Computation of the Diffraction of Plane Electromagnetic Waves at Certain Metal Bodies (I and II)," *Sov. Phys. Tech. Phys.*, **27**, 1957, 1708–1718; **28**, 1958, 2386–2396.
210. P. Ya Ufimtsev, *Method of Edge Waves in the Physical Theory of Diffraction*, Sovyetskoye Radio, Moscow, 1962. (Available in English from Nat. Tech. Inform. Serv., Springfield, VA 22161, AD 733203).
211. J. L. Volakis and L. Peters Jr., "Evaluation of Reflected Fields at Caustic Regions using a Set of GO Equivalent Line Currents," *IEEE Trans. Antennas and Propagat.*, **AP-33**(8), 1985, 869–876.
212. R. F. Millar, "An Approximate Theory of the Diffraction of an Electromagnetic Wave by an Aperture in a Plane Screen," *Proc. Inst. Elec. Eng., Part C*, **103**, 1956, 177–185. (First published as Monograph 152R, Oct. 1955.)
213. R. F. Millar, "The Diffraction of an Electromagnetic Wave by a Circular Aperture," *Proc. Inst. Elec. Eng., Part C*, **104**, 1957, 87–95.
214. R. F. Millar, "The Diffraction of an Electromagnetic Waves by a Large Aperture," *Proc. Inst. Elec. Eng., Part C*, **104**, 1957, 240–250. (First published as Monograph 213R, Dec. 1956.)
215. A. Michaeli, "Equivalent Edge Currents for Arbitrary Aspects of Observation," *IEEE Trans. Antennas and Propagat.*, **AP-32**, 1984, 252–258 (Correction: **AP-33**, 1985, 227).
216. E. F. Knott and T. B. A. Senior, "Comparison of Three High-Frequency Diffraction Techniques," *Proc. IEEE*, **62**, 1974, 1468–1474.
217. R. A. Shore and A. D. Yaghjian, "Incremental Diffraction Coefficients for Planar Surfaces," *IEEE Trans. Antennas and Propagat.*, **AP-36**(1), 1988, 55–70.

218. G. Pelosi, S. Maci, R. Tiberio, and A. Michaeli, "Incremental Length Diffraction Coefficients for an Impedance Wedge," *IEEE Trans. Antennas and Propagat.*, **AP-40**(10), 1992, 1201–1210.

219. R. G. Rojas, "Diffraction of EM Waves by a Dielectric/Ferrite Half-Plane and Related Configurations," *IEEE Trans. Antennas and Propagat.*, **AP-37**(6), 1987, 751–763.

220. R. G. Rojas, H. C. Ly, and P. H. Pathak, "Electronmagnetic Plane Wave Diffraction by a Planar Junction of Two Thin Dielectric/Ferrite Half Plane," *Radio Science*, **26**(3), 1991, 641–660.

221. T. B. A. Senior, "Scattering From Imperfect Half-Planes," In *Electromagnetic Scattering*, edited by P. L. E. Uslenghi, Academic Press, New York, 1978.

222. V. Daniele, "On the Solution of Two Coupled Weiner–Hopf Equations," *SIAM J. Appld Math.*, **44**, 1984, 667–680.

223. A. A. Khrapkov, "Certain Cases of the Elastic Equilibrium of an Infinite Wedge with a Nonsymmetric Notch at the Vertex Subjected to Concentrated Forces," *Prikl. Mat. Mekh.*, **35**, 1971, 652–663.

224. R. A. Hurd and E. Luneberg, "Diffraction by an Anisotropic Impedance Half Plane," *Can J. Phys.*, **63**, 1965, 1135–1140.

225. A. H. Serbest and A. Yazici, "Plane Wave Diffraction by an Anisotropic Dielectric Half Plane," *IEEE Antenna and Propagat. Soc. Conf.*, *London, Canada*, June 23–26, 1991, Session 32, Paper 7, pp. 562–565.

226. A. H. Serbest, A. Buyukaksoy, and G. Uzgoren, "Diffraction at a Discontinuity Formed by Two Anisotropic Impedance Half Planes," *IEICE Trans.*, **E74**(5), 1991, 1283–1287.

227. J. M. L. Bernard, "Diffraction by a Metallic Wedge Covered with a Dielectric Material," *Wave Motion*, **9**, 1987, 543–561.

228. H. J. Bilow, "Scattering by an Infinite Wedge with Tensor Impedance Boundary Conditions—A Moment Method/Physical Optics Solution for the Currents," *IEEE Trans. Antennas and Propagat.*, **AP-39**(6), 1991, 767–773.

229. W. D. Burnside and K. W. Burgener, "High Frequency Scattering by a Thin Lossless Dielectric Slab," *IEEE Trans. Antennas and Proagat.*, **AP-31**(1), 1983, 104–110.

230. R. J. Luebbers, "Finite Conductivity Uniform GTD versus Knife Edge Diffraction in Prediction of Propagation Path Loss," *IEEE Trans. Antennas and Propagat.*, **AP-32**(1), 1984, 70–76.

231. R. J. Luebbers, "A Heuristic UTD Slope Diffraction Coefficient for Rough Lossy Wedges," *IEEE Trans. Antennas and Propagat.*, **AP-37**(2), 1989, 206–211.

232. R. Mittra and M. Tew, "Accuracy Test for High-Frequency Asymptotic Solutions," *IEEE Trans. Antennas and Propagat.*, **AP-37**(1), 1979, 62–68.

233. M. Tew and R. Mittra, "An Integral E-Field Accuracy Test for High Frequency Asymptotic Solutions," *IEEE Trans. Antennas and Propagat.*, **AP-28**(4), 1980, 513–518.

234. J. A. Aas, "On the Accuracy of the Uniform Geometrical Theory of Diffraction Close to a 90° Wedge," *IEEE Trans. Antennas and Propagat.*, **AP-27**(5), 1979, 704–705.

235. J. Bach Andersen and V. V. Solodukhov, "Field Behaviour Near a Dielectric Wedge," **AP-26**(4), 1978, 598–602.

236. K. R. Jacobsen, "An Alternative Diffraction Coefficient for the Wedge," *IEEE Trans. Antennas and Propagat.*, **AP-32**(2), 1984, 175–177.

237. "Bridging the Gap Between the Low and High Frequency Regimes," *IEEE APS/URSI Symp.*, Chicago, IL, July 20–25, 1992, Session MP09, pp. 80–87.

238. P. H. Pathak and R. J. Marhefka, "On the Behaviour of Uniform Ray Solutions at Lower Frequencies," *IEEE/URSI Symp., Chicago, IL*, July 20–25, 1992, Session MP09, p. 97.

239. L. K. Warne and K. C. Chen, "A Bound on EMP Coupling," *IEEE Trans. Electromag. Compat.*, **EMC-32**(3), 1990, 217–222.

240. W. A. Davis, "Bounding EMP Interaction and Coupling," *IEEE Trans. Electromag. Compat.*, **EMC-29**(6), 1981, 842–846.

241. M. K. Sistanizadeh and W. A. Davis, "Time-Domain Upper Bound on Signals on a Multi-Conductor Transmission Line Behind An Aperture," *IEEE Trans. Antennas and Propagat.*, **AP-30**(6), 1982, 1247–1250.

242. A. Taflove and K. Umashankar, "A Hybrid Moment Method/Finite Difference Time-Domain Approach to Electromagnetic Coupling and Aperture Penetration," *IEEE Trans. Antennas and Propagat.* **AP-30**(4), 1982, 617–627.

243. J. B. Keller and A. Blank, "Diffraction and Refraction of Pulses by Wedges and Corners," *Commun. Pure Appld Math.*, **4**, 1951, 75–94.

244. F. Oberhettinger, "On the Diffraction and Refraction of Waves and Pulses by Wedges and Corners," *J. Res. Nat. Bur. Stand.*, **61**(5), 1958, 343–365.

245. L. B. Felsen, "Diffraction from Pulse Field from an Arbitrarily Oriented Electric or Magnetic Dipole by a Perfectly Conducting Wedge," *SIAM Jr. Appld Math.*, **26**(2), 1974, 306–312.

246. D. E. Merewether, "Transient Current Induced on a Body of Revolution by an Electromagnetic Pulse," *IEEE Trans. Electromag. Compat.*, **EMC-13**(2), 1971, 41–44.

247. C. Eftimiu and P. L. Huddleson, "The Transient Response of Finite Open Cylinders," *IEEE Trans. Antennas and Propagat.*, **AP-32**(4), 1984, 356–363.

248. T. W. Verrutipong, "Time Domain Version of the Uniform GTD," *IEEE Trans. Antennas and Propagat.*, **AP-38**(11), 1990, 1757–1764.

249. A. K. Dominek, "Transient Scattering Analysis for a Circular Disk," *IEEE Trans. Antennas and Propagat.*, **AP-39**(6), 1991, 815–819.

250. P. R. Barnes and F. M. Tesche, "On the Direct Calculation of a Transient Plane Wave Reflected from a Finitely Conducting Half Space," *IEEE Trans. Electromag. Compat.*, **EMC-33**(2), 1991, 90–96.

251. B. D. Kahler and A. K. Bhattacharyya, "Physical Optics and GTD Time-Domain Scattering from a Finite PEC Cylinder with One End Open and One End Closed," *Micro. Opt. Tech. Lett.*, **5**(10), 1992, 524–529.

252. A. K. Bhattacharyya, "Time-Domain UTD Coefficient for Scattering from a Wedge with Two Face Impedances and Some Applications," 1992 *IEEE EMC Conference, Anaheim CA*, August 1992, pp. 499–502.

1.31 ADDITIONAL SOURCES

H. Bach and H.-H. Viskum, "The SNFGTD Method and Its Accuracy," *IEEE Trans.* **AP-35**(2), 1987, 169–175.

D. V. Giri, *Canonical Examples of High Power Microwave (HPM) Radiation Systems for the Case of One Feeding Waveguide*, Sensor and Simulation Note, 326, April 10, 1991, Pro. Tech Lafayette, CA.

S. I. Hariharan, *Absorbing Boundary Conditions for Exterior Problems*, Inst. Comput. Appln. Sci. and Eng. (ICASE), Nasa Langley Research Center, Hampton, VA, Nasa Contractor Rep. 177944.

V. Kerdemelidis, G. L. James and B. A. Smith, "Some Methods of Reducing the Sidelobe Radiation of Existing Reflector and Horn Antennas," ch. 8 in Y. T. Lo and S. W. Lee (Eds.), (Supply book title) Van Nostrand Reinhold, New York, 1988.

A. Michaeli, "Equivalent Edge Currents For Arbitrary Aspects of Observation," *IEEE Trans, Antennas and Propagat.*, **AP-32**(3), 1984, 252–258. (Correction: *IEEE Trans. Antennas and Propagat.*, **AP-33**(2), 1985, 227.)

R. Paknys, "On the Accuracy of the UTD for the Scattering by a Cylinder," *IEEE Trans. Antennas and Propagat.*, **AP-42**(5), 1994, 757–760.

N. D. Taket and R. E. Burge, "A Physical Optics Version of the Geometrical Theory of Diffraction," *IEEE Trans. Antennas and Propagat.*, **AP-39**, 1991, 719–731.

P. Ya Ufitmtsev, Y. Rahmat-Samii and K. M. Mitzner, "The Physical Theory of Slope Diffraction," *PIERS Conference Special Session on Current Topics in Scattering and Diffraction, JPL*, July 12–16, 1993.

CHAPTER TWO

Near and Far Fields of Electromagnetic Horn Antennas

2.1 INTRODUCTION

The use of different types of electromagnetic horns as primary feeds is very common in reflector antenna analysis, design, and applications. Hence, one needs to accurately know the near- and far-field patterns of horn antennas; how accurately depends on the practical applications and the set priorities. For ordinary applications a PO method of pattern calculation may be sufficient. But for a situation where interference to neighboring equipment or where the secrecy of communication is important one might need to avoid strong edge illumination and spillovers to suppress large-angle sidelobes and also to increase the front-to-back lobe ratio of the reflector system. This calls for a more complete analysis.

2.2 THE DIFFRACTION PROBLEM OF ELECTROMAGNETIC HORNS

Application of the GTD to horn antennas for accurate prediction of near and far sidelobes has been investigated for quite some time. Perhaps the earliest work on horn antennas using Keller's GTD was due to Russo, Rudduck, and Peters [1]. In this work, they approximated a dihedral corner reflector excited by a magnetic line source and then obtained its E-plane far-field radiation patterns. Yu, Rudduck, and Peters [2] carried out a comprehensive study of the E-plane radiation pattern considering multiply reflected images. Yu and Rudduck [3] obtained the H-plane pattern using Keller's GTD and parallel-plate waveguide mode approximation. This was refined by Mentzner, Peters, and Rudduck [4] and by Narasimhan and Rao [5] using UTD slope diffraction correction. The near- and far-field E- and H-plane patterns were found in [5]

176   NEAR AND FAR FIELDS OF ELECTROMAGNETIC HORN ANTENNAS

using a spherical source excitation. UAT has been used by Menendez and Lee [6] on rectangular horns considering up to $k^{-1/2}$ terms. Rectangular waveguide modes were approximated by parallel-plate modes and the resulting plane waves were used to give a clear insight to the horn problem. The computed E-plane patterns in subsequent investigations [4–6] showed no significant improvement over that obtained in [2]. A fundamental limitation in [2] was the approximation (up to $k^{-3/2}$ terms) which in general differ from that of UAT [7–9]. This limitation does not exist in UAT. The near- and far-field E- and H-plane patterns for uncorrugated pyramidal horns using UTD have been described in a recent text [10].

The steps of the basic approach for the analytical formulation for the field due to an electromagnetic horns at any observation point are: (1) identify the dominant scattering centers in the plane of interest; (2) postulate a source inside the horn, possibly at the phase center; (3) obtain the scattered field from the scattering centers by vectorially adding the contributions; each contribution is expressed as the product of the incident field from the source at the scattering center, its diffraction coefficient, divergence (spreading) factor, and the phase of the diffracted field to the observation point; and (4) for near-field calculations, the slope diffraction correction (Appendix A.1) should be incorporated as the edge cannot be considered uniformly illuminated in near-field estimates.

2.3 PYRAMIDAL HORNS: UAT ANALYSIS

Since the analysis and applications of E- and H-plane patterns of pyramidal horns using UTD have been described in a recent text [10, sections 5.3 and 5.4] they will not be discussed here. Hence we describe here the UAT treatment of the same problem.

2.3.1 *E*-Plane Diffraction Problem

First-Order Fields. The pyramidal horn can be approximated by a dihedral corner reflector formed by two perfectly conducting half planes and excited by an isotropic magnetic line current source at the vertex S (Fig. 2.1). The omnidirectional ray-optical field (magnetic field normal to the E-plane) due to this source is given by

$$H^i(r_1) \sim e^{i(kr_1 - \pi/4)} \left(\frac{2}{\pi k r_1}\right)^{1/2} \cdot \left\{1 + \frac{1}{i8kr_1}\right\} + O(k^{-2}) \qquad (2.1)$$

This ray is incident tangentially on each half plane. At the observation point P, the resulting total field H^t_A due to diffraction at the edge A is the sum of the geometrical optics (GO) field H^G and the diffracted field H^d [11] and is given by

$$H^t_A = H^G(r, \phi) + H^d(r, \phi) \qquad (2.2a)$$

2.3 PYRAMIDAL HORNS: UAT ANALYSIS

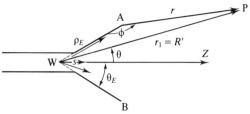

FIGURE 2.1 Geometry of the corner reflector in the E-plane. (From [11] © 1983 IEEE.)

where
$$H^G(r, \phi) = H^{Gi}(r, \phi) + H^{Gr}(r, \phi) \tag{2.2b}$$
$$H^d(r, \phi) = H^{di}(r, \phi) + H^{dr}(r, \phi) \tag{2.2c}$$
where
$$H^{Gr}(r, \phi) = R \cdot H^{Gi}(r, 4\pi - \phi) \tag{2.2d}$$
$$H^{dr}(r, \phi) = R \cdot H^{di}(r, 4\pi - \phi) \tag{2.2e}$$

R is the reflection coefficient, which is equal to unity as the incident magnetic field is parallel to the edge. The GO and the diffracted fields are given by

$$H^G(r, \phi) \sim 2\{F(k^{1/2}\xi) - \hat{F}(k^{1/2}\xi)\} H^i(r_1) \tag{2.3a}$$

$$H^d(r, \phi) \sim -\frac{e^{ik(r+\rho_E)}}{\pi k(\tau \rho_E)^{1/2}} \sec\frac{\phi}{2}\left[1 + \sec^2\frac{\phi}{2}\right] \cdot \frac{1}{4}\left(\frac{1}{\rho_E} + \frac{1}{R}\right) \bigg/ ik + O(k^{-2}) \tag{2.3b}$$

The expressions for the Fresnel integral $F(X)$ and its dominant term $\hat{F}(X)$ are given in eqns. (A.9b) and (A.19) in Appendix A.2. The detour parameter (DP) ξ is given by

$$\xi = (r + \rho_E - r_1)^{1/2} \cdot \text{sgn}\left(\cos\frac{\phi}{2}\right) \tag{2.3c}$$

At the near shadow/reflection boundary (SB/RB) ($\phi = \pi$), the asymptotic expansions of the GO and the diffracted fields given in eqn. (2.3a,b) become very large. The total field given by eqn. (2.2) then involves the subtraction of two large numbers. Exactly at the SB/RB boundary singularities in H^d cancels out with those of H^G. Then eqn. (2.2) reduces to

$$H_A^t(SB/RB) \sim H^i(r_1 = r + \rho_E) \tag{2.4}$$

Similarly, the total field H_B^t due to diffraction at edge B may also be found.

The E-plane field is then obtained by first summing the contributions H_A^t and H_B^t. However, for angles $|\theta|\{< \theta_E$, this results in the incident field being computed twice [11, 12]. Thus on subtracting the incident field (2.1) from this sum we obtain the first-order field from the horn.

$$H^t(r_1, \phi) = H_A^t + H_B^t - H^i(r_1)|_{|\theta| < \theta_E} \tag{2.5}$$

It may be noted here that for $\phi > 3\pi/2 - \theta_E$ and $\theta < \pi - \theta_E$ (see Fig. 2.1) edge B does not contribute and therefore $H_B^t = 0$ in this region.

Second-Order Fields Using Images

The first-order diffracted line sources at edges A and B (Fig. 2.2) give rise to images due to reflections from the walls [2], each of which can be considered as a nonisotropic line source giving rise to the incident field at the other.

$$H^i(r_{1I}, \phi_{1I}) \sim -\frac{\exp ik(r_{1I} + \rho_E)}{\pi k(r_{1I}\rho_E)^{1/2}} \cdot \sec\left(\frac{\phi_{1I}}{2}\right) \cdot \frac{1}{4}$$

$$\cdot \sec^3 \frac{\phi_{1I}}{2} \left(\frac{1}{D} + \frac{1}{r_{1I}}\right) \bigg/ ik + O(k^{-2}) \quad (2.6)$$

where

$$\phi_{1I} = \Omega_I + \Delta\phi_{1I} \quad \text{(Fig. 2.2)} \quad (2.7)$$

Here, the first subscript 1 corresponds to source coordinate (Fig. 2.2), the second subscript I corresponds to secondary image sources due to reflections. The second-order fields due to this source can be computed using eqn. (2.5).

Multiple Diffractions. The multiple diffractions become significant for θ around $\pi/2$ (Figs 2.3 and 2.4). The multiply diffracted field from edge B (far away from the SB at $\phi_1' = 3\pi/2 - \theta_E$) is given by

$$H_B^i(r_1', \phi_1') = \exp(ikr_1) \cdot Z^i(r_1', \phi_1') \quad (2.8)$$

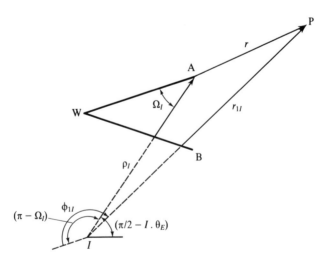

FIGURE 2.2 Image source. (From [11] © 1983 IEEE.)

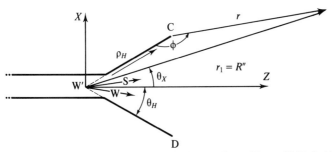

FIGURE 2.3 Geometry of the corner reflector in the H-plane. (From [11] © 1983 IEEE.)

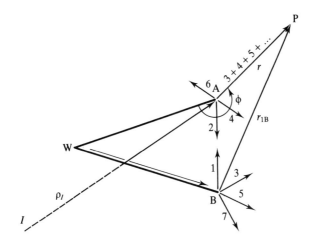

FIGURE 2.4 Multiple reflections. (From [11] © 1983 IEEE.)

where

$$Z^i(r'_1, \phi'_1) = \frac{\exp(\pi/4 + kD)}{(2\pi k)^{3/2}(r'_1 D)^{1/2}} \sum_{I=0}^{IH-1} \frac{\exp(ik\rho_I)}{\rho_I^{1/2}} \cdot \sec\frac{\Omega_I}{2} \cdot \left[\sec\frac{\phi'_1}{2} + \sec\frac{\phi'_1 + \Omega_I}{2}\right]$$
$$+ C \cdot \left[\sec\frac{\phi'_1 - \Omega_0}{2} + \sec\frac{\phi'_1 + \Omega_0}{2}\right] + O(k^{3/2}) \quad (2.9)$$

and

$$C = -\frac{\exp i(\pi/4 + k\rho_0) \sum_{I=0}^{IH-1} \frac{\exp(ik\rho_I)}{\rho_I^{1/2}} \cdot \sec\frac{\Omega_I}{2} \cdot \left\{\sec\frac{\Omega_0 - \Omega_I}{2} + \sec\frac{\Omega_0 - \Omega_I}{2}\right\}}{\pi k 2(2\pi k \rho_0)^{1/2} + \exp i(\pi/4 + k\rho_0) \cdot (1 + \sec\Omega_0)}$$
$$(2.10)$$

where IH is the number of images formed at each wall [2]. This field from edge B gives rise to multiple diffraction from edge A and results in

$$H^t_{\text{multiple}} = \exp(r' + \rho_0)[F(k^{1/2}\xi_1) - \hat{F}(k^{1/2}\xi_1)] \cdot Z^t(r'_1, \phi'_1)$$

$$+ \frac{\exp i[\pi/4 + k(r' + \rho_0)]}{(2\pi k)^{3/2}(r'D)^{1/2}} \cdot C \cdot \left[\sec \frac{\phi' - \Omega_0}{2} + \sec \frac{\phi' + \Omega_0}{2} \right]$$

$$+ O(k^{-5/2}) \tag{2.11}$$

where

$$\phi' = 3\pi/2 - \theta_E \quad \text{or, in the far field case, } \theta \neq \pi/2$$

$$\xi_1 = (r' + \rho_0 - r'_1)^{1/2} \cdot \operatorname{sgn} \cos \frac{\phi' - \Omega_0}{2} \tag{2.12}$$

The functions $F(X)$ and $\hat{F}(X)$ are defined in Appendix A.2.

At the ISB of the far-field incident from edge B (at $\phi' = 3\pi/2 - \theta_E$) and in the far-field at $\theta = \pi/2$

$$H^t_{\text{multiple}} = \frac{H^I_B(r'_1, \phi'_1)}{2} + \frac{\exp i[\pi/4 + k(r' + D)]}{(2\pi k)^{3/2}(r'D)^{1/2}}$$

$$\cdot C \cdot \sec \frac{\phi' + \Omega_0}{2} + O(k^{-5/2}) \tag{2.13}$$

Around $\theta = 0$ equal to $(\pi - \theta_E)$ wedge diffractions become important [2] and these can be computed using UAT formulation for a curved wedge.

Results and Discussions. Numerical computations were performed in [11] for near- and far-field patterns of some horns. One important issue associated with UAT computations involves the subtraction of two large numbers. Near the SB the large term due to $\hat{F}(X)$ in the GO field (2.3a), should be separately computed and added to the diffracted field. Finally, the total field can be computed by adding the remaining part of the GO field, consisting of $F(X)$. Using a Burroughs-6700 computer the typical processing time for a data file was 39 sec for a radiation pattern with $0 \leq \theta \leq 180°$ in steps of 1°. E-plane near- and far-field patterns for a typical pyramidal horn are shown in Figures 2.5 and 2.6 at 9.8 GHz. It is found that UAT with all factors taken into account shows better agreement than before [2] by at least 1.0 dB at 35 dB below reference. UAT predicts even the ripples at wide angles between 90° and 160°, as shown in Figure 2.6. This is due to the diffracted field from the edge, nonuniformity of the cylindrical sources and images, and the inclusion of $k^{-3/2}$ terms. This does not seem to be due to the presence of the feeder waveguide, attenuator, detector, and so on, as explained in [2]. The presence of ripples in the near-field pattern (Fig. 2.6) is predicted unlike in [5]. The inclusion of multiple diffraction improves the agreement (from $\theta = 75°$ to 90° in Fig. 2.5).

2.3 PYRAMIDAL HORNS: UAT ANALYSIS

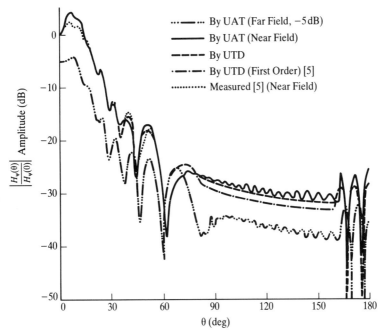

FIGURE 2.5 E-plane near-field patterns, $R = 47.5\lambda$. (From [11] © 1983 IEEE.)

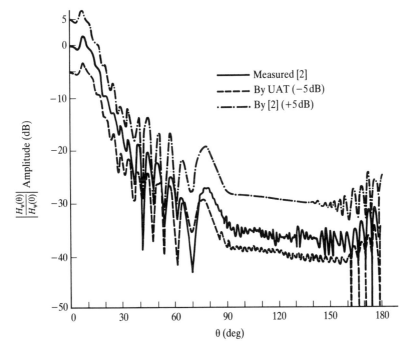

FIGURE 2.6 E-plane far-field patterns of a pyramidal horn, $\rho_E = 24.8\lambda$, $\theta_E = 17.5°$. (From [11] © 1983 IEEE.)

182 NEAR AND FAR FIELDS OF ELECTROMAGNETIC HORN ANTENNAS

A 2 dB difference in the main lobe appears between UAT and near-field experimental results (Fig. 2.6). This is explained on the same lines as [11]. The magnetic line source is not an exact representation of the real nature of the exciting source. The incident fields on edges A and B are shadow boundary fields hence they are not true ray-optical fields. It is quite likely this has contributed to the disagreement. The discrepancy can also be accounted for partly by probe–aperture interaction. In the case of small horns the agreement between the theory and experiments is of the same order as in [2]. This analysis is therefore less accurate and is again attributed to the actual source being nonray-optical. Also, higher-order modes should be considered in this case. The theoretical results are strictly applicable to the E-plane pattern of two-dimensional horns with E-plane flares. This is true as long as edges in the E-plane are not very close to each other. In practice it works well even when the separation is a few wavelengths [4, 5]. The omnidirectional source at the vertex is an adequate approximation to the actual field for dominant mode excitation of the waveguide. This representation is valid for all conditions of waveguide–horn junction as long as higher modes are not excited. A quantitative estimation of the above calls for a large number of accurate experimental near- and far-field patterns compared with theoretical results for a wide variety of horn dimensions.

This study concludes the uniform theory of diffraction gives good agreement for near- and far-field patterns compared with theoretical results for electromagnetic horns. A better agreement is obtained when the terms up to $k^{-3/2}$ are included, particularly beyond an observation angle equal to 90°.

2.3.2 *H*-Plane Patterns

The pyramidal horn is again approximated by a dihedral corner reflector formed by, two perfectly electrically conducting (PEC), inclined, half planes, Figure 2.5(a). However, unlike the E-plane case described earlier, the corner reflector is now excited by an anisotropic electric line current source at the vertex S_H. Rays from this source are assumed to have their electric vector parallel to the corner reflector edges (i.e. normal to the H-plane). These are incident tangentially on the reflector walls. The ray-optical expansion of this source is given by

$$E^i(r_1, \theta_x) \sim e^{i(kr_1 - \pi/4)} \cdot \left(\frac{2}{\pi k r_1}\right)^{1/2} \cdot \cos\left(\frac{\pi}{2}\frac{\theta_x}{\theta_H}\right) \qquad (2.14)$$

Consider now the corner reflector wall W'C (Fig. 2.7) which forms a part of the upper half plane. Using UAT, the field incident on this half plane may be reproduced below.

$$E^i(r_1, \phi_1) \sim e^{i(kr_1 - \pi/4)} \cdot \left(\frac{2}{\pi k r_1}\right)^{1/2} \cdot \cos\left(\frac{\pi}{2}\frac{\pi - \phi_1 + \theta_H}{\theta_H}\right) \qquad (2.15)$$

where $\phi_1 = \pi - \theta_X + \theta_H$. The reflected field in this case of grazing incidence is then given by

$$E^r(r_{-1}, \phi_{-1}) = -E^i(r_1, \phi_{-1}) - e^{i(kr_1 - \pi/4)}\left(\frac{2}{\pi k r_1}\right)^{1/2} \cos\left(\frac{\pi}{2} \cdot \frac{\pi - \phi_{-1} + \theta_H}{\theta_H}\right)$$

(2.16)

where $\phi_{-1} = \pi + \theta_X - \theta_H$. Then the GO field is given by

$$E^G \sim \frac{e^{-\pi/4}}{\sqrt{\pi}} \int_{-\infty}^{k^{1/2}\xi} e^{it^2} + \frac{e^{i(k\xi^2 + \pi/4)}}{2\sqrt{\pi}k^{1/2}\xi}(1 + 1/2ik\xi^2 + O(k^{-2}))$$
$$\cdot \{E^i(r_1, \phi_1) + E^r(r_{-1}, \phi_{-1})\}$$

(2.17)

The detour parameter is given by

$$\xi = (r + \rho_H - r_1)^{1/2} \cdot \text{sgn}\left(\cos\frac{\phi}{2}\right)$$

(2.18)

The rays emanating from the source S_H are diffracted by edge C (Fig. 2.7). The diffracted field is of the form

$$E^d \sim e^{ik(r + \rho_H)} \cdot k^{-1/2}\hat{v}_0(r, \phi) - \hat{v}_0(r, 4\pi - \phi)$$
$$+ (\hat{v}_1(r, \phi) - \hat{v}_1(r, 4\pi - \phi))/ik + O(k^{-2})$$

(2.19)

Since this is a case of grazing incidence with an E-vector parallel to the edge,

$$\hat{v}_0(r, \phi) - \hat{v}_0(r, 4\pi - \phi) = 0$$

(2.20)

Moreover, as the incident field vanishes along the walls of the corner reflector

$$\hat{v}_0(r, \phi) = \hat{v}_0(r, 4\pi - \phi) = 0$$

(2.21)

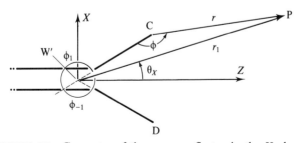

FIGURE 2.7 Geometry of the corner reflector in the H-plane.

Therefore, considering the next-higher-order terms we obtain the diffracted field as

$$E^d \sim \frac{e^{ik(r+\rho_H)}}{4ik^2\rho_H\theta_H} \cdot \frac{1}{(r\rho^H)^{1/2}} \cdot \frac{\sin\phi/2}{\cos^2\phi/2} + O(k^{-3}) \tag{2.22}$$

Then the UAT field E_C^t due to diffraction at edge C is obtained as the sum of the GO field (2.17) and the diffracted field (2.19), i.e.

$$E_C^t = E^G + E^d \tag{2.23}$$

At the SB/RB boundary, corresponding to $\phi = \pi$ (Fig. 2.7), $\xi = 0$. The singularities present in E^G and E^d cancel out. Then the total field reduces to

$$E_C^t(SB/RB) \sim \frac{e^{ik(r+\rho_H)}}{\theta_H \cdot k(r\rho_H)} \cdot \left(\frac{r}{\rho_H}\right)^{1/2} \tag{2.24}$$

Proceeding as above, the total field E_D^t due to diffraction at edge D can now be similarly obtained.

The H-plane horn field E^t is then obtained by first summing the contributions E_C^t and E_D^t. However, this results in the incident field being computed twice for angles $|\theta_X| < \theta_H$. Thus on subtracting the incident field (2.14) from this sum we obtain

$$E^t = E_C^t + E_D^t - E^i(r_1, \theta_x) \tag{2.25a}$$

where

$$E^i(r_1, \theta_x) = 0 \quad \text{for } |\theta_X| < \theta_H \tag{2.25b}$$

It may be noted that for observation positions between $\phi < 3\pi/2 - \theta_H$ and $\theta_X < \pi - \theta_H$ (Fig. 2.7), edge D remains invisible and we have $E_D^t = 0$.

It should be mentioned here that eqn. (2.25a) does not completely represent the H-plane radiation patterns, especially in the backlobe region. In the backlobe region the diffraction from the two edges perpendicular to the E-plane should be considered. Referring to Figure 2.8, the rays incident at the midpoint of this edge result in conical diffraction because of the skewed angle of incidence. These rays significantly contribute to the H-plane radiation pattern in the backlobe region. The diffracted field at the observation point P due to the top edge in the H-plane is given by

$$E_T^d = -\frac{e^{i(k\sigma + \pi/4)}}{2(2\sigma)^{1/2}} \cdot \frac{1}{\sin\beta} \cdot \frac{2}{\cos\frac{\pi + \psi^i}{2}} E^i(Q) \tag{2.26}$$

where $E^i(Q)$ is the value of the incident electric field at the point of diffraction and $\psi^i = \pi - \theta_E + v$. This electric field makes an angle $(\pi - v)$ with the y-axis

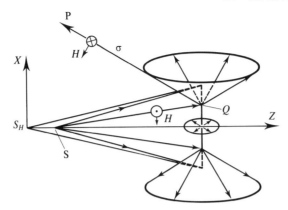

FIGURE 2.8 Conical diffraction.

at the observation point P. Keeping this in view and taking into account the bottom edge in the H-plane, we have the contribution of the source S (Fig. 2.8) to the H-plane field at P as

$$E_S^d \sim \frac{e^{|i(k\sigma + \pi/4)|}}{(2\pi k\sigma)^{1/2} \sin \beta} \frac{2E^i(Q)}{\sin \psi^i/2} \cos \nu \qquad (2.27)$$

The total field in the H-plane is given by

$$E^t = E_A^t + E_B^t - E^i(r_1, \theta_X) + E_S^d \qquad (2.28)$$

It may be noted that the expression for the field of a pyramidal horn using UAT is valid for all distances from the edges greater than $\rho = \lambda/4$.

2.4 CONICAL HORNS

In this section we describe how the pattern at any point due to a conical horn can be determined using the uniform theory of diffraction. Conical horns have been treated using high-frequency techniques in a number of references [13–21].

The field $u(R, \theta)$ at any observation point due to a scattering center on an edge is given by

$$u(R, \theta) = E^i|_{edge} \cdot D_{s,h} \cdot \Gamma \cdot \Psi(R) \qquad (2.29)$$

where $E^i|_{edge}$ = the incident field at the scattering center on the edge
$D_{s,h}$ = the diffraction coefficient of the edge for soft and hard polarizations
Γ = the divergence factor for the diffracted wave
$\Psi(R)$ = the phase factor

186 NEAR AND FAR FIELDS OF ELECTROMAGNETIC HORN ANTENNAS

In situations where more than one edge contributes to the field at the observation point, the resultant field is the vector sum of the individual contributions determined by eqn. (2.28).

2.4.1 Uncorrugated Conical Horns

E-Plane Patterns. The GTD analysis of the near-field patterns of uncorrugated and corrugated conical horns using UTD is available in [16]. Figure 2.9 shows the geometry and coordinate system of the conical horn in the $\phi = 90°$ plane. α_0 is the half angle of the horn. The apex of the horn is at the center of the coordinate system. The total field at any point is the vector sum of the diffracted field from the scattering centers on edges A and B. The horn is assumed to support only the dominant TE_{11} mode. This is true as long as the horn flare angle is small.

Near-Field Patterns. To calculate near-field patterns, since the edges are nonuniformly illuminated one needs to use the slope diffraction correction (Appendix A.1) in the UTD coefficient.

The slope-corrected diffracted fields from the two edges are given by [16]

$$E_\theta^{d,A} = E_\theta^G(r_0, \alpha_0) D_h^s + \frac{1}{jkr_0} \left.\frac{dE_\theta^g}{d\theta}\right|_{r=r_0, |\theta|=\alpha_0} \cdot \frac{dD_h^A}{d\xi} \sqrt{\frac{\rho_{1,2}}{s_{1,2}(s_{1,2}+\rho_{1,2})}}$$
$$\cdot e^{-jks_{1,2}} \cdot e^{j(\alpha_0+\theta_1-\theta-\pi/2)} \tag{2.30}$$

$$D_h^{A,B} = -e^{-j\pi/4}\sqrt{8\pi k}\, \frac{F[kL_{1,2}a(\theta_{1,2})]}{\cos\dfrac{\theta_{1,2}}{2}} \tag{2.30a}$$

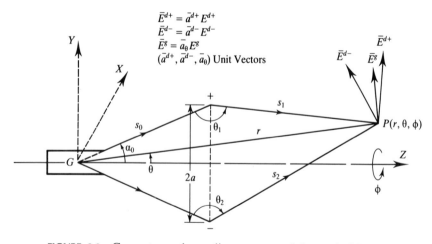

FIGURE 2.9 Geometry and coordinate system of the conical horn.

where the distance parameter $L_{1,2}$ is given by

$$L_{1,2} = \frac{s_{1,2} r_0}{r_0 + s_{1,2}} \tag{2.30b}$$

and

$$a(\theta_{1,2}) = 2\cos^2 \frac{\theta_{1,2}}{2} \tag{2.30c}$$

$$F(u) = j2\sqrt{u}\, e^{ju} \int_{\sqrt{u}}^{\infty} e^{-jz^2}\, dz \tag{2.30d}$$

$$\frac{dD_h}{d\xi} = \frac{dD_s}{d\xi} = je^{j\pi/4}\sqrt{\frac{x}{\pi}}\, kx \cdot \sin\xi \cdot \operatorname{sgn}\left\{\cos\frac{\xi}{2}\right\} \cdot e^{jkx(1+\cos\xi)} \int_{\sqrt{kx(1+\cos\xi)}}^{\infty} \frac{e^{-jz^2}}{\sqrt{kx(1+\cos\xi)}}\, dz$$

$$+ \frac{j}{2\sqrt{kx(1+\cos\xi)}} \tag{2.30e}$$

$$\frac{dD_h^1}{d\xi} = \left.\frac{dD_h}{d\xi}\right|_{\xi=\theta_{1,2},\, x=L_{1,2}} \tag{2.30f}$$

$$dE_\theta^g\big|_{r=r_0,\,|\theta|=\alpha_0} = A\,\frac{e^{-jkr_0}}{|r_0|}\sin\theta_0\,\frac{d}{d\alpha_0}P_S^1(\cos\alpha_0)$$

$$- \cos\alpha_0 P_S^1(\cos\alpha_0) \sin^2\alpha_0 \tag{2.30g}$$

$$\rho_{1,2} = -r_0 \sin\alpha_0 / \sin(\theta_{1,2} + \alpha_0) \tag{2.30h}$$

$$s_{1,2} = \sqrt{r^2 + r_0^2 - 2rr_0 \cos(\alpha_0 \mp \theta)} \tag{2.30i}$$

From the geometry, $\theta_{1,2}$ is given by

$$\theta_{1,2} = \begin{cases} \cos^{-1}\left(\dfrac{r_0^2 + s_{1,2}^2 - r^2}{2r_0 s_{1,2}}\right) & \text{for } |\theta| \leq \alpha_0 \text{ and for} \\ & 0 \leq |\theta| \leq (\pi - \alpha_0) \text{ respectively} \\ 2\pi - \cos^{-1}\left(\dfrac{r_0^2 + s_{1,2}^2 - r^2}{2r_0 s_{1,2}}\right) & \text{for } \alpha_0|\theta| \leq \pi \text{ and for} \\ & (\alpha - \alpha_0) < |\theta| \leq \pi \text{ respectively} \end{cases} \tag{2.31}$$

The GO field is given by

$$E_\theta^G = AP_S^1(\cos\theta)e^{-jkr\sin\theta_r} \cdot 3e^{j\pi/2} \quad \text{for } \alpha_0 \leq |\theta| \leq \pi$$

$$= 0 \quad \text{for } |\theta| \leq \alpha_0 \tag{2.32}$$

The total nonaxial near field of the horn is given by

$$E_T = E_\theta^G + E_\theta^{D,A} + E_\theta^{D,B} \quad \text{for } 0 < |\theta| \leq \alpha_0$$

$$= E_\theta^{D,A} + E_\theta^{D,B} \quad \text{for } \alpha_0|\theta| \leq \theta_3 \quad \text{for } \pi - \alpha_0 \leq |\theta| < \pi \tag{2.33}$$

where

$$\theta_3 = \cos^{-1}\left(\frac{r_0 \cos \theta_0}{r}\right) \quad (2.34)$$

The directions $\theta = 0$ and π are the caustic directions where ray techniques fail and the fields can be determined using either PO (Section 1.8) or an equivalent edge current method (Section 1.19.1).

Radiation Patterns. In order to arrive at an expression of the radiation pattern [17], accurate expressions for fields in the uncorrugated horn (Fig. 2.9) are needed. These are available in [19]. Simplified asymptotic expressions [16–19] for the field components in TE_{11} in the radial and transverse directions

$$E_\theta = \frac{B_1}{\sin \theta} J_1\left(1.841 \frac{\theta}{\theta_0}\right) \frac{\hat{H}_\nu(kr)}{r} \sin \phi \quad (2.35a)$$

$$E_\phi = \frac{1.841}{\alpha_0} B_1 J_1'\left(1.841 \frac{\theta}{\alpha_0}\right) \frac{\hat{H}_\nu(kr)}{r} \cos \phi \quad (2.35b)$$

$$H_r = \frac{1}{j\omega\mu} \left(\frac{1.841}{\alpha_0}\right)^2 B_1 J_1\left(\frac{1.841}{\alpha_0} \theta\right) \frac{\hat{H}_\nu(kr)}{r^2} \cos \phi \quad (2.35c)$$

where

$$\nu = \left(-0.5 + 0.25 + \left(\frac{1.841}{\alpha_0}\right)^2\right)^{1/2} \quad (2.35d)$$

For long horns where $kr_0 > 15$, the modified Hankel function is given by

$$\hat{H}_\nu(kr) + (j)^{\nu+1} e^{-jkr} \quad (2.35e)$$

To formulate the expression for the radiation field, the phase reference is taken at edge A. The radiated field diffracted by edge A is given by

$$E_\theta^{D,A} = E_\theta(r_0, \alpha_0, \pi/2) \, v(r_0, \theta_{2A}) \frac{\sqrt{a}}{\sqrt{\frac{a}{r} + \sin \theta}} \cdot \frac{e^{-jkr}}{r}$$

$$\text{for } -90° < \theta < (180° + \alpha_0) \quad (2.36a)$$

$$v(r_0, x) = -\frac{e^{-j\pi/4}}{2\sqrt{2\pi k}} \frac{F[kr_0 a(x)]}{\cos \frac{x}{2}} \quad (2.36b)$$

$$a(x) = 2 \cos^2 \frac{x}{2} \quad (2.36c)$$

$$F(p) = j2\sqrt{p}e^{jp}\int_{\sqrt{x}}^{\infty}e^{-jz^2}\,dz \qquad (2.36d)$$

$$\theta_{2A,B} = \pi \pm \theta - \alpha_0$$

The nonaxial edge-diffracted field in the E-plane of the horn from edge B is given by

$$E_\theta^{d,B} = -E_\theta(r_0, \alpha_0, -\pi/2) \cdot v(r_0, \theta_{2B})e^{\pi/2} \cdot \sqrt{\frac{a}{\sin\theta - \frac{a}{r}}} \cdot e^{-j2ka\sin\theta}\frac{e^{-jkr}}{r}$$

$$\text{for } -180° - \alpha_0 < \theta < 90° \qquad (2.37)$$

For horns with $kr_0 > 15$, the GO field as an asymptotic expansion of Hankel functions is given by

$$E_\theta^g = \begin{cases} B_1 \dfrac{(j)^{v+1}}{\sin\theta} J_1\dfrac{1.841}{\alpha_0}\theta \dfrac{e^{-jkr}}{r} e^{-jkr_0\cos(\alpha_0-\theta)} & \text{for } |\theta| \leq \alpha_0 \\ 0 & \text{for } \alpha_0 |\theta| \leq 180° \end{cases} \qquad (2.38)$$

The nonaxial far field of the horn is given by

$$E_\theta^T = E_\theta^G + E_\theta^{d,A} + E_\theta^{d,B} \qquad (2.39)$$

H-Plane Patterns. For the H-plane in the grazing direction (e.g. for $\theta = 0$) H_r is nonvanishing and, as before, a slope diffraction correction is required.

The diffracted field in the nonaxial direction of the long conical horn is expressed as [17]

$$E_\phi^{D,A} = jkZ_0 H_r(r_0, \alpha_0, 0) r_0^{3/2} e^{+jkr_0} \frac{e^{-jkr}}{\sqrt{r}} v_s(r_0, \theta_{2,B}) \cdot \sqrt{\frac{\dfrac{a}{\sin\theta}}{a + \dfrac{a}{\sin\theta}}} \qquad (2.40a)$$

$$= \left(\frac{1.841}{\alpha}\right)^2 B_1(j)^{v+1} J_1(1.841) \sqrt{\frac{a}{r_0\sin\theta}} \cdot \frac{v_s(r_0, \theta_{2,A})}{\sqrt{1 + \dfrac{a}{r\sin\theta}}} \cdot \frac{e^{-jkr}}{r}$$

$$-90° < \theta < (180° + \alpha_0) \qquad (2.40b)$$

where

$$v_s(x, p) = \frac{\sin p \, \text{sgn}[-\cos(p/2)] \exp[-j(kx - \pi/4)]}{\sqrt{\pi}} \cdot \exp(jkx(1 + \cos p))$$

$$\cdot \int_{\sqrt{kx(1+\cos p)}}^{\infty} e^{-jz^2} \, dz + \frac{j}{2[kx(1 + \cos p)]^{1/2}} \qquad (2.40c)$$

and

$$\text{sgn}[-\cos(p/2)] = \text{sgn}[-\cos(p/2)]$$

For the edge B,

$$E_\phi^{D,B} = \left(\frac{1.841}{\alpha_0}\right)^2 B_1(j)^{v+1} J_1(1.841) \sqrt{\frac{a}{r_0 \sin \theta}} \cdot \frac{v_s(r_0, \theta_{2,B})}{\sqrt{1 - \frac{a}{r \sin \theta}}}$$

$$\cdot e^{j\pi/2 - j2ka \sin \theta} \frac{e^{-jkr}}{r} \quad \text{for } 90° > \theta > -(180° + \alpha_0) \qquad (2.40d)$$

Thus the total pattern is the vector superposition of the different contributions and is expressed as [17]

$$E_\phi^T = E_\phi^G + E_\phi^{D,A} + E_\phi^{D,B} \qquad (2.41)$$

where

$$E_\phi^g = \begin{cases} \dfrac{1.841}{\alpha_0} B_1(j)^{v+1} J_1' \, 1.841 \dfrac{\theta}{\alpha_0} \cdot e^{-jkr_0 \cos(\theta - \alpha_0)} \dfrac{e^{-jkr}}{r} & \text{for } |\theta| \leq \alpha_0 \\ 0 & \text{for } \alpha_0 \leq |\theta| \leq 180° \end{cases}$$

Caustic Region Fields. Once again using the equivalent edge current (Section 1.19.1), the caustic region fields are expressed as [17]

$$E(\theta = 0) = bfa_y B_1(j)^{v+1} \frac{1.841}{\alpha_0} \frac{e^{-jkr}}{r} \frac{a}{2} \sqrt{\frac{2\pi k}{r_0}} \cdot \frac{1.841}{\alpha_0} J_1(1.841) \cdot v_s(r_0, \pi - \alpha_0) e^{j\pi/4}$$

$$+ \frac{a}{2r_0} \sqrt{2\pi k} \frac{\alpha_0}{1.841} \frac{J_1(1.841)}{\sin \alpha_0} v(r_0, \pi - \alpha_0) \cdot e^{-jkr_0 + j\pi/4} \qquad (2.42)$$

$$E(\theta = \pi) = a_y B_1(j)^{v+1} \frac{1.841}{\alpha_0} \frac{e^{-kr}}{r} \frac{a}{2} \sqrt{\frac{2\pi k}{r_0}} \cdot \frac{1.841}{\alpha_0} J_1(1.841) \cdot v_s(r_0, 2\pi - \alpha_0) e^{j\pi/4}$$

$$- \frac{a}{2r_0} \sqrt{2\pi k} \frac{\alpha_0}{1.841} \frac{j_1(1.841)}{\sin \alpha_0} v(r_0, 2\pi - \alpha_0) \cdot e^{-jkr_0 + j\pi/4} \qquad (2.43)$$

2.4 CONICAL HORNS

Near-Field Patterns. The procedure to arrive at expressions for near-field and radiation patterns in the H-plane is similar to the procedure for the E-plane in Section 2.4.1.1 except the field component diffracted is E_ϕ^g and the hard diffraction coefficient D_h to be used instead in eqns. (2.35a–d) and (2.36).

Radiation Patterns. H-plane radiation patterns for uncorrugated conical horns have been presented in [17]. The component to be used for the diffraction problem is H_r since it is nonvanishing at $\theta = \alpha_0$ and $E_\theta = E_\phi = 0$. Since H_r varies inversely as r^2, the edges for H-plane calculations are nonuniformly illuminated, hence a slope diffraction correction is needed.

The radiated fields from the two edges are given by

For edge A

$$E_\phi^{D,A} = jkZ_0 H_r(r_0, \alpha_0, 0) r_0^{3/2} e^{jkr_0} \frac{e^{-jkr}}{\sqrt{r}} v_s(r_0, \theta_{2,A}) \cdot \sqrt{\frac{\frac{a}{\sin\theta}}{r + \frac{a}{\sin\theta}}}$$

$$= \frac{1.814}{\alpha_0} B_1(j)^{v+1} J_1(1.841) \sqrt{\frac{a}{r_0 \sin\theta}} \cdot \frac{v_s(r_0, \theta_{2,A})}{\sqrt{1 + \frac{a}{r \sin\theta}}} \frac{e^{-jkr}}{r}$$

$$\text{for } -90° < \theta < (180° + \alpha_0) \quad (2.44)$$

where

$$v_s(x, p) = \frac{\sin p \, \text{sgn}[-\cos(p/2)] \exp(-j(kx - \pi/4))}{\sqrt{\pi}} \cdot \exp[jkx(1 + \cos p)]$$

$$\cdot \int_{\sqrt{k\pi(1 + \cos p)}}^{\infty} e^{-jz^2} dz + \frac{j}{2[kx(1 + \cos p)]^{1/2}} \quad (2.45)$$

For edge B

$$E_\phi^{D,B} = \left(\frac{1.841}{\alpha_0}\right)^2 B_1(j)^{v+1} J_1(1.841) \sqrt{\frac{a}{r_0 \sin\theta}} \cdot \frac{v_s(r_0, \theta_{2,B})}{\sqrt{1 - \frac{a}{r \sin\theta}}}$$

$$\cdot e^{j\pi/2 - j2ka \sin\theta} \cdot \frac{e^{-jkr}}{r} \quad \text{for } 90° < \theta < -(180° + \alpha_0) \quad (2.46)$$

The nonaxial radiated field at any observation point is given by

$$E_\phi^T = E_\phi^g + E_\phi^{D,A} + E_\phi^{D,B} \quad (2.47)$$

The GO field is given by

$$E_\phi^g = \begin{cases} \left(\dfrac{1.841}{\alpha_0}\right) B_1(j)^{v+1} J_1' \; 1.841 \; \dfrac{\theta}{\alpha_0} \cdot e^{-jkr_0 \cos(\theta-\alpha_0)} \dfrac{e^{-jkr}}{r} & \text{for } |\theta| < \alpha_0 \\ 0 & \text{for } \alpha_0 \leq |\theta| \leq 180° \end{cases}$$

(2.48)

The radiated fields along the front and back directions are found by using the equivalent edge current method since these directions are caustic directions and ray techniques will give singular results.

$$E(\theta = 0) = \hat{a}_y B_1(j)^{v+1} \frac{1.841}{\alpha_0} \frac{e^{-jkr}}{r} \frac{a}{2} \sqrt{\frac{2\pi k}{r_0}} \cdot 1.841 \{\alpha_0 J_1(1.841) \cdot v_s(r_0, \pi - \alpha_0)\} e^{j\pi/4}$$

$$+ \frac{a}{2r_0} \sqrt{2\pi k} \frac{\alpha_0}{1.841} \frac{J_1(1.841)}{\sin \alpha_0} v(r_0, \pi - \alpha_0) \cdot \exp(-jkr + \pi/4) \quad (2.49a)$$

$$E(\theta = \pi) = \hat{a}_y B_1(j)^{v+1} \frac{1.841}{\alpha_0} \frac{e^{-jkr}}{r} \frac{a}{2} \sqrt{\frac{2\pi k}{r_0}} \cdot \frac{1.841}{\alpha_0} J_1(1.841) \cdot v_s(r_0, 2\pi - \alpha_0) e^{j\pi/4}$$

$$- \frac{a}{2r_0} \sqrt{2\pi k} \frac{\alpha_0}{1.841} \frac{J_1(1.841)}{\sin \alpha_0} v(r_0, 2\pi - \alpha_0) \cdot e^{-jkr_0 + j\pi/4} \quad (2.49b)$$

Results and Discussion. Some numerical computations and comparisons were made for near and far patterns. Figure 2.10 shows comparison of theoretical and experimental results for near-field patterns of uncorrugated conical horns. The agreement is shown to be good up to a maximum of 40° since the dynamic range of the receiver was limited. Typical *E*- and *H*-plane radiation patterns are shown in Figure 2.11 and, using eqns. (2.39) and (2.40), agreement between theory and experiment is fair.

2.5 CORRUGATED CONICAL HORNS

The corrugated electromagnetic horn antenna has been established as an antenna with low sidelobes and backlobe, broad-band performance and rotationally symmetric patterns (for square pyramidal and conical horns). These properties make the corrugated horn antenna useful for many practical applications. Corrugated horns were being studied long ago [13]. Hamid [14] considered the problem of diffraction by conical horn using a Brillouin ray technique; though a good initial attempt to analyze conical horns, it has many shortcomings. Mentzer and Peters Jr. [15] analyzed patterns for corrugated horn antennas. The pattern in the main lobe region is computed using conventional aperture integration, the contribution of the *H*-plane edges as

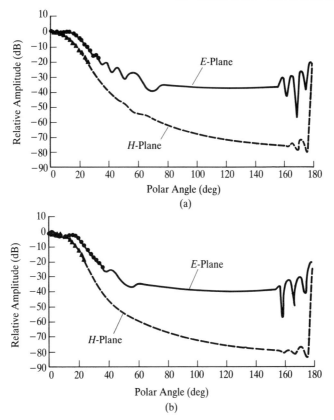

FIGURE 2.10 E- and H-plane near-field patterns of conical horns [16]. (a) $r = 46.23\lambda$; (———), (– – –) calculated; (● ●), (▲ ▲) measured. (b) $r = 23.52\lambda$; (———), (– – –) calculated; (● ●), (▲ ▲) measured.

determined using slope diffraction analysis, and the contribution of the E-plane edges was found by duality. The asymptotic solution for the field components in the corrugated horn antenna are given by [17]

$$E_\theta = A_1 \frac{2.405}{\alpha_0} J_0\left(\frac{2.405}{\alpha_0}\theta\right) \frac{\hat{H}_s(kr)}{r} \sin\phi \qquad (2.50a)$$

$$E_\phi = A_1 \frac{2.405}{\alpha_0} J_0\left(\frac{2.405}{\alpha_0}\theta\right) \frac{\hat{H}_s(kr)}{r^2} \cos\phi \qquad (2.50b)$$

$$H_r = \frac{A_1}{j\omega\mu} \frac{2.405^2}{\alpha_0} J_1\left(\frac{2.405}{\alpha_0}\theta\right) \frac{\hat{H}_s(kr)}{r^2} \cos\phi \qquad (2.50c)$$

where $s = \left[-0.5 + \left(0.25 + \frac{2.405}{\alpha_0}\right)^2\right]^{1/2}$

a = radius of the horn aperture
α_0 = half angle of the horn

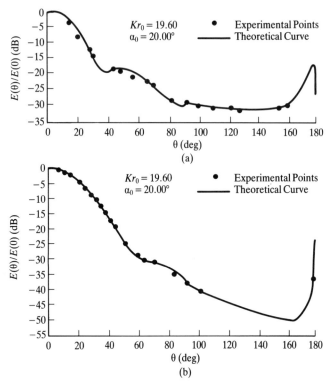

FIGURE 2.11 Radiation patterns for an uncorrugated conical horn; $kr_0 = 19.6$; $\alpha_0 = 20°$: (a) E-plane; (b) H-plane. (From [17] © 1978 IEEE.)

2.5.1 Near-Field Patterns

The near-field patterns for the E-plane characteristics of the corrugated horns (Fig. 2.12) can be formulated on the same lines as for uncorrugated horns. As a first approximation the corrugations can be considered to be infinitely thin. The near field patterns of corrugated conical horns [16] are shown in Fig. 2.13 compared with the experimental results.

2.5.2 Far-Field Patterns

The radiation pattern for corrugated horns [17] can be computed in exactly the same way as the uncorrugated horns with two assumptions: (1) the edges have small thickness compared to the wavelength and the corrugated walls are replaced by conducting surfaces; and (2) the excited dominant mode in the corrugated horn is HE_{11}, the field component H_r exists, and $E_\theta = E_\phi = 0$. The radiated fields from the edges A and B are given by [17]

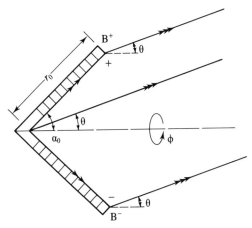

FIGURE 2.12 Geometry and coordinate system of the corrugated horn. (From [16] © 1979 IEEE.)

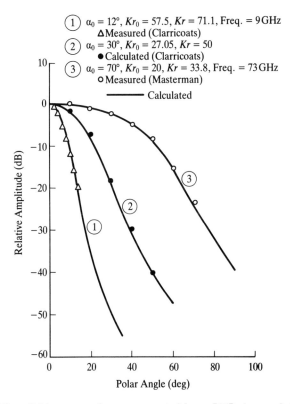

FIGURE 2.13 Near-field patterns for a corrugated horn [16]: (———) theory.

From edge A

$$E_\phi^{D,A} = \frac{2.405^2}{\alpha_0} A_1(j)^{s+1} J_1(2.405) \sqrt{\frac{a}{r_0}} \cdot \sqrt{\frac{1}{\sin\theta}} \frac{v_s(r_0, \theta_{2a})}{\sqrt{1 + \frac{a}{r\sin\theta}}} \frac{e^{-jkr}}{r}$$

$$\text{for } -90° < \theta < (90° + \alpha_0) \quad (2.51a)$$

From edge B

$$E_\phi^{D,B} = \frac{2.405^2}{\alpha_0} A_1(j)^{s+1} J_1(2.405) \sqrt{\frac{a}{r_0}} \cdot \sqrt{\frac{1}{\sin\theta}} \frac{v_s(r_0, \theta_{2,B})}{\sqrt{1 - \frac{a}{r\sin\theta}}}$$

$$\cdot e^{j\pi/2 - j2ka\sin\theta} \frac{e^{-jkr}}{r} \quad \text{for } 90 > \theta > -(90 + \alpha_0) \quad (2.51b)$$

The total nonaxial H-plane radiation pattern is given by

$$E_\phi^T(\theta) = E_\phi^G + E_\phi^{D,A} + E_\phi^{D,B} \quad (2.52)$$

where

$$E_\phi^g = \begin{cases} \dfrac{2.405}{\alpha_0} A_1(j)^{s+1} J_0 \dfrac{2.405\theta}{\alpha_0} \cdot e^{-jkr_0 \cos(\theta - \alpha_0)} \dfrac{e^{-jkr}}{r} & \text{for } |\theta| < \alpha_0 \\ 0 & \text{for } \alpha_0 \leq |\theta| \leq 180° \end{cases} \quad (2.53)$$

The field in the axial direction ($\theta = 0$), which is a caustic, is calculated by the equivalent edge current method

$$E = a_y \frac{2.405}{\alpha_0} A_1(j)^{s+1} \sqrt{\frac{a}{r_0}} \sqrt{2\pi k a} \cdot e^{j\pi/4} J_1(2.405) v_s(r_0, \pi - \alpha_0) \frac{e^{-jkr}}{r} \quad (2.54)$$

Figure 2.14 [17] shows the comparison of H-plane theoretical and experimental radiation patterns for a corrugated horn with HE_{11} mode. The eigenvalues for a class of spherical wave function are available in [22].

2.6 SPECIAL HORNS

There are situations where the user needs horns with low sidelobes and low cross-polar levels and also large bandwidth. These special requirements need

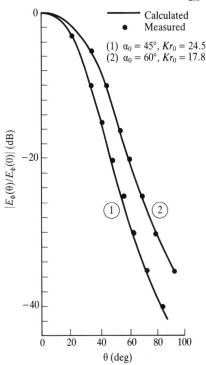

FIGURE 2.14 H-plane radiation pattern for a corrugated conical horn [17]: (1) half angle $\alpha_0 = 45°$, $kr_0 = 24.5$; (2) half angle $\alpha_0 = 60°$, $kr_0 = 17.8$.

special attention in designing the horns. Since the generation of sidelobes and backlobe is due to the fringing field of edge illumination, the basic need for suppressing the sidelobes is to considerably reduce the edge illumination. An early paper [28] discusses significant reduction of sidelobe levels over a 2-to-1 frequency band. Two methods based on the reduction of edge illumination were used. One used a series of quarter-wavelength chokes in the form of slots on the wall; another in the form of cutoff corrugated guides on the wall. Two horns were investigated. The aperture of the choke-slot horn is a square of side 3.5 in. and a flare angle of 92°; the second horn has the same aperture but a flare angle of 50°. Another antenna with plane inner walls was used for comparison. Figure 2.15 shows the E-plane patterns of the small and large corrugated horns at 10 GHz compared with the control horn.

2.6.1 Broad-Band Flared Horns with Low Sidelobes

Flared horns (Fig. 2.16) with small flare angles can be designed to give sidelobes as low as -75 dB.

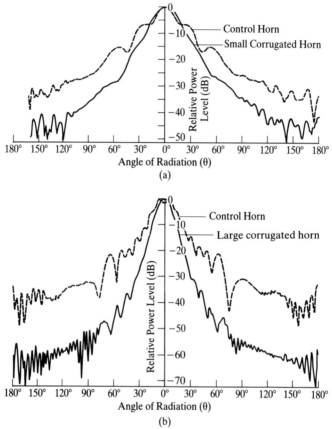

FIGURE 2.15 Normalized E-plane patterns of corrugated horns compared with control horn at 10 GHz [28]: (a) comparison with small horn; (b) comparison with large horn.

The diffracted field $u_d(P)$ at any point is given by

$$u_d(P) = A_i(Q_1) \exp ik[\phi_i(Q_1) + t + s] \cdot \left[\frac{d\sigma(Q_1)}{d\sigma(P_1)}\right]^{1/2} \cdot \left[\frac{\rho_1}{s(\rho_1 + s)}\right]^{1/2}$$

$$\cdot \sum_m D_m(P_1) D_m(Q_1) \exp\left[-\int_0^t \alpha_m(\tau)\, d\tau\right] \tag{2.55}$$

In eqn. (2.49), the incident field amplitude at Q_1 is $A_i(Q_1) = \exp(ik\phi_i(Q_1))$ normalized to $(\pi r_1^2)^{-1/2}$.

The two coefficients $D_m(Q_1)$ and $D_m(P_1)$ are the surface wave coefficients generated by radiation at Q_1. $|D_m|^4$ is the effective thickness of the mode. The factor $\exp[-\int_0^t \alpha_m(\tau)\, d\tau]$ is the attenuation of the mode where α_m is a function of the local radii of curvature and the material. There are two cases of polarization. Case I is where the electric vector is parallel to the surface with

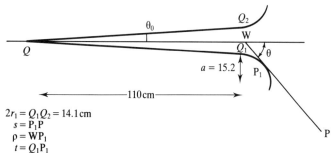

FIGURE 2.16 Small flared long horn.

$u = 0$. This corresponds to the H-plane pattern. Case II is where the electric vector is perpendicular to the surface with $\partial u/\partial n = 0$, corresponding to the E-plane pattern. The attachment coefficients D_m are given by

$$D_m^2 = \tfrac{1}{2}\lambda^{1/6}a^{1/3}(\pi/3)^{4/3}e^{j\pi/12} \times \begin{cases} [A'(q_m)]^{-2}, & \text{case I} \\ 3q_m^{-1}[A(q_m)]^{-2}, & \text{case II} \end{cases}$$

The forward gain is given by

$$g(\theta) = \lim_{s \to \infty} (2\pi r_1/\lambda)^2$$

Figure 2.17 compares theoretical and measured H- and E-plane patterns. The horn had a 90° flare. The frequency used was 31.4 GHz and the mode excited was TE_{11}. The disagreement is due to the simplifying assumptions made in the theory. The horn just described can be used in very broad band applications [23]. The aperture size required is larger than that in [24]. The pattern of the horn in the sidelobe region improves rapidly.

2.7 CONTROL OF SIDELOBES

As indicated earlier, the control of sidelobes is of great practical importance since the presence of sidelobes has the potential to create considerable inconvenience. First, a system loses energy through radiation in undesired directions. Second, it may communicate in unwanted directions. And third, radar using such an antenna may get jammed through the sidelobes. The essence of sidelobe control is to find ways and means of reducing the edge illumination in the case of aperture antennas, such as horns and reflectors. A corrugated conical horn of the scalar feed horn antenna type was designed and constructed at the Jet Propulsion Laboratory [24]. The reader is referred to the experimental paper [24] for further details. Figure 2.18 compares theoretical and experimental E- and H-plane patterns of the low sidelobe horn.

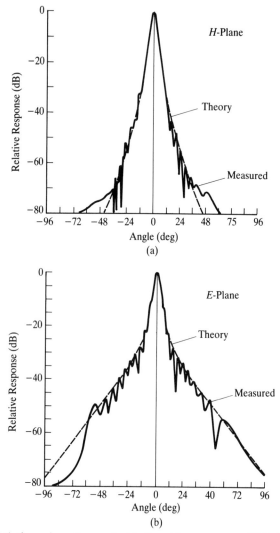

FIGURE 2.17 Relative gain patterns of wide-flared horn antenna [23], flare angle = 90°: (a) H-plane; (b) E-plane.

Recently, it has been demonstrated that low sidelobe levels of horns can be obtained by various means, such as a metal-strip-loaded dielectric substrate [25, 26] and by curving [27] the edges of the aperture. Fabrication of corrugated metal horns is expensive and tedious. It has been found [25, 26] that periodic strips of dielectric substrate are much easier to fabricate by photolithography and they fulfill the same function as a metal corrugated horn.

The geometry of the strip-loaded feedhorn is shown in Figure 2.19 and the E-plane pattern is shown in Figure 2.20. For details of the design of the horn,

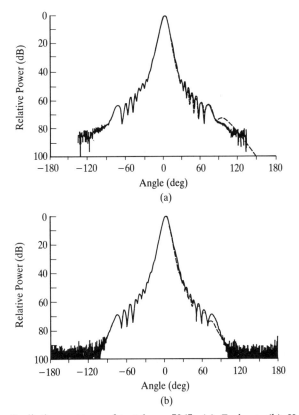

FIGURE 2.18 Radiation pattern of test horn [24]: (a) E-plane; (b) H-plane. (———) experiment; (- - - - -) calculated.

the reader is referred to [25]. Table 2.1 shows the co- and cross-polar characteristics and beamwidths with frequencies in the X-band.

It has been established [27] that the far-out sidelobes of a horn antenna can be suppressed significantly by curving the edges of the horn (Fig. 2.21). A three-dimensional UTD E-plane analysis was recently carried out in [27]. It differs from conventional models in that the fields in the E- and H-planes are generated by line sources which are displaced from one another: one source in the E-plane at the apex of the horn; the other in the H-plane at a position δ (Fig. 2.22 [27]) where $\delta = \frac{1}{2} l_e \cos(\theta_e)$. The following are the expressions for the different components of the fields [27]:

Incident Field. The field due to the line source excitation [10] is given by

$$E_i(s) = \begin{cases} E_0 \sqrt{\dfrac{\rho_1 \rho_2}{(\rho_1 + s)(\rho_2 + s)}} e^{-jks}, & 0 \leq \theta \leq \theta_e \\ 0, & \text{elsewhere} \end{cases} \qquad (2.56)$$

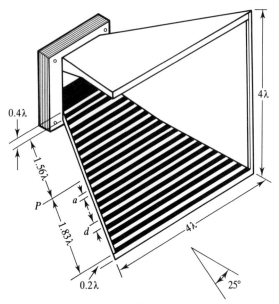

FIGURE 2.19 Strip-loaded feedhorn [25].

In the far field ($s \to \infty$), the incident field is approximated as

$$E^i(\theta) = \begin{cases} \sqrt{\rho_1 \rho_2}\, e^{jkl_e \cos(\theta_e)}\, \dfrac{e^{-jkr}}{r}\, \hat{\theta}, & 0 \leq \theta \leq \hat{\theta}_e \\ 0, & \text{elsewhere} \end{cases}$$

where (Fig. 2.21)

$$\rho_1 = l_e \cos(\theta_e)$$

$$\rho_2 = l_e \cos(\theta_e) - \frac{\delta}{\cos(\theta)}$$

First-Order Diffracted Field. The diffracted fields are from edges A and B, as shown in Figure 2.23. The soft and hard diffraction coefficients D_s and D_h are for the infinite wedge, and the effect of curvature of the edge is taken care of in the radius of curvature of the diffracted wavefront. The diffracted field from edge A is given by

$$E_1^d(\theta) = \begin{cases} -E^i(A)\sqrt{\rho}\,\dfrac{D_h}{2}\, e^{jkl_e \cos(\theta_e - \theta)}\, \dfrac{e^{-jkr}}{r}\, \hat{\theta}, & 0° \leq \theta \leq 180° \\ 0, & \text{elsewhere} \end{cases} \quad (2.57)$$

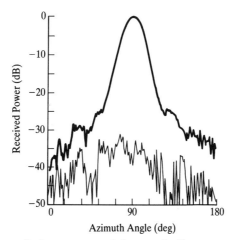

FIGURE 2.20 E-plane radiation patterns of the new feedhorn antenna at 10 GHz [25]. (———) copolar; (———) cross-polar.

TABLE 2.1 E-Plane Radiation Characteristics of the New Feedhorn Antenna [25]

Frequency (GHz)	Maximum Sidelobe Level (−dB)	Maximum Cross-Polar Level (−dB)	3 dB Beamwidth (deg)	10 dB Beamwidth (deg)
8.4	29.30	33.73	21.60	41.72
8.6	27.34	26.00	22.76	44.23
8.8	26.50	25.00	23.09	44.35
9.0	30.10	35.00	22.76	42.16
9.2	30.73	26.82	22.49	41.48
9.4	28.54	30.82	20.93	39.16
9.6	26.24	28.26	20.52	38.34
9.8	26.00	26.00	20.47	36.73
10.0	25.00	32.00	20.00	34.89
10.2	27.50	28.81	18.32	33.61
10.4	31.46	33.00	17.44	32.97
10.6	26.71	35.88	17.25	32.41
10.8	34.61	36.48	16.63	32.55

where

$$\frac{1}{\rho} = \frac{1}{\rho_e^i} + \frac{1 - \cos(\theta_e)\cos(\theta) - \sin(\theta_e)\sin(\theta)}{a_e} \quad (2.58)$$

The factor $\frac{1}{2}$ in $\frac{1}{2}D_h$ is because of the grazing incidence of the incident field at the edge.

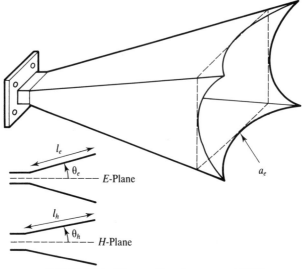

FIGURE 2.21 Horn with edges curved [27].

The diffracted field from edge B is given by

$$E_2^d(\theta) = \begin{cases} 0, & 90° \leq \theta \leq 180° - \theta_e + \theta' \\ E^i(B)\sqrt{\rho}\,\dfrac{D_h}{2}\,e^{jkj_e \cos(\theta_e + \theta)}\,\dfrac{e^{-jkr}}{r}\,\theta, & \text{elsewhere} \end{cases} \quad (2.59)$$

θ' is shown in Figure 2.22 along with the geometry of the second-order diffracted fields. Derivation and discussion of the second-order fields is not be

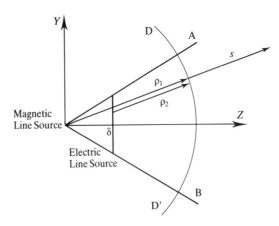

FIGURE 2.22 A two-line source model for the incident field in the aperture of the curved edge horn antenna [27].

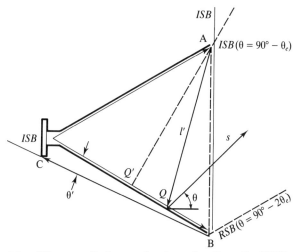

FIGURE 2.23 The different radiating mechanisms that form the UTD model for a horn antenna [27].

TABLE 2.2 The Dimensions of the Horn Antennas [27]

	Small Horn	Large Horn
l_e(cm)	17.0	27.2
θ_e(degrees)	13.2	14.0
a_e(cm)	4.5	9.0
l_h(cm)	22.7	31.6
θ_h(deg)	10.2	12.3

reproduced here; instead the reader is referred to [27]. Table 2.2 gives the dimensions of the two horns with experimental edges that were used in [27] to compare theoretical results.

Figure 2.24 shows the E-plane radiation patterns of large ($\delta = 140$ mm) and small horns ($\delta = 84$ mm) at 11 GHz. Figure 2.25 shows the variation in the sidelobe levels in the E-plane radiation patterns with different radii of curvature of the edges at 9 GHz for the small horn.

2.8 CROSS-POLARIZATION CHARACTERISTICS OF HORNS

A horn may be designed to radiate in single mode or in dual mode. Along with the question of mode comes another important point as to whether the polarization is pure or not. So if a horn is radiating with a certain polarization

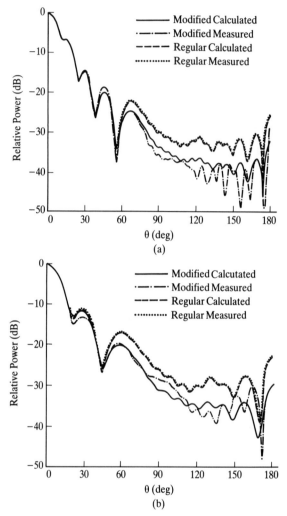

FIGURE 2.24 E-plane radiation pattern [27]: (a) large horn at 11 GHz, $\delta = 140$ mm; (b) small horn at 11 GHz, $\delta = 84$ mm.

it is important to know the strength of the cross-polarization or the orthogonal polarization relative to the desired polarization. The presence of significant cross-polarization leads to energy loss, potential interference to equipment, and possible detection and jamming.

Several references [25, 27, 29–31] discuss cross-polarization effects in horns. We summarize the results here even though it looks as though cross-polarization cannot be predicted by high-frequency techniques. It appears that most of the results on cross-polarization of horns are experimental. And a MOM formulation may well be more appropriate than a high-frequency technique.

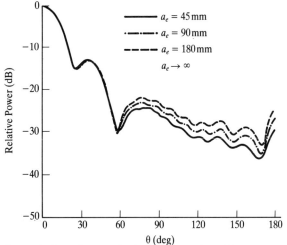

FIGURE 2.25 Theoretical E-plane radiation patterns for different radii of curvature at 9 GHz (small horn) [27].

We summarize here the published results on cross-polarization of electromagnetic horns. The cross-polar pattern in the E-plane of the horn in [25] is shown in Figure 2.26. Cross-polarization from horn antennas operating in dual mode and supporting the TE_{11} and TM_{11} modes has been treated both analytically and experimentally in [29]. Figure 2.26 shows the calculated and experimental cross-polarization for the dual mode horns used in [29] where R is the straight length and α is the half angle of the cone. Figure 2.27 shows the measured cross-polarization in contrast to the main beam for the magnetically coated square horn designed for low cross-polarization. The wall thickness was 1 mm and with no serration. Another simplified design of dual-band corrugated conical horns with cross-polarization levels better than -34 dB is discussed [31]. For design details of the corrugated horn, the reader is referred to [31]; the final design has the following parameters: the aperture diameter $D = 6.372$ in., the axial length of the horn $L = 25.22$ in., and $\Theta_0 = 7.2°$. Figure 2.28 [31] shows the co- and cross-polar properties of the dual corrugated horns.

2.9 SUMMARY

This chapter dealt with the diffraction problem of electromagnetic horns. Both UAT treatment of pyramidal horns and UTD treatment of uncorrugated and corrugated conical horns were described. And both near and far fields were addressed. The problem of low sidelobes and cross-polarization were discussed.

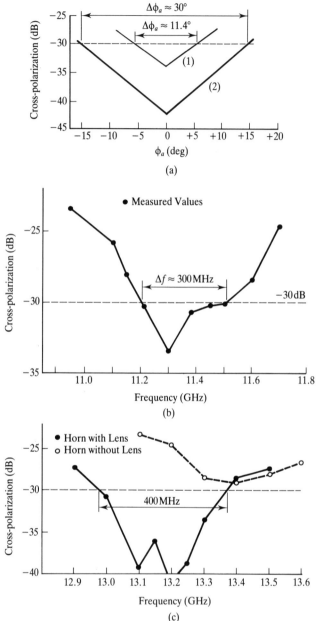

FIGURE 2.26 Calculated and measured cross-polarization with phase and frequency [29]. (a) Calculated cross-pol versus the relative phase between the TM_{11} and TE_{11} modes in the horn aperture: (1) $\alpha = 11°$, $R/\lambda = 21.7$; (2) plane aperture, $D \gg \lambda$. (b) Measured cross-polarization versus the frequency from a dual mode horn antenna ($\alpha = 11°$, $R/\lambda = 21.7$, $2b = 40$ mm, $a/b = 0.79$, $L = 10$ mm). (c) Measured cross-polarization versus the frequency from a dual mode horn antenna ($\alpha = 18.5°$, $R/\lambda = 13.9$, $2b = 34$ mm, $a/b = 0.79$, $L \approx 30$ mm).

2.9 SUMMARY   209

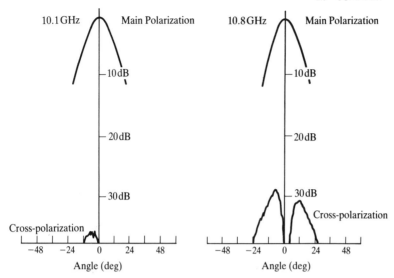

FIGURE 2.27 Measured cross-polarization pattern in comparison with the main beam of the magnetic horn; square horn with 1 mm coating on all walls, no serration [30].

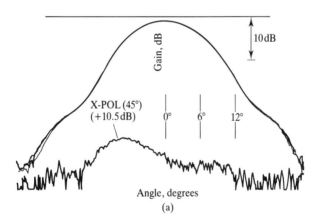

FIGURE 2.28 Far-field patterns of the dual-band corrugated horn. (a) Measured far-field patterns of the corrugated horn at 11.94 GHz frequency (center of the transmit band). Co-pol patterns are measured in E-, H-, and 45° planes, while the cross-pol patterns are measured in the 45° plane. (b) Measured far-field patterns of the corrugated horn at 17.54 GHz frequency (center of the receive band). Co-pol patterns are measured in E-, H-, and 45° planes while the cross-pol patterns are measured in the 45° plane. (c) Comparison between the computed and measured far-field patterns of the horn in 45° plane. Frequency is 11.94 GHz. □-□—copal pattern (computed); ×-×—co-pol pattern (measured); ○-○—×-pol pattern (computed). (d) Comparison between the computed and measured far-field patterns of the horn in the 45° plane. Frequency is 17.54 GHz. □-□—co-pal pattern (computed); ×-×—co-pol pattern (measured); ○-○— ×-pol pattern (computed). (From [30] © 1989 IEEE.) (*continued on next page*)

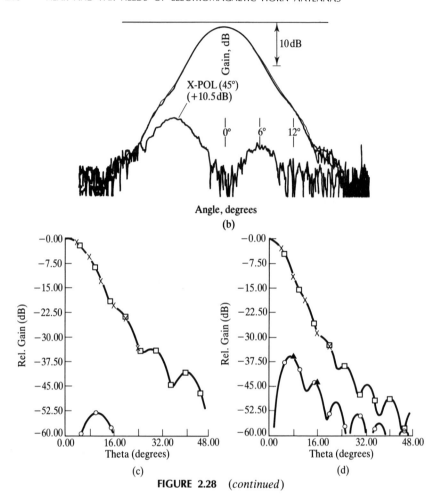

FIGURE 2.28 (*continued*)

2.10 REFERENCES

1. P. M. Russo, R. C. Rudduck, and L. Peters Jr., "A Method for Computing E-Plane Patterns of Horn Antennas," *IEEE Trans. Antennas and Propagat.*, **AP-13**(2), 1965, 219–224.

2. J. S. Yu, R. C. Rudduck, and L. Peters Jr., "Comprehensive Analysis for E-Plane Patterns of Horn Antennas by Edge Diffraction Theory," *IEEE Trans. Antennas and Propagat.*, **AP-14**(2), 1966, 138–149.

3. J. S. Yu and R. C. Rudduck, "H-Plane Pattern of a Pyramidal Horn," *IEEE Trans. Antennas and Propagat.*, **AP-17**(5), 1969, 651–652.

4. C. A. Mentzner, L. Peters Jr., and R. C. Rudduck, "Slope Diffraction and Its Application to Horns," *IEEE Trans. Antennas and Propagat.*, **AP-23**(2), 1975, 153–159.

5. M. S. Narsimhan and K. S. Rao, "GTD Analysis of the Near Field Patterns of Pyramidal Horns," *Proc. IEE*, **126**(12), 1979, 1223–1226.
6. R. C. Menendex and S. W. Lee, "On the Role of the Geometrical Optics Field in Aperture Diffractions," *IEEE Trans. Antennas and Propagat.*, **AP-25**(2), 1977, 187–193.
7. J. Boersma and S. W. Lee, "High-Frequency Diffraction of a Line-Source Field by a Half-Plane: Solution by Ray Techniques," *IEEE Trans. Antennas and Propagat.*, **AP-25**(2), 1977, 171–179.
8. J. Boersma and Y. Rahmat-Samii, "Comparison of Two Leading Uniform Theories of Edge Diffraction with the Exact Uniform Asymptotic Solution," *Radio Science*, **15**, 1980, 1179–1194.
9. Y. Rahmat-Samii and R. Mittra, "A Spectral Domain Interpretation of High Frequency Diffraction Phenomena," *IEEE Trans. Antennas and Propagat.*, **AP-25**, 1977, 676–687.
10. D. A. McNamara, C. W. I. Pistorius, and J. A. G. Malherbe, "*Introduction to the Uniform Geometrical Theory of Diffraction*," Artech House, Boston, MA, 1990, Sections 5.3 and 5.4.
11. S. Sanyal and A. K. Bhatacharyya, "UAT Analysis of E-Plane Near and Far-Field Patterns of Electromagnetic Horn Antennas," *IEEE Trans. Antennas and Propagat.*, **AP-31**(5), 1983, 817–819.
12. S. Sanyal, "Some Investigations on Electromagnetic Radiation and Scattering Problems using Uniform Asymptotic Theory of Diffraction," PhD thesis, Electronics and Electrical Communication Eng., Indian Inst. Tech., Kharagpur, 1985.
13. A. W. Love, *Electromagnetic Horn Antennas, Parts II and VI*, IEEE Press, New York, 1976.
14. M. A. K. Hamid, "Diffraction by a Conical Horn," *IEEE Trans. Antennas and Propagat.*, **AP-16**(5), 1968, pp. 520–528.
15. C. A. Mentzer and L. Peters Jr., "Pattern Analysis of Corrugated Horn Antennas," *IEEE Trans. Antennas and Propagat.*, **AP-24**(3), 1971, 304–309.
16. M. S. Narsimhan and K. Sudhakar Rao, "GTD Analysis of the Near-Field Patterns of Corrugated Conical Horns," *IEEE Trans. Antennas and Propagat.*, **AP-27**(5), 1979, 705–708.
17. M. S. Narsimhan and M. S. Shesadri, "GTD Analysis of the Radiation Patterns of Conical Horns, *IEEE Trans. Antennas and Propagat.*, **AP-26**(6), 1978, 774–778.
18. R. F. Harrington, *Time-Harmonic Electromagnetic Fields*, McGraw-Hill, New York, 1961, Ch. 6, 264–316.
19. M. S. Narsimhan and B. V. Rao, "Modes in a Conical Horn: New Approach," *Proc. Inst. Elec. Eng.*, **118**(2), 1971, 287–292.
20. M. S. Narsimhan and B. V. Rao, "Hybrid Modes in Corrugated Conical Horns," *Electron Lett.*, **6**, 1970, 32–34.
21. M. S. Narsimhan and B. V. Rao, "Diffraction by Wide-Flare-Angle Corrugated Conical Horns," *Electron Lett.*, **6**, 1970, 469–471.
22. M. S. Narsimhan, "Eigenvalues of a Class of Spherical Wave Functions," *IEEE Trans. Antennas and Propagat.*, **AP-20**(1), 1973, 8–14.
23. J. C. Mather, "Broad-Band Flared Horn with Low Sidelobes," *IEEE Trans. Antennas and Propagat.*, **AP-29**(6), 1981, 967–969.

24. M. A. Janssen, S. M. Bednarczyk, S. Gulkis, H. W. Marlin, and G. F. Smoot, "Pattern Measurements of a Low-Sidelobe Horn Antenna," *IEEE Trans. Antennas and Propagat.*, **AP-27**(7), 1979, 551–555.

25. S. Rodrigues, P. Mohmanan, and K. G. Nair, "A Strip-Loaded Feed-Horn Antenna," *IEEE Microwave and Guided Wave Letters*," **1**(11), 1991, 318–319.

26. E. Lier and T. S. Petterson, "The Strip-loaded Hybrid Mode Feed Horn," *IEEE Trans. Antennas and Propagat.*, **AP-35**, 1987, 1086–1089.

27. J. W. Odendaal and C. W. I. Pistorious, "E-Plane Analysis of a Modified Horn Antenna with Suppressed Far-Out Sidelobe Level," *IEEE Trans. Antennas and Propagat.*, **AP-40**(6), 1992, 620–627.

28. R. E. Lawrie and I. Peters Jr., "Modifications of Horn Antennas for Low Sidelobe Levels," *IEEE Trans. Antennas and Propagat.*, **AP-14**(5), 1966, 605–610.

29. E. Lier, "Cross Polarization From Dual Horn Antennas," *IEEE Trans. Antennas and Propagat.*, **AP-34**(1), 1986, 106–110.

30. J. J. H. Wang, V. K. Tripp, and R. P. Zimmer, "Magnetically Coated Horn for Low Sidelobes and Low Cross-Polarization," *Proc. IEE, Part H*, **136**(2), 1989, 132–138.

31. K. Sudhakar Rao, "A Simple Dual-Band Corrugated Horn with Low Cross Polarization," *IEEE Trans. Antennas and Propagat.*, **AP-38**(6), 1990, 946–951.

32. P. J. B. Clarricoats et al., "Near Field Radiation Characteristics of Corrugated Horns," *Electron Lett.*, **7**, 1971, 446–448.

CHAPTER THREE

Reflector Antennas and Rim Loading

3.1 INTRODUCTION

Reflector antennas for transmitting and receiving purposes have been around since a century and astronomers and scientists were aware of then useful properties. Optical astronomers used the fact that a line source illumination from the focus of a parabolic cylinder gives rise to parallel rays. Hertz [1] in 1888 used a zinc-sheet parabolic reflector supported by a wooden frame to receive electromagnetic waves from his spark-gap transmitter. Since then reflector antennas have undergone constant modifications in design, analytical modeling, and fabrication to meet specific requirements. Modern radar technology and availability of digital computer power made possible their applications in space vehicles, telecommunications, defense and deep-space explorations.

The subject of reflector antennas has been described in a number of review papers [2–5], textbooks [6–13] including book chapters on reflectors. Three good examples of practical reflector antennas are: (1) the high-gain paraboloid on the 1975 Viking Orbiter, which used a 1.5 m dish using graphite–epoxy face skins on a honeycomb backing; (2) the 9.1 m 48-rib, flexible radial wrap-rib umbralla reflector used on the ATS F and G spacecraft; (3) the 4 GHz feedhorn array for the Intelsat VI contour-beam spacecraft, which is a part of Intelsat's worldwide high-capacity commercial communication net; and (4) the NASA Deep Space Network (DSN) [14] is the largest and most sensitive scientific telecommunications and radionavigation network. It uses a Ka-band (32 GHz) reflector of physical diameter 70 ms. Offset feed, subreflectors and different shaped reflectors have been used to tailor the illumination of the main dish to minimize the spillover. This suppresses the large-angle sidelobes and backlobe. Instead of a continuous metal as the reflector surface, gold-plated molybdenum and wire grids have been in use. Deployable types have been in wide use in

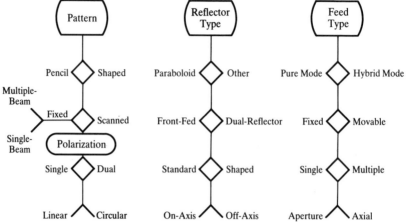

FIGURE 3.1 Reflector antenna classification based on pattern, reflector, and feed types [11].

satellite antennas. Numerous factors come into reflector antenna design, e.g. peak and average power, bandwidth, loss, input impedance and VSWR, cross-polarization level and cross-polar rejection, spillover, scan volume, scan time, pointing accuracy, size, weight, environment, and last but not the least

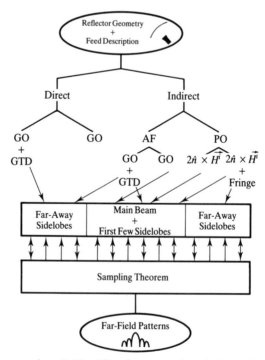

FIGURE 3.2 Summary of available diffraction analysis techniques for reflectors; AF = aperture field [20].

3.2 REFLECTOR ANALYSIS AS AN ELECTROMAGNETIC SCATTERING PROBLEM

the cost. Figure 3.1 shows how reflector antennas can be classified according to pattern, reflector, and feed types [11]. Figure 3.2 is a summary of the available diffraction analysis for reflectors [20]. We discuss the application of the high-frequency techniques described in Chapter 1 to the analysis and design of reflector antennas. High-frequency techniques have proved useful in pattern calculations of all types of reflector configurations. For example, in multibeam antennas the feed has to be offset, which gives rise to reflected rays in the field patterns. The prediction of isolation between beams and the deep minima of the patterns is very difficult and expensive using PO alone.

3.2 REFLECTOR ANALYSIS AS AN ELECTROMAGNETIC SCATTERING PROBLEM

Figures 3.3 and 3.4 respectively show focus-fed and cassegrainian reflector systems. The reflector antenna system, in fact for any aperture antenna, can be considered as an electromagnetic bistatic scattering problem. We have already discussed in Chapter 2 how an electromagnetic horn antenna can be analyzed as a bistatic scatterer using high-frequency techniques and PO, or an equivalent edge current method for the caustic directions. As shown in Figure 3.4, the main reflector now can be considered as scattering energy due to the illumination of the main reflector by the horn antenna. For observations in the plane $\phi = 0°$ points A and B are the two dominant scattering centers, and for $\phi = 90°$, points

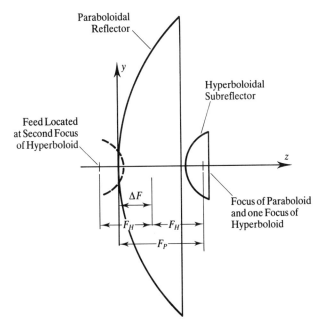

FIGURE 3.3 Geometry and coordinate system of the focus-fed reflector.

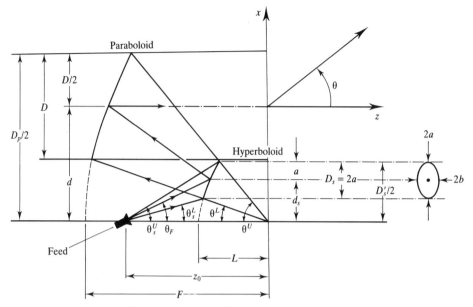

FIGURE 3.4 Geometry and coordinate system of cassegrainian system.

C and D. For any other observation plane with $\phi \neq 0°$, or $90°$ four dominant scattering centers will be contributing. For the cassegrainian system (Fig. 3.4), there is a main and a secondary, or subreflector. The illumination due to the primary feed and subreflector together can be analyzed as a scattering problem. The resulting beam then can be considered as being scattered by the main dish in the same way as described in the case of the focus-fed main dish. The analysis and design involves ray-tracing methods (Section 1.6.1), geometrical optics (Section 1.3), physical optics (Section 1.8), the geometrical theory of diffraction (Section 1.9), and the equivalent (edge) current method (Section 1.19.1). Both near- and far-field characteristics of the reflector systems are important. Near-field information is needed to design radomes and also because engineers are more and more inclined to predict far-field behavior of antennas and scatterers from the near-field information. This saves the necessity of having a far-field test range, either indoor or outdoor. But the near-field sampling facility is costly and does require correction for probe–field interactions.

3.3 FOCUS-FED REFLECTOR

Figure 3.3 shows the basic geometry of the parabolic reflector antenna having a focal length F with a feed at the focus. Using geometrical optics (GO), which is based on the fact that the wave travels in a straight line, a spherical wave from the feed is transformed to a plane or quasi-plane wave after reflection from the reflector. The equation of the parabolic surface with its vertex at the origin $(0, 0, 0)$ of the coordinate system is given by

$$z = (x^2 + y^2)/4F \tag{3.1}$$

In spherical polar coordinates (r, θ, ϕ) with the feed at the origin, the equation of the surface is given by

$$r = F/\cos^2(\phi/2) \tag{3.2}$$

The beam of the feed should have the angle of max $2\phi_0$, which is the aperture angle of the reflector. The angle can be found from

$$\tan(\phi_0/2) = D/(4F) = \frac{1}{4}\frac{1}{F/D} \tag{3.3}$$

The parameter F/D, which is the ratio of the focal length to the diameter of the dish, is a very important parameter in reflector antenna design.

Normally, given a structure, there are three steps to be carried out:

1. Ray tracing to determine the points of reflection and diffraction.
2. Determination of the geometrical and electrical parameters at the point of reflection and diffraction.
3. Calculation of the reflected and diffracted fields due to each ray at the observation point.

3.3.1 Near-Field Patterns of Paraboloids

The near-field patterns of paraboloids can be derived using the scattering center approach in the same manner as described in Chapter 2 of this book for electromagnetic horns. This has been presented in [22]. Figure 3.5 shows the paraboloid geometry and the geometries for H- and E-plane analysis of near-field patterns.

Incident Fields. The incident field on the paraboloid (Fig. 3.5) is determined by the pattern of the feed in the plane of interest. The feed radiation patterns can be described as

$$E_\theta^i(s, \theta, \phi) = \frac{e^{-jkx}}{r} A(\theta) \cos \phi \tag{3.4a}$$

$$E_\phi^i(s, \theta, \phi) = \frac{e^{-jks}}{s} B(\theta) \sin \phi \tag{3.4b}$$

218 REFLECTOR ANTENNAS AND RIM LOADING

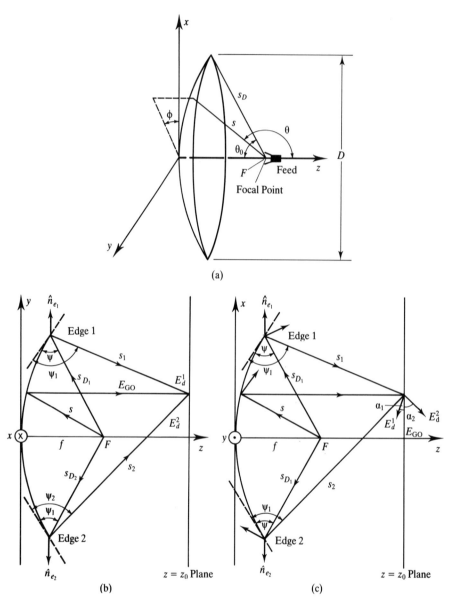

FIGURE 3.5 Near-Field analysis of a prime-focus paraboloid: (a) geometry of the paraboloid [22]; (b) geometry and coordinate system of H-plane analysis [22]; (c) E-plane and coordinate system of E-plane analysis [22].

3.3 FOCUS-FED REFLECTOR

where s = the distance from the feed phase center to the observation point
$A(\theta), B(\theta)$ = the feed pattern functions in the $\phi = 0°$ and $\phi = 90°$ planes, respectively.

The values of the two pattern functions in the different angular sectors of observation are given by

$$A(\theta) = B(\theta) = 1, \quad (\pi - \theta_D) \le |\theta| \le \pi$$
$$A(\theta) = B(\theta) = 0, \quad \text{otherwise}$$

Diffracted Fields. In the H-plane which is the yz-plane (Fig. 3.5(b)) for $\phi = \pi/2$, all fields are x-polarized. The total field at any point, irrespective of whether it is near or far, is a superposition of the GO field and the diffracted fields from the two pertinent scattering centers on the paraboloid. It is given by [22]

$$E_z(0, y, z_0) = E^{GO} + E_A^s + E_B^d \tag{3.5}$$

where

$$E^{GO} = -\frac{e^{-jk(f+z_0)}}{s(y)} B(\pi - \Theta(y)), \quad 0 \le |y| \le D/2 \tag{3.6a}$$

$$= 0, \qquad \text{otherwise} \tag{3.6b}$$

$$s = \left[\left(F - \frac{t^2}{4F}\right)^2 + t^2\right]^{1/2} \tag{3.6c}$$

$$\Theta(t) = \tan^{-1}\left(\frac{t}{F - \frac{t^2}{4F}}\right) \tag{3.6d}$$

In eqn. (3.6), $t = y$ and x respectively in the H- and E-planes.

$$E_{1,2}^d = E_i D_s^{1,2} \left[\frac{\rho_{1,2}}{s_{1,2}(\rho_{1,2} + s_{1,2})}\right]^{1/2} e^{-jks_{1,2}} \tag{3.7a}$$

$$E^i = \frac{e^{-jks_D}}{s_D} B(\pi - \theta_D) \tag{3.7b}$$

$$\rho_{1,2} = \frac{D/2}{\hat{n}_{e_{1,2}} \cdot \hat{s}_{1,2}} \tag{3.7c}$$

$$D(s, h_{1,2}(\psi_{1,2}, \psi) = -\frac{e^{-j\pi/4}}{\sqrt{8\pi k}} \left[\frac{F[kL_{1,2}^i a(\psi_{1,2} - \psi)]}{\cos[(\psi_{1,2} - \psi)/2]} \mp \frac{F[kL_{1,2}^r a(\psi_{1,2} + \psi)]}{\cos[(\psi_{1,2} + \psi)/2]}\right]$$

$$\tag{3.7d}$$

The distance parameters are given by the following expressions

$$L^i_{1,2} = \frac{s_D s_{1,2}}{(s_D + s_{1,2})}; \quad L^r_{1,2} = s_{1,2} \tag{3.7e}$$

$$a(\psi_{1,2} \pm \psi) = 2\cos^2[(\psi_{1,2} \pm \psi)/2] \tag{3.7f}$$

$$F(u) = j2\sqrt{u}\,e^{ju} \int_{\sqrt{u}}^{\infty} e^{-jz^2}\,dz \tag{3.7g}$$

E-Plane Near-Field Patterns. Unlike the H-plane, there are two field components in the E-plane (Fig. 3.5). It serves our purpose to consider only one component. The x-component of the total field is given by [22]

$$E_x(x, 0, z_0) = E^{GO} + E^d_A \cos \alpha_1 + E^d_B \cos \alpha_2 \tag{3.8}$$

where

$$E^{GO} = \begin{cases} \dfrac{e^{-jk(f+z_0)}}{s(x)} A(\pi - \Theta(x)), & 0 \le |x| \le \dfrac{D}{2} \tag{3.9a} \\ 0, & \text{otherwise} \tag{3.9b} \end{cases}$$

$$E^{1,2}_D = E_i D_h^{1,2} \left[\frac{\rho_{1,2}}{s_{1,2}(\rho_{1,2} + s_{1,2})}\right]^{1/2} e^{-jks_{1,2}} \tag{3.9c}$$

$$E_i = \frac{e^{-jks_D}}{s_D} A(\pi - \theta_D) \tag{3.9d}$$

Fields at the Caustic. The diffracted field in the near zone evaluated by the equivalent edge current method (Section 1.19.1) can be shown to be given by [22]

$$E_x = (E^e + E^m + E^{GO})\hat{x} \tag{3.10a}$$

where

$$E^{GO} = -\frac{e^{-jk(F+z_0)}}{F} \tag{3.10b}$$

$$E^{e,m} = \mp j \frac{D}{2} \frac{e^{-jks_D}}{s_D} \sqrt{\frac{8}{\pi k}} e^{-j\pi/4} \frac{e^{-jkq}}{4q} \cdot \begin{matrix} D_s(\psi_0, \psi) B(\pi - \theta_D) \\ D_h(\psi_0, \psi) A(\pi - \theta_D) \end{matrix} \tag{3.10c}$$

$$q = \left[\left(\frac{D}{2}\right)^2 + \left(z_0 - \frac{D^2}{16f}\right)^2\right]^{1/2} \tag{3.10d}$$

$$\psi_0 = \pi - \tan^{-1}\left(\frac{4F}{D}\right) + \tan^{-1}\left[\frac{D}{2(z_0 - D^2/16F)}\right] \tag{3.10e}$$

$$L^i = qs_D/(q + s_D); \quad L^r = q \tag{3.10f}$$

Figure 3.6 [20] shows the amplitude and phase of the H-plane near-field patterns; Figure 3.7 does the same for the E-plane near field. Results from Silver's aperture integration formula [23] are also presented for comparison. In this example, the feed uniformly illuminates the paraboloid without edge taper. It has been shown that it is possible to obtain the near-field patterns for all angles, including the caustic region, using GTD if the condition that the plane of observation is not too close to the reflector antenna ($z < 0.4(D^2)/\lambda$). It may not be out of place to mention here that the GTD method takes less computer time and also is capable of accurately predicting the far-out sidelobes in both near- and far-field patterns, whereas PO aperture integration can only give the main lobe and the near sidelobes without PTD corrections incorporated in the PO formulation. The effect of the thickness of the edge can be considered as in [24].

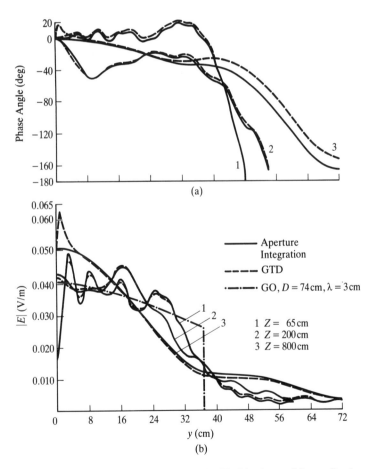

FIGURE 3.6 H-plane near-field patterns [22]: (a) phase; (b) amplitude.

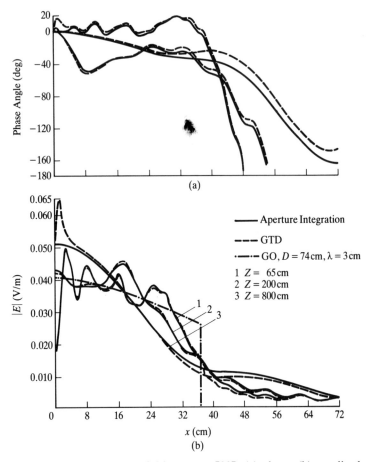

FIGURE 3.7 E-plane near-field patterns [22]: (a) phase; (b) amplitude.

3.3.2 Axially Symmetric Reflectors: GTD Analysis

The geometry and coordinate system of the axially symmetric reflector with feed [15] on axis is shown in Figure 3.8(a). Figure 3.8(b) and (c) shows the geometry of the rays reflected from the two scattering centers. The total SDR can be expressed as

$$E_d = [E_{d\theta}(\theta, \phi) a_\theta + E_{d\phi}(\theta, \phi) a_\phi] \frac{e^{-jkR}}{R} \qquad (3.11)$$

The diffracted fields in θ and ϕ are given by

$$E_{d\theta}(\theta, \phi) = E_{d\theta}^+(\theta, \phi) + E_{d\theta}^-(\theta, \phi) \qquad (3.12a)$$

$$E_{d\phi}(\theta, \phi) = E_{d\phi}^+(\theta, \phi) + E_{d\phi}^-(\theta, \phi) \qquad (3.12b)$$

3.3 FOCUS-FED REFLECTOR 223

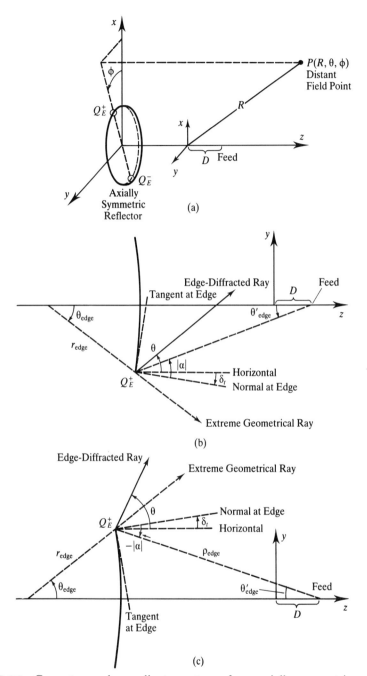

FIGURE 3.8 Geometry and coordinate system of an axially symmetric parabolic reflector with focal Length F: (a) a three-dimensional view; (b) a two-dimensional view (the xz-plane is the plane of a symmetry $\phi = 0°$); and (c) the yz-plane is the plane of asymmetry ($\phi = 90°$).

The diffracted fields from the + and − edges are given by

$$E^{\mp}_{d\theta,\phi}(\theta, \phi) = -E_{f\theta}(\pi - \theta'_{edge}, \phi) \frac{e^{-jk\rho_{edge}}}{\rho_{edge}} \cdot e^{j\pi/2} \sqrt{\frac{D_{refl}/2}{\sin \theta}}$$

$$\cdot \{\exp[jk[(2c + Z_{edge})\cos\theta \pm (D_{refl}/2)\sin\theta]]\} D^{+}_{h,s} \quad (3.13)$$

The symbols are illustrated in Figure 3.8. The diffraction coefficient (see Appendix A.1) is given by

$$D^{+}_{s,h} = -\frac{e^{-j\pi/4}}{2\sqrt{2\pi k}} \left[\frac{F[kK^i a(\theta - \delta_t - \alpha)]}{\cos(\theta - \delta_t + \alpha)/2} \mp \frac{F[kL' a(\pi + \theta - \delta_t + \alpha)]}{\sin(\theta + \delta_t - \alpha)/2} \right] \quad (3.14)$$

The diffraction coefficients given in Appendix A.1 are substituted for the geometry of the reflector.

The caustic fields on the Z-axis can again be computed using the equivalent edge current method of Section 1.19.1. Details are available in [15] for the GTD H-plane pattern of a hyperboloid with spherical wavefeed at the external focus.

3.4 REFLECTORS WITH OFFSET FEED

It is well known that for a focus-fed reflector, there is a considerable blockage due to the presence of the feed along the main beam axis. Therefore, offset feeds are commonly used to minimize aperture blockage. The offset reflectors in both single and dual mode configurations are used for low sidelobe applications and therefore it is necessary that their sidelobes and null locations and their levels need be predicted accurately. Figure 3.8(a) shows a three-dimensional view of a parabolic reflector with an offset feed and its coordinate system. The two-dimensional view is shown in Figure 3.8(b). The origin of the coordinate system is chosen at the center of the effective circular aperture σ of radius $D/2$. The effective aperture is described by the primed coordinate system. Many publications have appeared on the offset reflector and some of them are in [7, 15–21] discussing different aspects of the subject. A comparison has been made in [20] between GO/aperture field and physical optics methods for offset reflectors. The schematic diagram in Figure 3.9 shows the direct, reflected, and singly edge-diffracted rays.

3.4.1 Offset Reflector Analysis Using a PO Method

The expression for the far field of the offset reflector shown in Figure 3.8 using the PO method (see Section 1.8) is of the form [20]

$$\mathbf{E}(\theta, \phi) = -jk\eta \frac{e^{-jkr}}{4\pi r} (\hat{\mathbf{I}} - \hat{r}\hat{r}) \cdot \mathbf{T} \quad (3.15)$$

3.4 REFLECTORS WITH OFFSET FEED

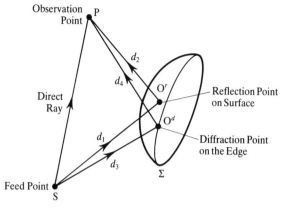

FIGURE 3.9 A schematic diagram showing the direct, reflected, and singly edge-diffracted rays.

where

$$\mathbf{T} = \iint_\Sigma \mathbf{J}(\mathbf{r}')e^{jk\hat{r}r}\,d\sigma' \tag{3.16a}$$

$$\mathbf{J}(\mathbf{r}') = 2(\hat{\mathbf{n}} \times \mathbf{H}^i) \tag{3.16b}$$

$j = \sqrt{-1}$; $k = 2\pi/\lambda$; $\eta = 120\pi$; $\mathbf{J}$ is the induced current, and $\mathbf{H}^i$ is the induced field.

The function $\mathbf{T}(\theta, \phi)$ is given by

$$\mathbf{T}(\theta, \phi) = \iint_S \mathbf{J}(\mathbf{r}')e^{jkz'\cos\theta}e^{jk\hat{r}\cdot\rho'}\,ds' \tag{3.16c}$$

$$\mathbf{J}(\mathbf{r}') = \sqrt{1 + \left(\frac{\partial f}{\partial x'}\right)^2 + \left(\frac{\partial f}{\partial y'}\right)^2}\cdot \mathbf{J}(\hat{\mathbf{r}}) \tag{3.16d}$$

where $z = f(x, y) =$ the reflector surface equation
$\rho' = x'\hat{x} + y'\hat{y} + z'\hat{z}$
$S =$ is the projected aperture area over which the integration is performed.

The incident field can be expressed as follows in terms of source coordinates (r_s, θ_s, ϕ_s) (Fig. 3.8)

$$\mathbf{H}^i = \frac{1}{\eta}\hat{\mathbf{r}}_s \times \hat{\mathbf{E}}^i \tag{3.17a}$$

$$\mathbf{E}^i(\mathbf{r}_s) = \frac{e^{-jkr_s}}{4\pi r_s}[U(\theta_s, \phi_s)\hat{\theta}_s + V(\theta_s, \phi_s)\hat{\phi}_s] \tag{3.17b}$$

Assuming the focal point is the unique phase center of the feed, the function $T(\theta, \phi)$ can be expressed as

$$T(\theta, \phi) = e^{-jkF} \iint_S |J(r')| e^{-jkz'(1-\cos\theta)} e^{jk\hat{r}\cdot\rho'} ds' \qquad (3.18)$$

where $F = r_s - z'$
$|J|$ = the absolute value of each component.

On expanding the second term of the integrand in a Taylor series, the function $T(\theta, \phi)$ becomes

$$T(\theta, \phi) = e^{-kF} \sum_{p=0}^{\infty} [-jk(1-\cos\theta)]^p J_P(\theta, \phi) \qquad (3.19)$$

where

$$T_P(\theta, \phi) = \iint_\sigma z'^P |J(r')| e^{jk\hat{r}\cdot\rho'} ds' \qquad (3.20)$$

The following conjugate relationship holds with function $T(\theta, \phi)$ above.

$$T_P(\theta, \phi) = T_P^*(\theta, \pi + \phi) \qquad (3.21)$$

From eqns. (3.14) and (3.17), one gets,

$$T(\theta, \phi) \neq T^*(\theta, \pi + \phi) \qquad (3.22)$$

As regards the pattern in the plane of offset (xz-plane), it follows from eqns. (3.17), (3.13) and (3.11) that for each component,

$$|E(\theta, \phi = 0)| \neq |E(\theta, \phi = 180°)| \qquad (3.23)$$

This indicates that the pattern is asymmetric. The degree of asymmetry depends on F/D, illumination taper, and so on. The pattern is approximately symmetrical for small values of θ.

3.4.2 GO and Aperture Field Formulation [20]

Calculating the reflected field at the aperture plane can be done using GO. This consists of launching and searching algorithms for the offset reflector. The tangential surface fields on the aperture are determined and they are Fourier transformed to get the far field at any observation point. The transform can use E- or H-fields on the surface [12]. Using the E-field formulation,

$$\mathbf{E}(\theta, \phi) = jk \frac{e^{-jkr}}{4\pi r} \hat{r} \times \mathbf{T} \qquad (3.24a)$$

$$\mathbf{T} = \iint_S -2\hat{z} \times \mathbf{E}_a e^{jk\hat{r}\cdot\rho'} \, d\sigma' \qquad (3.24b)$$

When the reflector is focus-fed, the phase of the integrand is constant; when offset-fed, the phase of the integrand varies. The focus-fed pattern is symmetric but not when diffracted fields are considered. However, an accurate estimate of the surface fields and hence the pattern does require inclusion of the diffracted fields.

Example 3.1 [20] A rectangular low tapered horn of aperture size $(2a \times 2b)$ is feeding the reflector as shown in Figure 3.8(a). The horn dominant mode is TE_{10}. Using the E-field formulation and the Fourier transform find the radiation pattern. The parameters are $F/\lambda = 22.7$, $\Psi = 44°$, $\Psi^* = 30°$. The aperture dimensions are $(1.57\lambda \times 2.14\lambda)$ tilted in the direction of Ψ_c.

Solution The aperture field is expressed in terms of a y-polarized field given by

$$\mathbf{E}_a(x', y') = \hat{y} E_0 \cos\left(\frac{\pi x'}{2a}\right) \qquad (3.25)$$

The Fourier transform of this aperture fields leads to an expression for the far field given by

$$U(\theta_s, \phi_s) = 4\pi ab E_0 \sin \phi_s \frac{\cos u}{(\pi/2)^2 - u^2} \frac{\sin u}{u} \qquad (3.26a)$$

$$V(\theta_s, \phi_s) = 4\pi ab E_0 \cos \theta_s \cos \phi_s \frac{\cos u}{(\pi/2)^2 - u^2} \frac{\sin u}{u} \qquad (3.26b)$$

where

$$u = ka \sin \theta_s \cos \phi_s; \quad v = kb \sin \theta_s \sin \phi_s \qquad (3.27a)$$

The parameters of the offset reflector result in

$$D/\lambda = 28.64; \quad d/\lambda = 19.89; \quad \Psi_c = 47.32° \qquad (3.27b)$$

Figure 3.10 shows the co- and cross-polar radiation pattern from the offset reflector antenna and Figure 3.11 shows the far-field pattern of the reflector in

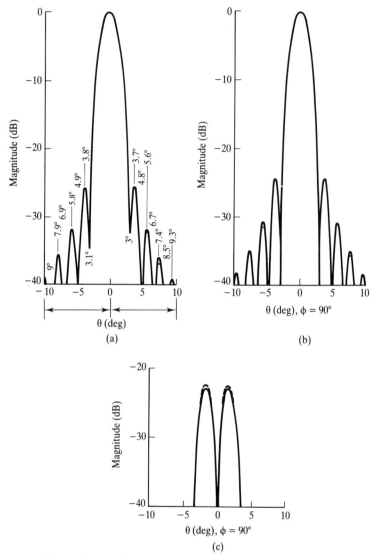

FIGURE 3.10 Normalized radiation fields from an offset reflector ($F = 22.7\lambda$, $\Psi_0 = 44°$, $\Psi^* = 30°$) fed by a linearly polarized rectangular horn with aperture dimensions $1.57\lambda \times 2.14\lambda$, measured [21]; symmetry predicted by PO. (a) Coplanar field in plane of symmetry (offset), $\phi = 0°$. (b) Copolar field in plane of asymmetry, $\phi = 90°$. (c) Cross-polar field in the plane of asymmetry.

the plane of offset with an edge taper of 10 dB. The differences between measured data and the GO/aperture-field method in the plane of offset ($\phi = 0°$) are small. But this difference may be important in many applications. The GO/aperture field theoretical results are symmetric but the experimental data

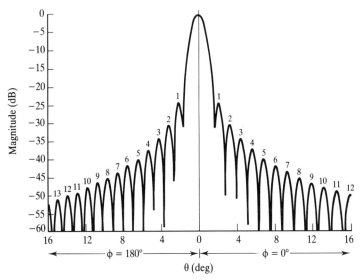

FIGURE 3.11 Far-field pattern of an offset parabolic reflector in the plane of offset ($\phi = 0°$). Reflector parameters are given in Figure 3.15b. Edge taper (ET) = -10 dB. (From [21] © 1977 IEEE.)

are asymmetric. Figure 3.10(c) shows good agreement and in particular the asymmetry in the pattern is predicted by the PO method. It is observed that the cross-polar patterns also agree very well with the PO method, better than the GO/aperture field method.

3.5 CASSEGRAINIAN REFLECTORS

A cassegrainian system of reflectors is shown in Figure 3.12. This system of reflectors has been developed to reduce the spillover and to increase the aperture efficiency. The schematic diagram in Figure 3.9 shows the direct, reflected, and singly edge-diffracted rays. The structure can be analyzed by PO and GTD methods, but the PO method is usually expensive in terms of computer time unless an efficient sampling method can be used. A summary of such sampling methods is given in Appendix A.3. The cassegrainian system can be considered as an electromagnetic scattering problem in two parts. One is the primary feed illuminating the subreflector; the other is the feed–subreflector combination illuminating the main reflector. A large number of papers, reports, and book chapters have been written on the cassegrainian system. A great deal of information is available in [27–34], which is a partial list. The analysis of the cassegrainian system of reflectors can be performed but not equally efficiently by GO (Section 1.3), PO (Section 1.8) and theories of diffraction (Section 1.9). In the following we present the analysis and design of the cassegrainian system using high-frequency methods.

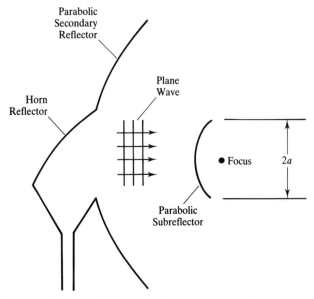

FIGURE 3.12 The near-field cassegrainian system and its geometry [34].

3.5.1 Near- and Far-Field Patterns of the Subreflector

The near-field cassegrainian system is different from a conventional far-field system; the near-field system uses a dual reflector not the conventional hyperboloidal reflector illuminated by a spherical wave. And it typically uses a dual reflector where a horn reflector is used to generate a plane wave. For details of a near-field cassegrainian system, the reader is referred to [35].

The analysis of the problem is available in [34]. The assumption as in [33, 34] is that the subreflector is being illuminated by a plane wave. The incident field on the parabolic subreflector is given by

$$E_\theta^i = KA(\theta)e^{jk\rho \cos\theta} \cos\phi \qquad (3.28a)$$

$$E_\phi^i = KB(\theta)e^{jk\rho \cos\theta} \sin\phi \qquad (3.28b)$$

where K = the complex amplitude of the electric field
$A(\theta), B(\theta)$ = the E- and H-plane radiation patterns
$\rho(r, \theta, \phi)$ = the position coordinate of a point on the subreflector.

Near-Field Analysis. The geometry and the coordinate system of the subreflector [34, 35] and near-field analysis are shown in Figures 3.13 and 3.14, respectively. Let us concentrate on the E_θ component. The GO field, as in eqn. (3.24a,b),

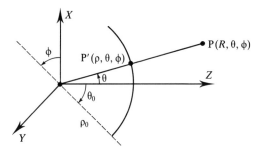

FIGURE 3.13 The coordinate system for a subreflector [34].

is given by

$$E_\theta^{GO} = +KA(\theta)e^{jk\rho\cos\theta}e^{jk\rho}\frac{\rho}{s}e^{-jks}\cos\phi F_1 \tag{3.29a}$$

$$E_\phi^{GO} = -KB(\theta)e^{jk\rho\cos\theta}e^{jk\rho}\frac{\rho}{s}e^{-jks}\sin\phi F_1 \tag{3.29b}$$

The edge-diffracted fields are given by [34]

$$E_\theta = -KA(\theta_0)e^{jk\rho_0\cos\theta_0}\frac{\exp(-j\pi/4)}{2\sqrt{2\pi k}}$$

$$\cdot \left[\frac{F\left(2kL_{1i}\cos^2\left(\frac{\phi_+ - \phi'_+}{2}\right)\right)}{\cos\left(\frac{\phi_+ - \phi'_+}{2}\right)} \pm \frac{F\left(2kL_{1r}\cos^2\left(\frac{\phi_+ + \phi'_+}{2}\right)\right)}{\cos\left(\frac{\phi_+ + \phi'_+}{2}\right)}\right]$$

$$\times \cos\alpha \cdot \sqrt{\frac{-a/\sin(\theta_1+\theta_0)}{s_1(s_1-a/\sin(\theta_1+\theta_0))}}\, e^{-jks_1}$$

$$+ \left[\frac{F\,2kL_{2i}\cos^2\left(\frac{\phi_- - \phi'_-}{2}\right)}{\cos\left(\frac{\phi_- - \phi'_-}{2}\right)} \pm \frac{F\left(2kL_{2r}\cos^2\left(\frac{\phi_- + \phi'_+}{2}\right)\right)}{\cos\left(\frac{\phi_- + \phi'_-}{2}\right)}\right]$$

$$\times \cos\alpha_2\sqrt{\frac{-a/\sin(\theta_2+\theta_0)}{s_2(s_2-a/\sin(\theta_2+\theta_0))}}\, e^{jks_2} F_3 \cos\phi \tag{3.30a}$$

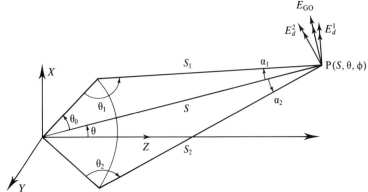

FIGURE 3.14 Near-field analysis of the subreflector. The GO and diffracted rays for the observation point [34].

The expression for E_ϕ is obtained by replacing $A(\theta_0)$ by $B(\theta_0)$, L_{Li} by L_{Lr}, $\alpha_1 = \alpha_2 = 0$, $\cos \phi$ by $\sin \phi$, where

$$\theta_{1,2} = \cos^{-1}\left(\frac{\rho_0^2 + s_{1,2}^2 - s^2}{2\rho_0 s_{1,2}}\right) \quad \text{for} \quad 0 \le \theta_0 \text{ and for } \theta \le (\pi - \theta_0), \text{ respectively}$$

$$360 - \cos^{-1}\left(\frac{\rho_0^2 + s_{1,2}^2 - s^2}{2\rho_0 s_{1,2}}\right) \quad \text{for} \quad \theta_0 \le 180$$

$$\text{and for} \quad (180 - \pi) \le 180, \quad \text{respectively} \quad (3.30b)$$

The parameters are defined below

$$s_{1,2} = \sqrt{s^2 + \rho_0^2 - 2s\rho_0 \cos(\theta_0 \mp \theta)}; \quad \alpha_{1,2} = \cos^{-1}\left(\frac{s_{1,2}^2 + s^2 - \rho_0^2}{2ss_{1,2}}\right)$$

$$L_{1,2i} = s_{1,2}; \quad L_{1,2r} = \rho_0 s_{1,2}/(\rho_0 + s_{1,2}) \quad (3.30c)$$

$$\phi'_+ = 90 - \theta_0/2 - \theta_0/2 = \phi'_-; \quad \phi_+ = \theta_1 - (90 - \theta_0/2)$$

$$\theta_2 - (90 - \theta_0/2)(90 - \theta_0/2) < \theta_2 \le 180 \quad (3.30d)$$

$$\phi_- = 360 + [\theta_2 - (90 - \theta_0/2)] \quad 0 \le \theta_2 \le (90 - \theta_0/2)$$

$$F_1 = \begin{cases} 0 & \text{for } 90 - \theta_0 < \theta_2 < 90 - \theta_0/2 \\ 1 & \text{otherwise} \end{cases}$$

The near fields along the two axial caustic directions $\theta = 0°$ and $180°$ are determined using the equivalent edge current method (Section 1.19). The total E_θ and E_ϕ are given by

$$E_\theta = E_\theta^e + E_\theta^m \quad (3.31a)$$

$$E_\phi = E_\phi^m + E_\phi^m \quad (3.31b)$$

where

$$E_\theta^e = KB(\theta_0)e^{jk\phi_0 \cos\theta_0} \frac{e^{-jks_p}}{s_p} F_4$$

$$\cdot \left[\frac{F\left(2kL_{1i}\cos^2\left(\frac{\phi_+ - \phi'_+}{2}\right)\right)}{\cos\left(\frac{\phi_+ - \phi'_+}{2}\right)} - \frac{F\left(2kL_{1r}\cos^2\left(\frac{\phi_+ + \phi'_+}{2}\right)\right)}{\cos\left(\frac{\phi'_+ + \phi'_+}{2}\right)} \right] \quad (3.32a)$$

$$E_\theta^m = -KA(\theta_0)e^{jk\rho_0 \cos\theta_0} \cos\alpha \frac{e^{-jks_p}}{s_p} \frac{a}{4}$$

$$\cdot \left[\frac{F\left(2kL_{1i}\cos^2\left(\frac{\phi_+ - \phi'_+}{2}\right)\right)}{\cos\left(\frac{\phi_+ - \phi'_+}{2}\right)} + \frac{F\left(2kL_{1r}\cos^2\left(\frac{\phi_+ + \phi'_+}{2}\right)\right)}{\cos\left(\frac{\phi_+ + \phi'_+}{2}\right)} \right] \quad (3.32b)$$

$$E_\phi^e = -KB(\theta_0)e^{jk\rho_0 \cos\theta_0} \frac{e^{-jks_p}}{s_p} \frac{a}{4}$$

$$\cdot \left[\frac{F\left(2kL_{1i}\cos^2\left(\frac{\phi_+ - \phi'_+}{2}\right)\right)}{\cos\left(\frac{\phi_+ - \phi'_+}{2}\right)} - \frac{F\left(2kL_{1r}\cos^2\left(\frac{\phi_+ + \phi'_+}{2}\right)\right)}{\cos\left(\frac{\phi_+ + \phi'_+}{2}\right)} \right] \quad (3.32c)$$

$$E_\phi^m = KA(\theta_0)e^{jkrh\sigma_0 \cos\theta_0} \cos\alpha_1 \frac{e^{-jks_p}}{s_p} \frac{a}{4} F_4$$

$$\cdot \left[\frac{F\left(2kl_{1i}\cos^2\left(\frac{\phi_+ - \phi'_+}{2}\right)\right)}{\cos\left(\frac{\phi_+ - \phi'_+}{2}\right)} + \frac{F\left(2kl_{1r}\cos^2\left(\frac{\phi_+ + \phi'_+}{2}\right)\right)}{\cos\left(\frac{\phi_+ + \phi'_+}{2}\right)} \right] \quad (3.32d)$$

where

$$s_p = \begin{cases} \sqrt{\rho_0^2 + s^2 - 2\rho_0 s \cos\theta_0} & \text{for } \theta = 0° \\ \sqrt{\rho_0^2 + s^2 + 2\rho_0 s \cos\theta_0} & \text{for } \theta = 180° \end{cases}$$

$$F_4 = \begin{cases} 1 & \text{for } \theta = 0 \\ -1 & \text{for } \theta = 180° \end{cases}$$

Figure 3.15 shows some results of near-field phase and amplitude patterns in the E- and H-plane of the parabolic subreflector of diameter 25λ and

234 REFLECTOR ANTENNAS AND RIM LOADING

$F/D = 0.33$ obtained using eqns. (2.31) and (3.32). The patterns in the figure are taken at a distance of 100λ from the focus of the paraboloid. GTD analysis of near-field patterns of prime-focus paraboloidal reflector antenna are discussed in [22].

Far-Field Analysis. The far field [34] is as usual a vector superposition of the GO reflected field and the edge-diffracted fields. With the incident fields remaining the same as given by eqn. (3.27) the GO reflected field components are given by

$$E_\theta^{r,\,GO} = KA(\theta)e^{jk\rho\cos\theta}\rho e^{jk\rho}\frac{e^{-jkR}}{R}\cos\phi F_1 \qquad (3.33a)$$

$$E_\phi^{r,\,GO} = -KB(\theta)e^{jk\rho\cos\theta}\rho\frac{e^{-jkR}}{R}\sin\phi F_1 \qquad (3.33b)$$

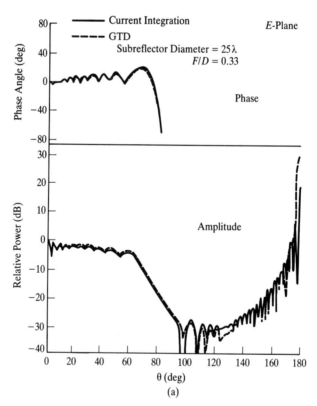

FIGURE 3.15 Far-field amplitude and phase patterns in the two principal planes [34]: (a) E-plane; (b) H-plane.

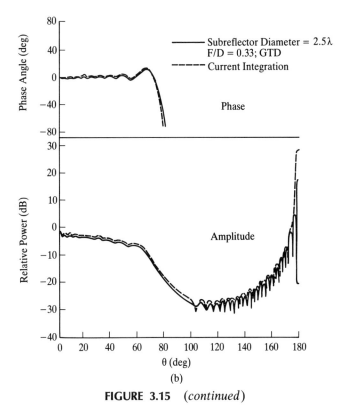

FIGURE 3.15 (*continued*)

where $\rho = F \sec^2 \theta/2$, F is the focal length of the parabolic reflector, and F_1 is given by

$$F_1 = \begin{cases} 1 & \text{for } 0° \leq \theta \leq \theta_0 \\ 0 & \text{for } \theta_0 < \theta \leq 180° \end{cases} \quad (3.34)$$

The edge-diffracted fields in the nonaxial directions ($\theta = 0°, 180°$) can be written as

$$E_\theta^d = KA(\theta_0) e^{jk\phi_0 \cos\theta} [D_h^+ e^{jk\rho_0 \cos(\theta_0 - \theta)} + D_h^- e^{jk\rho_0 \cos(\theta_0 + \theta)} e^{j\pi/2} F_2]$$

$$\times \sqrt{\frac{a}{\sin\theta}} \frac{e^{-jkR}}{R} \cos\phi \quad (3.35a)$$

$$E_\phi^d = KB(\theta_0) e^{jk\rho_0 \cos\theta_0} [D_s^+ e^{jk\rho_0 \cos(\theta_0 - \theta)} + D_s^- e^{jk\rho_0 \cos\theta_0}] \sqrt{\frac{a}{\sin\theta}} \frac{e^{-jkR}}{R} \sin\phi \quad (3.35b)$$

The fields in the axial directions and very close to axial directions, as usual are found by the equivalent edge current method. The fields are given by

$$E_\theta = E_\theta^e + E_\theta^m \quad (3.36a)$$

$$E_\phi = E_\phi^e + E_\phi^m \quad (3.36b)$$

where

$$E_{\theta e} = \frac{a \cos \theta}{4} e^{jk\rho_0 \cos(\theta_0 - \theta)} e^{jk\rho_0 \cos\theta - 0} D_s^+ \cdot P$$

$$\cdot [J_1(ka \sin \theta) + J_2(ka \sin \theta)] \frac{e^{-jkR}}{R} \cos \phi \quad (3.37a)$$

$$E_{\phi e} = -\frac{a}{4} e^{jk\rho_0 \cos(\theta_0 - \theta)} e^{jk\rho_0 \cos\theta_0} D_s^+ \cdot P$$

$$D[J_0(ka \sin \theta) - J_2(ka \sin \theta)] \frac{e^{-jkR}}{R} \sin \phi \quad (3.37b)$$

$$E_{\theta m} = -\frac{a}{4} e^{jk\rho_0 \cos(\theta_0 - \theta)} e^{jk\rho_0 \cos\theta_0} D_h^+ \cdot P$$

$$\cdot [J_0(ka \sin \theta) - J_2(ka \sin \theta)] \frac{e^{-jkR}}{R} \cos \phi \quad (3.37c)$$

$$E_{\phi m} = \frac{a \cos \theta}{4} e^{jk\phi_0 \cos(\theta_0 m - \theta)} e^{jk\phi_0 \cos\theta_0} \frac{e^{-jkR}}{R} \sin \phi \quad (3.37d)$$

where

$$P = -\frac{e^{-j\pi/4}}{2\sqrt{2\pi k}}$$

The far-field amplitude and phase patterns in the E- and H-planes have been computed [34] using eqns. (3.36) and (3.37) and are shown in Figure 3.15. The results obtained by PO integration are shown for comparison. The PO and GTD analysis compare very well. This justifies the validity of the analysis. The radiation pattern was computed at a distance of 10,000λ from the focus. The authors [34] claim this analysis is also valid for concave reflectors.

3.5.2 The Subreflector–Main Reflector System

The analysis of cassegrain antennas using different techniques and with different goals in view has been described in many references [25–48]. For the offset dual reflector it is advisable to have a shaping scheme [49–54] to avoid the improvement in the sidelobes at the cost of the aperture efficiency. Normally, it is wise to compute the pattern in the main beam region by conventional GO

3.5 CASSEGRAINIAN REFLECTORS   237

and PO (aperture integration) and GTD used to far-out sidelobes due to spillover and diffraction at the edges. The effects of curvature of the edge and the surface should be taken into account. The analysis is available in [38]. The geometry of the structure is shown in Figure 3.16(a) and the shadow boundaries and distances associated with the antenna system are shown in Figure 3.16(b).

GO and PO Analysis. The primary feed is placed at the second focus of the hyperboloid of revolution. The incident field is given by

$$F(\theta, s) = \begin{cases} (0.00316 + \cos^v \theta) \dfrac{e^{-jks}}{s} & \text{for } \theta < 90° \\ (0.00316) \dfrac{e^{-jks}}{s} & \text{for } \theta \geq 90° \end{cases} \quad (3.38)$$

where θ is the polar angle, s is the range, and the index m is chosen to provide the illumination with the desired amplitude taper on the subreflector.

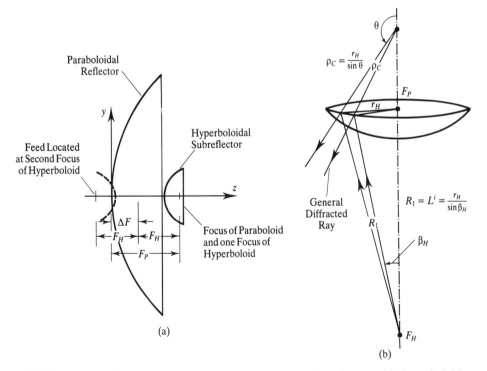

FIGURE 3.16 (a) Geometry of the cassegrain paraboloidal reflector with hyperboloid subreflector [38]. (b) Cassegrain geometry with shadow boundaries and distances [38]; direct feed spillover occurs for $\beta_H < \theta < \beta_p$.

GTD Analysis. The total diffracted field at any point is due to several sources: (1) the diffraction of the feed illumination by the scattering centers on the paraboloid; (2) the diffraction of the fields from the feed reflected from the hyperboloid; (3) the diffraction by the paraboloid rim of the fields diffracted by the hyperboloid; and (4) multiple diffraction between the two scattering centers on the paraboloid rim. The diffracted fields from the hyperboloid to be taken into consideration are: (1) the diffraction of the direct illumination from the feed; (2) the field from the feed reflected from the paraboloid; and (3) mutual coupling between the two scattering centers on the hyperboloid. Out of all these possible diffracted fields, only the dominant ones are considered. In the analysis to be presented here the effects of struts, general support structures, and the blockage of the diffracted fields from the rim of the paraboloid by the subreflector are not taken into consideration since they are small compared to the aperture blocking of the main beam. The diffracted field column vector is given by the following equation

$$\begin{bmatrix} E^d_{\parallel} \\ E^d_{\perp} \end{bmatrix} = \begin{bmatrix} D_s & 0 \\ 0 & D_h \end{bmatrix} \begin{bmatrix} E^i_{\parallel} \\ E^i_{\perp} \end{bmatrix} \sqrt{\frac{\rho_c}{s(\rho_c + s)}} e^{-jks} \tag{3.39}$$

where $E^i_{\parallel}, E^d_{\parallel}$ = incident and diffracted fields for parallel polarization
$E^i_{\perp}, E^d_{\perp}$ = incident and diffracted fields for perpendicular polarization
$\sqrt{\frac{\rho_c}{s(\rho_c + s)}}$ = the divergence factor of the diffracted field
ρ_c = distance from the diffracting edge to the caustic of the curved edge
s = distance from the point of diffraction to the observation point.

The hard and soft diffraction coefficients are given by

$$D_{s,h} = \frac{1}{\sin \beta_0} [d^+(\beta^-, n) F(ka^+(\beta^-)) + d^-(\beta^-, n) F[ka^-(\beta^-)]$$
$$\mp [d^+(\beta^+, n) F(ka^+(\beta^+)) + d^-(\beta^+, n) F[ka^-(\beta^+)]] \tag{3.40}$$

$$d^{\pm}(k, n) = -\frac{e^{j\pi/4}}{n\sqrt{2\pi k}} \frac{1}{2} \cot \frac{\pi \pm \beta}{2n}, \quad \text{in which } \beta = \beta^{\pm} = (\phi_d \pm \phi_i) \tag{3.41}$$

and

$$F[ka^{\mp}(\beta)] = j2|\sqrt{ka^{\pm}(\beta)}|e^{j(a\pm\beta)} \int_{|\sqrt{ka^{\pm}(\beta)}|}^{\infty} e^{-j\tau^2} d\tau, \tag{3.42}$$

The parameters are defined in Appendix A.3.

3.5 CASSEGRAINIAN REFLECTORS

For the special case of an infinitely thin half plane $n = 2$, the diffracted column vector is given by

$$\begin{bmatrix} E_\parallel^d \\ E_\perp^d \end{bmatrix} = \begin{bmatrix} v_B(L^i, \beta^-, 2) & 0 \\ 0 & v_B(L^i, \beta^-, 2) + v_B(L^r, \beta^+, 2) \end{bmatrix} \begin{bmatrix} E_\parallel^i \\ E_\perp^i \end{bmatrix} \sqrt{\frac{\rho_c}{s(\rho_c + s)}} e^{-jks} \quad (3.43)$$

where

$$v_B(L_{i,r}, \beta^\pm, 2) = -e^{j\pi/4} \frac{2}{\pi(1 + \cos \beta^\pm)} e^{jkL_{i,r}\cos\beta^\pm}$$

$$\cdot \left| \cos \frac{\beta^\mp}{2} \right| \int_{\sqrt{kL_{i,r}(1+\cos\beta^\pm)}}^\infty e^{-jr^2} dr \quad (3.43a)$$

The edge-diffracted field column from the subreflector is given by

$$\begin{bmatrix} E_\parallel^d \\ E_\perp^d \end{bmatrix} = \begin{bmatrix} D_s & 0 \\ D_H & 0 \end{bmatrix} \begin{bmatrix} \sin\phi \\ \cos\phi \end{bmatrix} E_x(\beta_H, r_1) \cdot e^{jk(z_H \cos\theta + r_H \sin\theta)} \frac{\sqrt{\rho_c}}{s} e^{-jks} \quad (3.44)$$

where $F_x(\beta_H, R_1)$ = incident field at the hyperboloid rim due to the feed illumination given by eqn. (3.39)

$R = \dfrac{r_H}{\sin\beta_H}$ = distance from the feedhorn to the hyperboloid

$\sqrt{\dfrac{\rho_c}{s}}$ = GO divergence factor when $s > \rho_c$

$\rho_c = \dfrac{r_H}{\sin\theta}$ = caustic distance associated with the circular rim of radius r_H of the hyperboloid (see Fig. 3.17).

The distance parameters L^i and L^r are given by

$$L^i = R_1 = r_H/\sin\beta_H; \quad L^r = R_2 = r_H/\sin\alpha_H$$

An expression similar to eqn. (3.41) can be obtained for the other edge of the hyperboloid. The feed radiation is reflected from the hyperboloid and subsequently by the paraboloid leading to a nonhomogeneous plane wave. This nonhomogeneous plane wave say, E_x^i is diffracted by the top edge of the hyperboloid and the diffracted column vector is expressed as (Fig. 3.18)

$$\begin{bmatrix} E_\parallel^d \\ E_\perp^d \end{bmatrix} = \begin{bmatrix} D_s & 0 \\ 0 & D_H \end{bmatrix} \begin{bmatrix} \sin\phi \\ \cos\phi \end{bmatrix} E_x^i e^{jk(z_H \cos\theta + r_H \sin\theta)} \frac{\sqrt{\rho^c}}{s} e^{-jks} \quad (3.45)$$

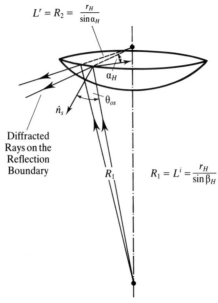

FIGURE 3.17 Ray geometry for diffraction of a source field by a hyperbola [38].

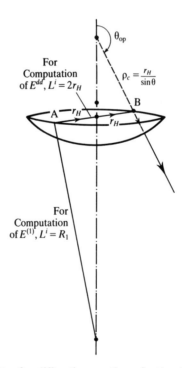

FIGURE 3.18 Ray geometry for diffraction on the reflection boundary of source fields by a parabola [38].

3.5 CASSEGRAINIAN REFLECTORS

The desired radius of curvature R_c is given by

$$R_2 = \frac{1}{R_1 + 2R_c \cos\theta_0} \tag{3.46a}$$

R_c turns out to be

$$\frac{1}{\rho^r} = \frac{1}{s'} + \frac{2}{R_c \cos\theta_{0p}} \tag{3.46b}$$

where S' is the source distance, assumed to be infinity, and θ_{0p} is the angle between the surface normal and the unit vector to the source point at the point of diffraction (Fig. 3.19).

The distance parameter L^r is given by

$$L^r = \rho^r = \frac{R_1 R_2}{R_1 - R_2} \frac{\cos\theta_{0p}}{\cos\theta_{0s}} \tag{3.46c}$$

Doubly Diffracted Rays (DDR). The doubly diffracted fields from the hyperboloid are given by

$$\begin{bmatrix} E_\parallel^{dd} \\ E_\parallel^{dd} \end{bmatrix} = \begin{bmatrix} D_s & 0 \\ 0 & D_H \end{bmatrix} \begin{bmatrix} E_\parallel^{(1)} + E_\parallel^{(2)} \\ E_\perp^{(1)} + E_\perp^{(2)} \end{bmatrix} e^{jk(z_H \cos\theta + r_H \sin\theta)} \frac{\sqrt{\rho_c}}{s} e^{-jks} \tag{3.47}$$

where

$$L^i = 2r_H \tag{3.48a}$$

$$L^r = \rho_1^r = \frac{2r_h R_1 R_2}{R_1 R_2 - 2r_H/R_1 - R_2 \dfrac{\cos\theta_0}{\cos\theta_s}} \tag{3.48b}$$

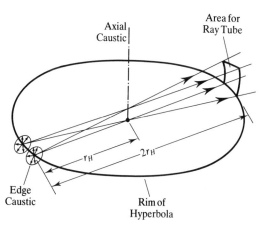

FIGURE 3.19 Ray geometry for second-order diffraction by the hyperbola rim [38].

242 REFLECTOR ANTENNAS AND RIM LOADING

These fields are given by

$$\begin{bmatrix} E_\parallel^{(1)} \\ E_\perp^{(1)} \end{bmatrix} = \begin{bmatrix} D_s & 0 \\ 0 & D_H \end{bmatrix} \begin{bmatrix} \sin \phi \\ \cos \phi \end{bmatrix} F_x(\beta_H, R_1) j \frac{1}{\sqrt{2r_H}} e^{-j2kr_H} \quad (3.49)$$

where $F_x(\beta_H, R_1)$ = the incident field at the bottom edge

$j \dfrac{1}{\sqrt{2r_H}} e^{-j2kr_H}$ = divergence factor multiplied by phase delay

$$L^i = R_1 = t_H / \sin \beta_H$$
$$L^r = R_2 = r_H / \sin \alpha_H$$

The plane wave fields originating from the paraboloid, diffracted by the hyperboloid, and incident on the opposite edge of the hyperboloid are given by

$$\begin{bmatrix} E_\parallel^{(2)} \\ E_\perp^{(2)} \end{bmatrix} = \begin{bmatrix} D_s & 0 \\ 0 & D_h \end{bmatrix} \begin{bmatrix} \sin \phi \\ \cos \phi \end{bmatrix} E_x^i j \frac{1}{\sqrt{2r_H}} e^{-j2kr_H} \quad (3.50)$$

where E_x^i = the amplitude of the nonhomogeneous plane wave incident on the hyperboloid rim (feed field after reflection from the hyperboloid and paraboloid).

The total diffracted field from the hyperboloid is a vector superposition of fields given by eqn. (3.44) from the two scattering centers.

Now the edge diffraction from the parabolic reflector needs be taken into account. For a feed design with efficient taper, perhaps a low-noise feed having sidelobes and backlobe, the rim of the paraboloid does not significantly diffract the direct field from the feed. Also included are the multiply diffracted terms for the illumination of the paraboloid rim due to diffraction from the hyperboloid rim.

The diffracted field column vector for the paraboloid rim due to the feed radiation reflected from the hyperboloid is given by

$$\begin{bmatrix} E_\parallel^d \\ E_\perp^d \end{bmatrix} = \begin{bmatrix} D_s & 0 \\ 0 & D_h \end{bmatrix} \begin{bmatrix} E_\parallel^{ri} \\ E_\perp^{ri} \end{bmatrix} e^{jk(x_p \cos \theta + r_p \sin \theta)} \sqrt{\frac{\rho_c}{s}} e^{-jks} \quad (3.51)$$

where $E_\parallel^{ri}, E_\perp^{ri}$ = fields from the feed reflected from the hyperboloid

$$\rho_c = r_p \sin \theta; \quad L^i = R_0 = \frac{r_p}{\sin \alpha_p}.$$

The DDR column vector from the paraboloid rim due to the incident fields as a result of diffraction by the hyperbola rim and the opposite rim of the paraboloid is given by (Fig. 3.20, [38])

$$\begin{bmatrix} E_\parallel^{dd} \\ E_\perp^{dd} \end{bmatrix} = \begin{bmatrix} D_s & 0 \\ 0 & D_h \end{bmatrix} \begin{bmatrix} E_\parallel^{(1)} + E_\parallel^{(2)} + E_\parallel^{(3)} \\ E_\perp^{(1)} + E_\perp^{(2)} + E_\perp^{(3)} \end{bmatrix} e^{jk(z_p \cos \theta + r_p \sin \theta)} \sqrt{\frac{\rho_c}{s}} e^{-jks} \quad (3.52)$$

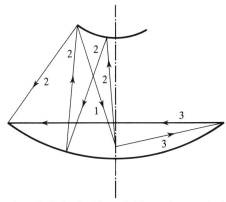

FIGURE 3.20 Rays used to find the incident field on the parabola edge. Ray system 1 used to obtain $E^{(1)}$. Ray system 2 used to obtain $E^{(2)}$. Ray system 3 used to obtain $E^{(3)}$ [38].

where $E_{\parallel}^{(1)}, E_{\perp}^{(1)}$ = incident fields due to the diffraction of the feed fields by the hyperboloid rim

$E_{\parallel}^{(2)}, E_{\perp}^{(2)}$ = incident fields due to the diffraction from the opposite edge of the paraboloid

$E_{\parallel}^{(2)}, E_{\perp}^{(2)}$ = incident fields due to the diffraction from the opposite edge of the paraboloid

$\rho_c = r_p/\sin\theta$, the caustic distance for the paraboloid rim of radius r_p.

Each of the fields $E_{\parallel}^{(1)}, E_{\parallel}^{(2)}, E_{\parallel}^{(3)}$ can be expressed in the same forms but with different divergence factors and length parameters. Further details are available in [38].

A typical plot of the E- and H-plane patterns of a cassegrainian antenna using the above analysis is shown in Figure 3.21 [38].

3.6 A TYPICAL SHAPING SCHEME

In this section a shaping based on GO will be described [46] and the effect of shaping on the performance of a typical cassegrain reflector will be presented. Shaped reflectors have been discussed in many references. Some of the research is available in [46, 49–54].

3.6.1 GO Shaping Scheme

The shaping scheme [46] is shown in Figure 3.22. The input parameters specified for shaping are the center locations of the two reflectors, horn angle and beam diameter, the feed pattern, and the desired aperture taper. The shaping scheme used is based on three optical laws: (1) the law of conservation

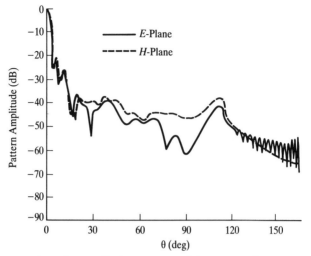

FIGURE 3.21 E- and H-plane radiation pattern of the cassegrain antenna; diameter = 20λ $F_p = 8\lambda$, $D_h = 2.67\lambda$, $F_H = 2.46\lambda$, and 15 dB feed taper over the subreflector shown in Figure 3.4. The direct feed spillover occurs for $17.5° \leq \theta \leq 89.8°$; reflected spillover also occurs. (From [38] © 1975 IEEE.)

of energy along a ray tube, (2) Snell's law, and (3) uniform phase and equal path length. The law of conservation of energy leads to a mapping function which transforms the coaxial cones of the horn (the primary feed) into concentric circles on the aperture with a prescribed radial power distribution. Use of Snell's law on the subreflector gives two first-order partial derivatives representing the slopes of the subreflector surface at each point. Since the

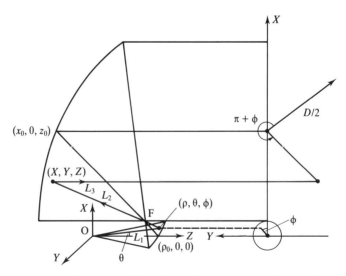

FIGURE 3.22 Geometry of the dual reflector antenna for shaping [46].

3.6 A TYPICAL SHAPING SCHEME

aperture has to be an equiphase surface, the path lengths of all rays emanating from the primary feed and reaching the aperture are equal.

We will quickly describe the analytical procedure for the shaping scheme. With the origin of the coordinate system at the phase center of the feed horn, the points on the subreflector and the main reflectors are given by $\rho = \rho(\theta, \phi)$ and $z = z(x, y)$ respectively. The feed and the radial power distribution are respectively given by $F(\theta)$ and $P(R)$. D is the diameter of the output beam and $2\theta_c$ is the total included angle of the horn.

The energy conservation law gives

$$\frac{\int_0^\theta F(\theta') \sin \theta' \, d\theta'}{\int_{c\,0}^\theta F(\theta') \sin \theta' \, d\theta'} = \frac{\int_0^R P(R') R' \, dR'}{\int_0^{D/2} P(R') R' \, dR'} \tag{3.53}$$

Assuming that $P(R')$ is uniform on the aperture, eqn. (3.53) turns out to be

$$\int_0^\theta F(\theta') \sin \theta' \, d\theta' = \frac{4S}{D^2} R^2 \tag{3.54a}$$

where

$$S = \int_0^{Q_c} F(\theta') \sin \theta' \, d\theta' \tag{3.54b}$$

The transformation between (x, y) and (θ, ϕ) is given by

$$x = x_0 \pm R \cos \phi \tag{3.55a}$$
$$y = \pm R \sin \phi \tag{3.55b}$$

where the plus sign is for a cassegrain reflector and the minus sign for a gregorian reflector.

The next task is find the corresponding z for (x, y) on the main reflector. It turns out that the relationships between different parameters are given by

$$\frac{\partial \rho}{\partial h} = \frac{QV}{Q^2 + U^2} \tag{3.56a}$$

and

$$\frac{\partial \rho}{\partial \phi} = \frac{U \cdot V \sin \theta}{Q^2 + U^2} \tag{3.56b}$$

The next step is to find a numerical solution of the reflector profiles.

Starting from the central ray, $\partial \rho / \partial \theta$ and $\partial \rho / \partial \phi$ at $(\rho_0, \theta_0, \phi_0) = (\rho, 0, 0)$ from eqn. (3.50) using the boundary condition that the central ray intersects the subreflector at $(\rho, \theta, \phi) = (\rho_0, 0, 0)$ and the main reflector at $(x, y, z) = (x_0, 0, z_0)$. Having determined the slope at the starting point, the differential increment in ρ can be found using

$$d\rho = \frac{\partial \rho}{\partial \theta} d\theta + \frac{\partial \rho}{\partial \phi} d\phi \tag{3.57}$$

The shaping is carried out by computing $\partial \rho / \partial \theta$ step by step in the θ direction only, since along each curve of these cuts on the subreflector $d\phi = 0$. The point (x, y) can be determined from eqns. (3.48) and (3.49) after the new intercept $(\rho = \rho_0 + d\rho, \theta = \theta_0 + d\theta, \phi = \phi_0 + d\phi)$.

If we use the reference point as $z = 0$, the optical length of the central ray is

$$OL = \rho_0 + l_0 - z_0 = \rho_0 + \sqrt{x_0^2 + (z_0 - \rho_0)^2} - z_0 \tag{3.58}$$

z can be determined from the relation

$$\rho + l - z = OL \tag{3.59}$$

This gives

$$z = \frac{a^2 + b^2}{2(\rho \cos \theta - \rho + OL)} + \tfrac{1}{2}(\rho \cos \theta - OL + \rho) \tag{3.60}$$

The step computation is repeated till the ranges of θ and ϕ are covered and as a result a complete shaping is achieved. Such a scheme generates the subreflector and the main reflector surfaces for an offset-fed antenna to achieve very low sidelobes and very high aperture efficiencies.

Example 3.2 Consider a shaping scheme with an unshaped baseline system. The main reflector is an offset (45°) portion of a parabola. The focal length = 3.5 meters. The focal point is at $(x, y, z) = (0.4, 0.0, 1.5)$ meters. The coordinates of the center of the subreflector are $(0.0, 0.0, 1.9)$ meters. The center of the main reflector is located at $(3.3, 0.0, -1.4)$. The total included angle of the horn is 28°.

Solution The feed pattern has an 18 dB taper at the edge as given in Table 3.1.
The parabola is used for reference purposes only, it has nothing to do with the shaping scheme.

TABLE 3.1 Feed Pattern $F(\theta)$ with Angle θ

θ(deg)	2	4	6	8	10	12	12
$F(\theta)$(dB)	0.5	1.8	4.2	7.1	10.3	14.1	18

3.7 EFFECT OF RIM LOADING AND SIDELOBE CONTROL

3.7.1 Introduction

One way of controlling the radiation characteristics of reflectors is to use radar-absorbing material (RAM) on the surface of the reflector. The effect of loading upon the radiation characteristics of reflectors has been a subject of research since the early 1980s [55–57]. An early experimental paper [58] described the possibility of generating low sidelobes using microwave absorbers. Some authors have characterized this loading as having a surface impedance. If a perfect conductor is coated with a RAM, then under certain conditions the coating can be represented by a surface impedance [59, 60]. Two canonical scattering problems have applications in loaded reflector analysis and design: (1) electromagnetic scattering by a half plane with two face impedances and (2) electromagnetic scattering by a straight wedge with two face impedances. These are discussed in Section 1.11 where all key references are available.

3.7.2 Rim-Loaded Focus-Fed Parabolic and Hyperbolic Dishes

Figures 3.23 and 3.24 respectively show the geometry and coordinate system of a parabolic and a hyperbolic dish. Each dish is divided into several annular regions with $\theta_i < \theta < \theta_{i-1}$ and each having a different surface impedance Z_i^+, $i = 1, \ldots, N$.

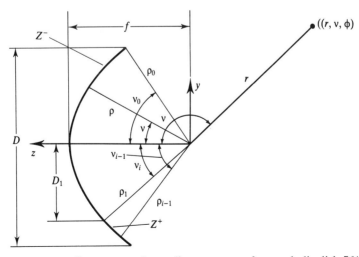

FIGURE 3.23 Geometry and coordinate system of a parabolic dish [61].

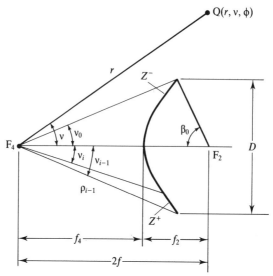

FIGURE 3.24 Geometry and coordinate system of a hyperbolic dish [61].

Incident Fields. The incident electric field [61] on the dish is given by

$$E_\theta^i = V \frac{\exp(-jk\rho)}{\rho} f_\theta(\theta, \phi) \sin \phi \qquad (3.61a)$$

$$E_\phi^i = V \frac{\exp(-jk\rho)}{\rho} f_\phi(\theta, \phi) \cos \phi \qquad (3.61b)$$

where f_θ, f_ϕ are the pattern functions in the θ and ϕ directions
$f_\theta = \cos \theta, f_\phi = 1$, for a y-directed dipole source
$f_\theta = f_\phi = (1 + \cos \theta)$, for a Huyghens source
V is a constant
The E- and H-plane cuts are given by $\phi = \pm \pi/2$ and $\phi = 0, \pi$ respectively.

Reflected Fields. The reflected field is given by [61]

For the parabolic dish

$$E^r = 0 \qquad (\theta \neq \pi) \qquad (3.62)$$

for the hyperbolic dish

$$E_\theta^r = \left(\frac{\sin \delta}{\sin \theta}\right) \exp\left(-j2kf \frac{1 - e \cos \theta}{e}\right) R_h(\theta) f_\theta(\delta, \phi) \sin \phi \qquad (3.63a)$$

$$E_\phi^r = \left(\frac{\sin \delta}{\sin \theta}\right) \exp\left(-j2kf \frac{1 - e \cos \theta}{e}\right) R_h(\theta) f_\theta(\delta, \phi) \sin \phi \qquad (3.63b)$$

3.7 EFFECT OF RIM LOADING AND SIDELOBE CONTROL

where

$$R(\theta) = \frac{\sin\left(\frac{\theta - \delta}{2}\right) - \sin \alpha^+}{\sin\left(\frac{\theta - \delta}{2}\right) + \sin \alpha^+} \tag{3.63c}$$

$$e = \frac{\sin(\beta_0 + \theta_0)}{\sin \beta_0 - \sin \theta_0} \tag{3.63d}$$

δ, the azimuthal angle of the point of reflection is zero for the parabolic dish and

$$\delta = \sin^{-1}\left[\frac{(e^2 - 1)\sin\theta}{e^2 + 1 - 2e\cos\theta}\right] \tag{3.63e}$$

Z^+ is the surface impedance at the point of reflection and α is given by

$$\alpha_e^+ = \sin^{-1}\left(\frac{\eta_0}{Z^+}\right) \quad \text{and} \quad \alpha_h^+ = \sin^{-1}\left(\frac{Z^+}{\eta_0}\right)$$

Diffracted Fields from the Rims. The fields diffracted by the rim are given by [61]

$$E_\theta^s = \sqrt{\frac{\sin\theta_0}{2\pi k \rho_0 \sin\theta}} \cdot D_h^u(\gamma_0, \gamma^u) f_\theta(\theta_0, \phi) \sin\phi$$

$$\cdot F_1 + GD_h^l(\gamma_0, \gamma^l) f_\theta(\theta_0, \phi + \pi) \sin\phi \cdot F_2 D_e^u(\gamma_0, \gamma^u) f_\theta(\theta, \phi) \cos\phi$$

$$\pm D_e^l(\gamma_0, \gamma^l) f_\phi(\theta - 0, \phi + \pi) \cos\phi \tag{3.64a}$$

$$E_\phi^s = \sqrt{\frac{\sin\theta_0}{2\rho_0 \sin\theta}} \cdot D_h^u(\gamma_0, \gamma^l) f_\phi(\theta_0, \phi + \pi) \cos\phi \cdot F_1$$

$$+ GD_e^l(\gamma_0, \gamma^l) f_\phi(\theta_0, \phi + \pi) \cos\phi \cdot F_2 D_{ei}^u(\gamma_i, \gamma_i^u) f_\phi(\theta_0, \phi) \cos\phi$$

$$\pm D_{ei}^l(\gamma_i, \gamma_i^l) f_\theta(\theta_i, \phi + \pi) \cos\phi \tag{3.64b}$$

For parabolic dish, the function G is given by

$$G = 1 - U\left(\theta - \frac{\pi - \theta_0}{2}\right) U\left(\frac{\pi}{2} - \theta\right) \tag{3.65a}$$

$$G = 1 - U\left(\frac{\pi}{2} + \frac{\beta_0 - \theta_0}{2} - \theta\right) U\left(\theta - \frac{\pi}{2}\right) \tag{3.65b}$$

$D^{u,l}$ are the diffraction coefficients for the upper and lower scattering centers of the dish, and F_1 and F_2 are given by

$$F_1 = \exp\left\{-j2k\rho_0 \sin^2\left[\left(\frac{\theta_0 - \theta}{2}\right) + \frac{\pi}{4}\right]\right\} \quad (3.65c)$$

$$F_2 = \exp\left\{-j2k\rho_0 \sin^2\left[\left(\frac{\theta_0 + \theta}{2}\right) - \frac{\pi}{4}\right]\right\} \quad (3.65d)$$

Diffracted Fields at the Junction of Annular Loaded Layers [61]. The diffracted field due to the interface between two regions having different surface impedances is given by [61]

$$E_{i\theta}^s = \sqrt{\frac{\sin\theta_i}{2\pi k\rho_i \sin\theta}} \{D_{hi}^u(\gamma_i, \gamma_i^u) \cdot f_0(\theta_0, \phi) \sin\phi \cdot F_3 U(\theta - \theta_i^u)$$

$$+ D_{hi}^l(\gamma_i, \gamma_i^l) f_\theta(\theta_i, \phi + \pi) \sin\phi \cdot F_4 U(\theta - \theta_i^l)\}$$

$$\cdot \begin{cases} D_{hi}^e(\gamma_i, \gamma_i^u) \cdot f_\phi(\theta_0, \phi) \cos\phi \cdot D_{ei}^l(\gamma_i, \gamma_i^l) \cdot f_\theta(\theta_i, \phi + \pi) \cos\phi \\ D_{hi}^e(\gamma_i, \gamma_i^u) \cdot f_\phi(\theta_0, \phi) \cos\phi \cdot D_{ei}^l(\gamma_i, \gamma_i^l) \cdot f_\theta(\theta_i, \phi + \pi) \cos\phi \end{cases}$$

$$\cdot \sqrt{\frac{\sin\theta_i}{2\pi k\rho_i \sin\theta}} D_{hi}^u(\gamma_i, \gamma_i^u) f_\phi(\theta_0, \phi) \cos\phi \cdot F_4 U(\theta - \theta_i^l) \quad (3.66a,b)$$

Total Field at any Observation Point. The total far field from the structure is given by [61]

$$\mathbf{E}(\mathbf{r}, \theta, \phi) = V \frac{\exp(-jkr)}{r} (\mathbf{E}^i + \mathbf{E}^r + \mathbf{E}^s) + \sum_{i=1}^{N} \mathbf{E}_i^s \quad (3.67)$$

The four terms respectively are the incident field, the reflected field, the scattered field and the contribution from the discontinuities at the junction of the loadings in the front part of the dish and the four different contributions are given by the above equations.

Results and Discussions. Figures 3.25 and 3.26 give the radiation patterns of the parabolic and hyperbolic reflector antennas. The parabolic dish has $\theta_0 = 60°$, and $\beta_0 = 64°$, $D/\lambda = 15$ and $2f/D = 1.2$. The dishes are annularly loaded and the loading tapers to the center of the dish. The important observations are: (1) the rear face loading has a moderate effect on the radiation characteristics; (2) the rear radiation is practically irrespective of the loading in the front or

3.7 EFFECT OF RIM LOADING AND SIDELOBE CONTROL 251

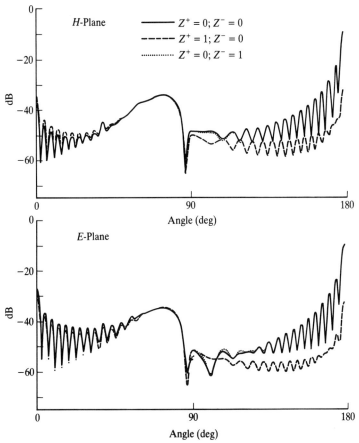

FIGURE 3.25 Radiation diagram of the parabolic dish in the two principal planes with resistive loading on the back side [61].

rear; and (3) the axial field in the forward direction is sensitive to the loading on the front. The last aspect is shown in Figure 3.27, where the normalized on-axis field of the rim-loaded dish is plotted versus $(D_0 - D_1)/D_0$, where D_0 is outer diameter of the dish and D_1 is the diameter of the PEC part. It may be mentioned that though this seems to be a consolidated study of the effect of rim loading, the analysis has been done without the consideration of aperture blockage. Hence the results are inaccurate to that extent, particularly in the forward or near-forward directions.

A separate study [61] was made on the control of reflector antenna performance by only rim loading. The theory is the same as above. Radiation patterns of rim-loaded parabolic antennas for various widths of tapering zone with loading impedances (a) $0.5\eta_0$ and (b) $0.5(1+j)\eta_0$ are shown in Figures 3.27 and 3.28.

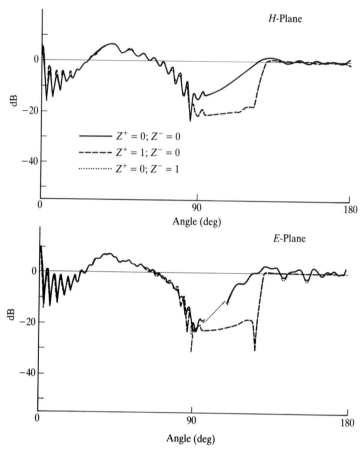

FIGURE 3.26 Radiation diagram of the hyperbolic dish in the two principal planes with resistive loading on the back side [61].

3.8 CROSS-POLARIZATION CHARACTERISTICS

The orthogonal, or cross-polarization [62], characteristics of reflector antennas have been of interest to researchers for quite some time. This is truer at high frequencies (above 10 GHz) where a small departure from regularity may easily cause cross-polarization. Nevertheless, using both co- and cross-polarization doubles the channel capacity. The cross-polarization is caused by any kind of irregularity or discontinuity. It may be due to offset illumination, unbalanced or asymmetric feed patterns, surface roughness, departure of a surface from a desired shape, longitudinal current distribution, poor pointing accuracy, and so on. To the best knowledge of the author, the first reported analysis of cross-polarization in cassegrain reflectors was by Condon [63]. It is also mentioned in the book by Silver [64]. Cutler [65] and Jones [66] investigated the paraboloidal reflector with a dipole at its focus. It has been shown [67] that if an electric and magnetic dipole pair at the focus is used to illuminate a paraboloid, a zero cross-polarization system can be designed in principle. The cross-polarization properties of a reflector excited by an arbitrary feed were studied by Afifi [68]. Potter [69] also found a similar expression for a cassegrain antenna. The cross-polarization isolation at off-axis incidence for front-end paraboloidal reflectors indicates that the cassegrain system is superior to the front-end antenna in terms of the off-axis cross-polarization properties. A brief account of offset dual-reflector systems having zero cross-polarization is given in [36]. Figure 3.29(a) plots the cross-polar diagrams of a cassegrain antenna with unloaded (solid line) and loaded (dotted line) subreflector. Figure 3.29(b)

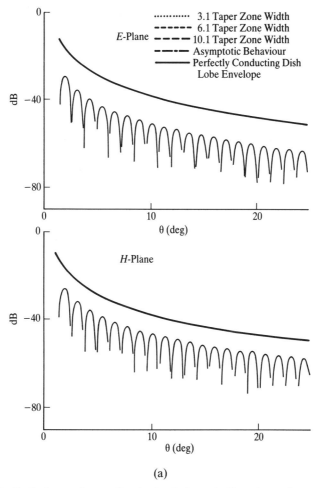

(a)

FIGURE 3.27 Radiation patterns of a rim-loaded parabolic antenna for various widths of tapering zone for E- and H-planes: (a) loading impedance $Z = 0.5\ \eta_0$. (b) $Z = 0.5(1 + j)\eta_0$. *(continued on next page)*

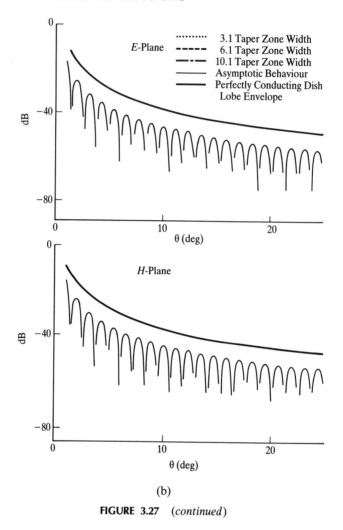

FIGURE 3.27 (continued)

compares the co- and cross-polar radiation diagrams of a cassegrain antenna. The parameters are $Z = 0.5\xi_0$, loading width $= 1.5\lambda$ and plane of cut is $\Phi = 45°$. The theory and structure configurations are given in Section 3.7.

3.8.1 Cross-Polarization Properties of Front-Fed and Offset Reflectors

The analysis of offset paraboloidal antenna from the point of view of de- or cross-polarization has been presented in a number of references. Some examples are [47, 71–75]. In this section, following [72], we present the formulation for the estimation of off-axis cross-polarization and polarization efficiencies of reflector antennas. Two important parameters in this connection, namely the peak polarization level and polarization efficiency, are defined in two ways in the literature. One is the ratio [47] of the antenna gain including the effects of

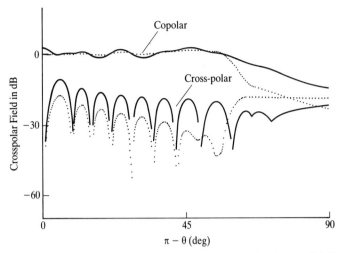

FIGURE 3.28 Radiation patterns of a hyperbolic dish: unloaded (solid line); loaded (dotted line). Loading impedance $Z = 0.5\eta_0$; loading width $= 1.5\lambda$; $\phi = 45°$ cut.

cross-polarization to the antenna gain if the cross-polarized energy were zero everywhere. Another definition [72] is the ratio of the radiated power in the copolar field to the total radiated power.

Let P_t be the total power fed to the reflector by the primary feed, then the incident field on the reflector surface is given by

$$\mathbf{E}^i = \sqrt{P_t Z_0 G_t} \, 2\pi \, \frac{e^{-jk\rho}}{\rho} \, \hat{e}^i \tag{3.68}$$

where e^i = unit vector describing the polarization of the incident field
 Z_0 = intrinsic impedance of the medium in which the antenna system is embedded
 ρ = distance from the focus to the reflecting surface
 G_t = gain function of the primary feed.

The reflected field with the phase reference at the aperture is given by

$$\mathbf{E}^r = \sqrt{\frac{P_t Z_0 G_t}{2\pi}} \, \frac{1}{\rho} \, e^{jk\rho} \tag{3.69}$$

The polarization vectors e^i and e^r for the incident and reflected are related by

$$(e^i + e^r) \times \hat{n} \tag{3.70}$$

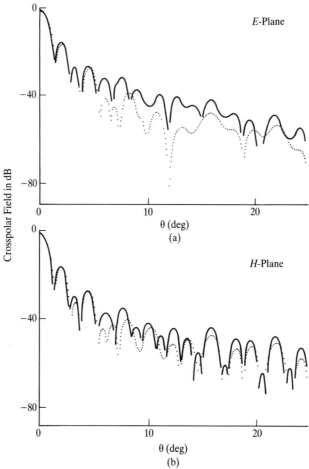

FIGURE 3.29 (a) Cross-polar diagrams of cassegrain antenna. (b) Copolar and cross-polar radiation diagrams of a cassegrain antenna. Unloaded (solid line); loaded (dotted line); $Z = 0.5\zeta_0$; loading width = 1.5λ; plane of cut $\Phi = 45°$.

The radiated field due to such a reflector is given by

$$\mathbf{E}(\zeta, \xi) = \frac{jk}{2\pi R} E^{jkR}(1 + \cos \zeta) \cdot \iint_S \mathbf{E}^r e^{jk\hat{r} \cdot \mathbf{R}} \, dS \tag{3.71a}$$

$$\mathbf{E}(\zeta, \xi) = \frac{jke^{jkR}}{2\pi R} \iint_S e^r A e^{jk\hat{r} \cdot \mathbf{R}} \, dS \tag{3.71b}$$

where

$$A = 2\sqrt{\frac{P_t Z_0 G_t}{2\pi} \cdot \frac{1}{\rho}}$$

3.8 CROSS-POLARIZATION CHARACTERISTICS

(R, ζ, ξ) are the polar coordinates of the point of observation.

Let the polarizations of the co- and cross-polarizations be in the directions u^c and u^x

$$e^r = (e^r \cdot e^c)u^c + (e^r \cdot u^x)u^x \tag{3.72}$$

The co- and cross-polar fields are given by

$$E^{co}(\zeta, \xi) = \frac{j}{\lambda R} \iint_S A(e^r \cdot u^c) e^{jk\hat{r} \cdot \mathbf{R}} \, dS \tag{3.73a}$$

$$E^x(\zeta, \xi) = \frac{j}{\lambda R} \iint_S A(e^r \cdot u^x) e^{jk\hat{r} \cdot \mathbf{R}} \, dS \tag{3.73b}$$

where $A(\hat{e}^r \cdot \hat{u}^c) e^{jk\hat{r} \cdot \mathbf{R}}$ are the co-polar and cross-polar field distributions over the aperture.

In this analysis, since the cross-polar field is much weaker than the copolar field, it is assumed that $E^r \cdot \hat{u}^c \approx 1$ without introducing any serious error. Also, the dot product $\hat{e}^r \cdot \hat{u}^c$ is a function of position at which it is computed. This product is unity and decrease towards the edge. The assumption is that the illumination is uniform. This assumption has negligible effect on the cross-polarization characteristics. For a conventional and axisymmetric feed, the dot product $\hat{e}^r \cdot \hat{u}^x$ is given by

$$\hat{e}^r \cdot \hat{u}^x = a \sin 2\phi \tag{3.74}$$

where, a is a constant to be determined.

The eqn. (3.73a,b) is given by

$$E_{co} = \frac{jD^2}{4\lambda R} \int_0^{2\pi} \int_0^1 A \exp[jur \cos(\phi - \xi)] r \, dr \, d\phi \tag{3.75a}$$

$$E_x = \frac{jD^2}{4\lambda R} \int_0^{2\pi} \int_0^1 Aa \sin 2\phi \exp[jur \cos(\phi - \xi)] r \, dr \, d\phi \tag{3.75b}$$

where

$$u = \frac{\pi D}{\lambda} \sin \zeta \tag{3.75c}$$

After integration it can be shown that the eqn. (3.75a,b) takes the forms given by

$$E_{co} = jA\lambda R \frac{\pi d^2}{4} 2J_1(u)/u \tag{3.76a}$$

$$E_x(\zeta, \xi) = \frac{jAa}{\lambda R} \frac{\pi D^2}{4} \sin 2\xi \cdot \sum_{k=0}^{\infty} \frac{2(3 + 2k)n}{(k + 2)} J_{3+2k}(u)/u \tag{3.76b}$$

From the expressions in eqn. (3.71a,b), the following conclusions are in order: (1) the cross-polar far field is zero in the boresight ($\zeta = 0$, $u = 0$) and in the two principal planes, i.e. $\xi = 0$ and $\xi = 90°$ (2) the peak value of the cross-polarized field occurs at $\xi = \pm 45°$.

The cross-polarization normalized with respect to copolarization in dB is given by

$$\text{Cross-polarization (dB)} = 20 \log_{10} |E_x(\zeta, \xi)/E_{co}(0)| \qquad (3.77)$$

The peak polarization ratio is defined by

$$\text{Cross-polarization (dB, } |\xi| = 45°) = 20 \log_{10} |aG(u)| \qquad (3.78)$$

The polarization efficiency of the antenna is defined as

$$\eta = \frac{\text{Power radiated in the Copolar Field}}{\text{Total radiated Power}}$$

$$= \frac{P_{co}}{P_{co} + P_x} \qquad (3.79)$$

Hence

$$\frac{P_{co}}{P_x} = \frac{\eta}{1-\eta} \qquad (3.80)$$

The co- and cross-polar field distributions on the aperture are

$$E_{co} = A \qquad (3.81a)$$

$$E_x(\text{aperture}) = Aa \sin 2\phi \qquad (3.81b)$$

with these, it turns out that the power radiated in the co- and cross-radiated field is given by

$$P_{co} = \frac{a^2}{Z_0} \frac{\pi D^2}{4} \qquad (3.82a)$$

$$P_x = \frac{A^2 a^2}{2Z_0} \frac{\pi D^2}{4} \qquad (3.82b)$$

Substituting eqn. (3.82a,b) in eqn. (3.79) one obtains

$$\frac{P_{co}}{P_x} = \frac{\eta}{1-\eta} = \frac{2}{a^2} \qquad (3.83)$$

Hence

$$A = \sqrt{\frac{2}{\eta - 1}} \qquad (3.84)$$

Then, the relative cross-polarization ratio in dB is given by

$$\text{Cross-polarization (dB)} = 10 \log_{10} |2(1/\eta - 1) G^2(u)| \qquad (3.85a)$$

The peak cross-polarization in dB is

$$10 \log_{10} |0.29(1/\eta - 1)| \qquad (3.85b)$$

Some comments on the limitations of eqn. (3.77b) are in order: (1) the equation ignores the r-dependence of the aperture field, the error can be shown to be less than 1 dB; (2) eqn. (3.82b) is applicable to both front-end and cassegrain antennas with axial symmetry. For offset cases the equation should be modified.

Example 3.3 This example [72] discusses the results using the above expressions for front-fed, cassegrainian antennas. The cross-polar radiation pattern of a 200λ diameter reflector compared with the exact method has been plotted in Figure 3.30 for a 200λ antenna with $f/D = 0.25$ with a short electric dipole

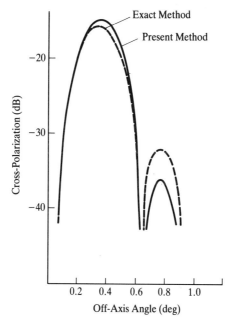

FIGURE 3.30 Cross-polar pattern for a 200λ antenna: $f/D = 0.25$; the primary feed is a short dipole [72].

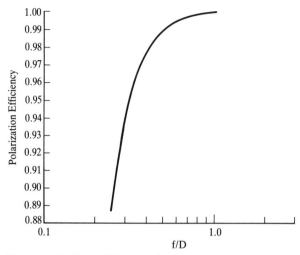

FIGURE 3.31 The polarization efficiency of axisymmetric antennas versus f/D ratio [72].

as the primary feed and Figure 3.31 shows the plot of polarization efficiency versus f/D ratio. This shows the present [72] method gives results close to the exact method [76]. The peak polarization in dB of front-fed reflectors is tabulated in Table 3.2 for different values of f/D and efficiency η. The results are more evident from the Table 3.2.

The polarization loss efficiency factor versus subtended angle ψ_1 in degrees for a classical cassegrain antenna is shown in Figure 3.32. Figure 3.33 gives the polarization efficiency of axisymmetric antennas as a function of f/D for both front-end and cassegrain antennas. As an example of the power of the technique described here, the peak cross-polarization of a cassegrain antenna of angular semiaperture $60°$ and magnification factor $M = 5$ was found using an exact method (ignoring diffraction) to be -50 dB, whereas using the present method the peak cross-polarization was found to be better than -48 dB with a polarization efficiency of 0.99995.

TABLE 3.2 Peak Cross-Polarization of Front-End Reflectors

f/D	Peak η	Cross-Polarization (dB)	
		Exact Method [76]	Present Method
0.25	0.893	15.8	14.6
0.30	0.942	18.1	17.5
0.40	0.979	22.2	22.1
0.46	0.987	24.3	24.2
0.60	0.995	28.0	28.4

3.8 CROSS-POLARIZATION CHARACTERISTICS   **261**

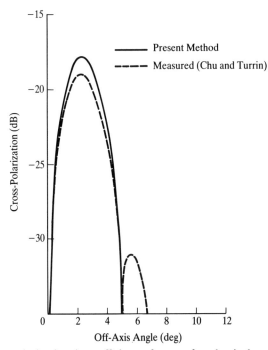

FIGURE 3.32 The polarization loss efficiency factor of a classical cassegrain antenna [47].

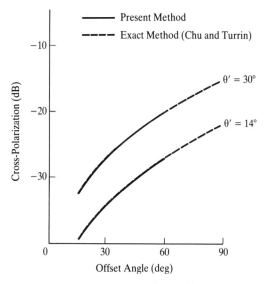

FIGURE 3.33 Cross-polar radiation pattern of an offset antenna: offset angle $= 45°$; angular semiaperture $= 45°$; $D/\lambda = 18.8$; a Huyghens' source is used as the primary feed.

3.8.2 Cross-Polarization Characteristics of Offset Antennas

We discuss here the polarization characteristics of an offset reflector. Figure 3.31 shows the polarization efficiency of axisymmetric antennas as a function of f/D ratio. This figure is equally appropriate for both front-end and cassegrain antennas. The cross-polarized field is given by

$$E_x(\zeta, \xi) \approx j \frac{\pi D^2}{4} \frac{jA}{\lambda R} a_1 H_1(u) \cos \xi \qquad (3.86\text{a})$$

where

$$H_1(u) = \sum_{k=0}^{\infty} \frac{8(k+1)}{(2k+1)(2k+3)} J_{2k+2}(u)/u \qquad (3.86\text{b})$$

$$a_1 = \sqrt{2(1/\eta - 1)} \qquad (3.86\text{c})$$

The cross-polarization is zero along the boresight and in the plane $\xi = 90°$ and the maximum is in the plane $\xi = 0°$.

The peak cross-polarization level occurs at $u \approx 2.4$, which corresponds to an off-axis angle of $\sin^{-1}(0.76\lambda/D)$.

The cross-polarization (dB) in $\xi = 0°$ is

$$10 \log_{10} |2(1/\eta - 1) H_1^2(u)| \qquad (3.87\text{a})$$

The peak cross-polarization (dB) is

$$10 \log_{10} |0.5(1/\eta - 1)| \qquad (3.87\text{b})$$

3.9 REFLECTOR ANTENNAS FOR HIGH-POWER MICROWAVE (HPM) APPLICATIONS

With the increased use of high-resolution radar, which involves large frequency bandwidth, it is necessary to design antenna to handle a large frequency bandwidth without much change of input impedance and radiation characteristics. A recent publication [77] reviews the state of the art, an art in which parabolic reflectors play their part; indeed a reflector with a TEM-feed is being considered as a candidate for generating wide-band signals. Several application notes addressing this aspect of the problem have appeared recently [78–80] in the literature. Since this is a relatively current topic and still under investigations, we will only touch upon some key points.

An impulse radiating antenna (IRA), which is a conical TEM wave launcher feeding reflector, is shown in Figure 3.34, (a) side view and (b) front view. It

3.9 REFLECTOR ANTENNAS FOR HIGH-POWER MICROWAVE (HPM) APPLICATIONS

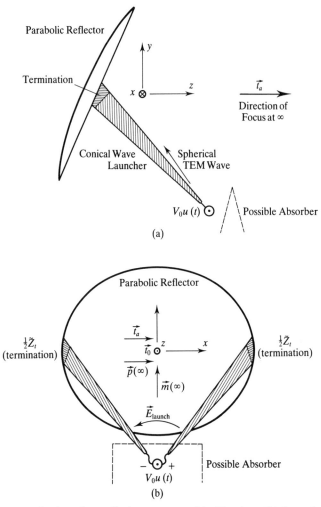

FIGURE 3.34 An impulse radiating antenna. (a) side view, (b) first view [79].

was first pointed out in [81] that a TEM feed can be used to feed a paraboloidal reflector. The TEM feed could be designed to provide a compromise between low and high frequencies. An expression for aperture efficiency [81] has been found in a closed form. Different types of TEM structures (e.g. two-arm and four-arm feeds, TEM horn feed) [79, 80, 81] have been proposed to achieve directive broad-band antennas and the effect of aperture blockage studied [80, 81]. Launching a fast-rising (step-like) TEM wave on the conical wave launcher, the wave is reflected and approximately time differentiated in the focal direction at large distances. A resistive termination is used at the reflector end of the conic to reduce multiple reflections.

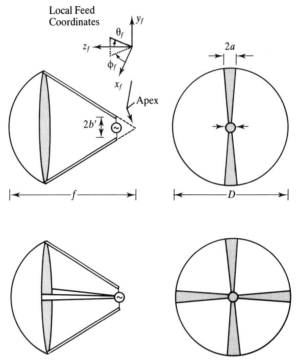

FIGURE 3.35 Antenna configurations for two-arm and four-arm TEM-fed reflectors [81]. Typical parameters: reflector diameter $D = 5$ m, focal length $f = 2$ m, plate width $2a = 0.72$ m, plate separation $2b = 5$ m, TEM Impedance $Z_c = 400\ \Omega$; $f/D = 0.4$; width/separation = 6.99.

The antenna configurations for two-arm and four-arm TEM-fed reflectors are shown in Figure 3.35. The antenna patterns with and without consideration of aperture blockage are shown in Figure 3.36.

3.10 CONTROL OF RADIATION CHARACTERISTICS OF A REFLECTOR ANTENNA WITH LOADED AND UNLOADED SHROUD AND FLANGE

It is expected the front-to-back lobe ratio would be enhanced and large-angle sidelobes suppressed with the use of shroud and flange. They may be loaded or unloaded. Research into this aspect of reflector analysis is available in some of the key references [82–87]. Figure 3.37 shows the H- and E-plane radiation of a parabolic dish having a right-angled flange with $W/D = 0.2$ (dotted lines) and unflanged (solid lines); $D/\lambda = 20$; $f/D = 0.5$. W is the width of the flange and D is the diameter of the dish without flange. The effect of a shroud on the radiation pattern of the prime focus paraboloid is shown in Figure 3.38 as

3.10 CONTROL OF RADIATION CHARACTERISTICS OF A REFLECTOR ANTENNA

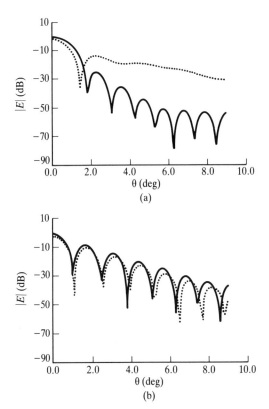

FIGURE 3.36 Normalized far-field pattern without (solid line) and with (dashed line) aperture blockage consideration, $f = 3$ GHZ; (a) $\phi = 0$, horizontal plane, H-plane, xz-plane; (b) $\phi = 90°$, vertical plane, E-plane, yz-plane.

discussed in [85]. It was observed [85] that the use of a shroud suppresses the far-out sidelobe levels and there is a considerable enhancement of F/B ratio, typically 20 dB and the sidelobes in far angles ($>90°$) increase but they can easily be suppressed by loading the rim of the shroud by radar-absorbing materials. It was also found in [85] that there is an optimum length of the shroud beyond which there is no appreciable improvement in F/B ratio. This optimum length for 12 ft ($\approx 73\lambda$) diameter dish at $\lambda = 5$ cm was found [85] to be 100 cm (20λ). In [86], the parabolic antenna has been analyzed with and without a loaded flange. The flange was loaded with a perfect conductor, corrugations, and isotropic material. Figure 3.39 shows the E- and H-plane radiation patterns of a parabolic dish with and without a corrugated flange. Figure 3.40 shows the plot of the E- and H-plane rear fields of a parabolic dish without flange and with flange loaded differently. The source is a Huyghens source with a $\cos \theta$ law of taper. The conclusions that can be arrived at from Figures 3.39 and 3.40 are: (1) the flange does not have much effect on the field

266 REFLECTOR ANTENNAS AND RIM LOADING

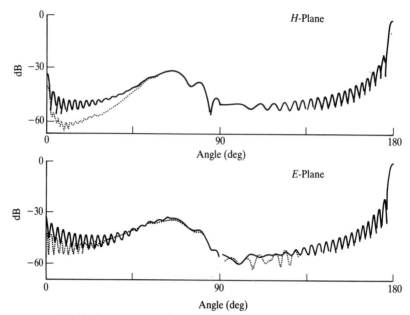

FIGURE 3.37 Radiation patterns of a parabolic dish having a right-angled flange with $W/D = 0.2$ (dotted lines) and unflanged (solid lines); $D/\lambda = 20$; $f/D = 0.5$; W is the width of the flange [84].

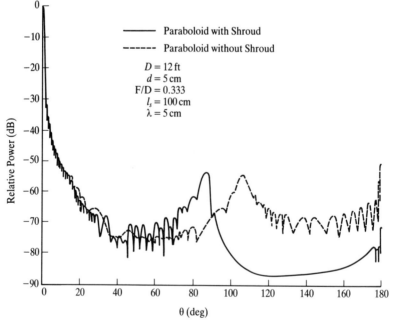

FIGURE 3.38 The comparison of radiation patterns of the paraboloid with and without shroud; $D = 12$ ft; $d = 5$ cm; $F/D = 0.333$; optimum length of the shroud = 100 cm; $\lambda = 5$ cm [85].

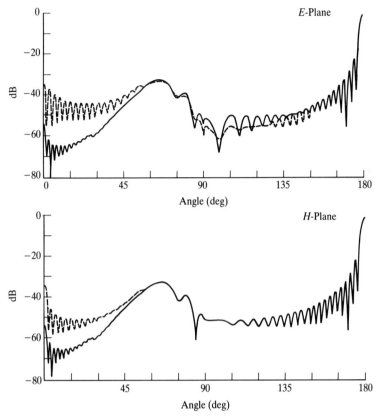

FIGURE 3.39 E- and H-plane radiation patterns of the parabolic dish with a corrugated flange (solid lines) and unflanged (dashed lines). $D/\lambda = 20$; $F/D = 0.5$; $W/\lambda = 5$ [86].

in the forward direction but causes a considerable reduction in the rear angular sector particularly with "optimum" isotropic and anisotropic loadings; (2) it is recommended that a corrugation is the best choice unless the bandwidth requirement is too wide, such as for an antenna in a high-resolution radar. Polarization properties of edge-loaded high-gain reflector antenna is described in [88].

3.11 SUMMARY

This chapter dealt with aspects of analysis and design of parabolic reflectors using high-frequency techniques. It discussed the effect of rim loading on the patterns of the reflectors and the cross-polarization characteristics.

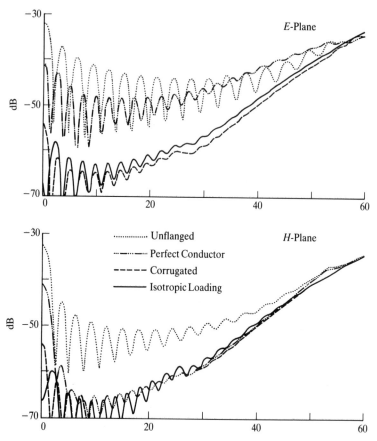

FIGURE 3.40 E- and H-plane rear fields of a parabolic dish with a loaded flange compared with the unloaded case. $D/\lambda = 20$; $F/D = 0.5$; $W/\lambda = 5$ [86].

3.12 REFERENCES

1. H. Hertz, *Electric Waves*, Macmillan, London, 1983.
2. A. W. Love, "Some Highlights in Reflector Antenna Development," *Radio Science*, **11**(8, 9), 1976, 671–684.
3. P. J. B. Clarricoats and G. T. Poulton, "High Efficiency Microwave Antennas—A Review," *Proc. IEEE*, **65**, 1977, 1470–1504.
4. W. V. T. Rusch, "The Current State of the Reflector Antenna Art," *IEEE Trans. Antennas and Propagat.*, **AP-32**, 1984, 313–329.
5. G. Franceschetti and A. Mohsen, "Recent Developments in the Analysis of Reflector Antennas—A Review," *Proc. IEE, Part H*, **133**(1), 1986, 65–77.
6. W. V. T. Rusch and P. D. Potter, *Analysis of Reflector Antennas*, Academic Press, New York, 1970.
7. A. W. Love (ed.), *Reflector Antennas*, IEEE Press, Piscataway, NJ, 1978.

8. H. E. Scharank, G. E. Evans, and D. Davis, "Reflector Antennas," in *Radar Handbook* edited by M. Skolnik, McGraw-Hill, New York, 1990.
9. W. L. Stutzman and G. S. Thiele, *Antenna Theory and Design*, John Wiley, New York, 1981, Ch. 9, pp. 446–512.
10. C. A. Balanis, *Antenna Theory—Analysis and Design*, John Wiley, New York, 1982, Ch. 11, pp. 446–529.
11. Y. Rahmat-Samii, "Reflector Antennas," in *Antenna Handbook—Theory, Applications and Design*, edited by Y. T. Lo and S. W. Lee, Van Nostrand Reinhold, New York, 1988, pp. 15-1 to 15-123.
12. R. E. Collin and F. J. Zucker (eds.), *Antenna Theory, Part 1 and 2*, New York, McGraw-Hill, 1969.
13. P. J. Wood, *Reflector Antenna Analysis and Design*, Electromagnetic Wave Series, Peter Peregrinus, 1980.
14. W. A. Imbriale, "Evolution of the Deep Space Network Antennas," *IEEE Trans. Antennas and Propagat, Society Magazine*, **33**(6), 1991, 7–19.
15. W. V. T. Tusch and O. Sorensen, "The Geometrical Theory of Diffraction for Axially Symmetric Reflector," *IEEE Trans. Antennas and Propagat.*, **AP-23**(2), 1975, 414–419.
16. A. W. Rudge, "Multiple-Beam Antennas: Offset Reflectors with Offset Feeds," *IEEE Trans. Antennas and Propagat.*, **AP-23**(3), 1975, 317–322.
17. J. F. Kauffman, W. F. Croswell and L. J. Jowers, "Analysis of the Radiation Patterns of Reflector Antennas," *IEEE Trans. Antennas and Propagat.*, **AP-28**(24), 1976, 53–65.
18. Y. Rahmat-Samii and V. Galindo-Israel, "Shaped Reflector Analysis using the Jacobi–Bessel Series," *IEEE Trans. Antennas and Propagat.*, **AP-28**(4), 1980, 425–435.
19. Y. Rahmat-Samii and V. Galindo-Israel, "Scan Performance of Dual Offset Reflector Antennas for Satellite Communications," *Radio Science*, **16**(6), 1981, 1093–1099.
20. Y. Rahmat-Samii, "A Comparison Between GO/Aperture Field and Physical-Optics Methods for Offset Reflectors," *IEEE Trans. Antennas and Propagat.*, **AP-32**(3), 1984, 301–306.
21. P. J. B. Clarricoats and G. T. Poulton, "High Efficiency Microwave Reflector Antennas—A Review," *Proc. IEEE*, **65**(10), 1977, 1470–1504.
22. M. S. Narasimhan and K. M. Prasad, "GTD Analysis of the Near-Field Patterns of a Prime-Focus Symmetric Paraboloidal Reflector Antennas," *IEEE Trans. Antennas and Propagat.*, **AP-29**(6), 1981, 959–961.
23. S. Silver, *Microwave Antenna Theory and Design*, MIT Radiation Laboratory Series, Cambridge, MA, Ch. 6, 1950, pp. 169–199.
24. G. Millington, R. Hewitt, and F. S. Immirzi, "Double Knife-Edge Diffraction in Field-Strength Prediction," *IEE Monograph 507E*, Mar. 1962, pp. 419–429.
25. H. E. Schrank, G. E. Evans, and D. Davis, "Reflector Antenna," in *Radar Handbook*, edited by M. L. Skolnik, Van Nostrand Reinhold, New York, 1988, pp. 6.1 to 6.64.
26. V. Jamnejad and Y. Rahmat-Samii, "A Modified Subreflector Design and Diffraction Analysis of Voyager's High Gain Antenna for Cassini Spacecraft," *IEEE Antenna and Propagat., London, Ontario*, 1991, Session 37, pp. 670–673.

27. Y. Rahmat-Samii, "Subreflector Extension for Improved Efficiencies In Cassegrain Antenna—GTD/PO Analysis," *IEEE Trans. Antennas and Propagat.*, **AP-34**(10), 1986, 1266–1269.
28. P.-S. Kildal, "The Effects of Subreflector Diffraction on the Aperture Efficiency of a Conventional Cassegrain Antenna—An Analytical Approach," *IEEE Trans. Antennas and Propagat.*, **AP-31**, 1983, 903–909.
29. B. Houshmand, S. W. Lee, Y. Rahmat-Samii, and P. T. Lam, "Analysis of Near-Field Cassegrain reflector: Plane Wave versus Element-by-Element Approach, *IEEE Trans. Antennas and Propagat.*, **AP-38**(7), 1990, 1010–1017.
30. R. J. Pogorzelski, "A New Integration Algorithm and Its Applications to the Analysis of Symmetrical Cassegrain Microwave Antennas," *IEEE Trans. Antennas and Propagat.*, **AP-31**(4), 1983, 748–755.
31. S. W. Lee, P. Cramer, Jr., K. Woo, and Y. Rahmat-Samii, "Diffraction by an Arbitrary Subreflector: GTD Solution," *IEEE Trans. Antennas and Propagat.*, **AP-27**(3), 1979, 305–316.
32. M. S. Narasimhan, P. Ramanujam, and K. Raghavan, "GTD Analysis of the Radiation Patterns of a Shaped Subreflector," *IEEE Trans. Antennas and Propagat.*, **AP-29**(5), 1981, 792–795.
33. C. A. Mentzner and L. Peters Jr., "A GTD Analysis of the Far-Out Sidelobes of Cassegrain Antennas," *IEEE Trans. Antennas and Propagat.*, **AP-23**, 1975, 702–709.
34. M. S. Narasimhan and P. Ramanujam, and K. Raghavan, "GTD Analysis of Near-Field and Far-Field Patterns of a Parabolic Subreflector Illuminated by a Plane Wave," *IEEE Trans. Antennas and Propagat.*, **AP-29**(4), 1981, 654–659.
35. D. C. Hogg and R. A. Samplak, "An Experimental Study of Near-Field Cassegrainian Antennas," *Bell. Syst. Tech. J.*, 1964, 2677–2704.
36. C. Scott, *Modern Methods of Reflector Antenna Analysis and Design*, Artech House, Norwood, NA, 1990, Ch. 9.
37. P. J. Wood, *Reflector Antenna Analysis and Design*, Peter Peregrinus, 1980, Ch. 7.
38. C. A. Mentzer and L. Peters Jr, "A GTD Analysis of the Far-Out Sidelobes of Cassegrain Antennas," *IEEE Trans. Antennas and Propagat.*, **AP-23**(5), 1975, 702–709.
39. R. J. Pogorzelski, "A New Integration Algorithm and Its Application to the Analysis of Symmetrical Cassegrain Microwave Antennas," *IEEE Trans. Antennas and Propagat.*, **AP-31**(5), 1983, 748–755.
40. V. Jamnejad and Y. Rahmat-Samii, "A Modified Subreflector Design and Diffraction Analysis of Voyager's High Gain Antenna for Cassini Spacecraft," *IEEE Antennas and Propagat. Symp. London, Ontario*, 1991, Paper 37.3, pp. 670–673.
41. Y. Rahmat-Samii, "Subreflector Extension for Improved Efficiencies in Cassegrain Antennas—GTD/PO Analysis," *IEEE Trans. Antennas and Propagat.*, **AP-34**(10), 1986, 1266–1269.
42. P.-S. Kildal, "The Effects of Subreflector Diffraction on the Aperture Efficiency of a Conventional Cassegrain Antenna—An Analytical Approach," *IEEE Trans. Antennas and Propagat.*, **AP-31**(6), 1983, 903–909.
43. B. Houshmand, S. W. Lee, Y. Rahmat-Samii, and P. T. Lam, "Analysis of Near-Field Cassegrain Reflector: Plane Wave Versus Element-by-Element Approach," *IEEE Trans. Antennas and Propagat.*, **AP-38**(7), 1990, 1010–1017.

44. W. V. T. Rusch, "Scattering From a Hyperboloidal Reflector in a Cassegrainian Feed System," *IEEE Trans. Antennas and Propagat.*, **AP-11**(4), 1963, 414–421.
45. P. D. Potter, "Aperture Illumination and Gain of a Cassegrainian System," *IEEE Trans. Antennas and Propagat.*, **AP-11**(3), 1963, 373–375.
46. J. J. Lee, L. I. Parad, and R. S. Chu, "A Shaped Offset-Fed Dual-Reflector Antenna," *IEEE Trans. Antennas and Propagat.*, **AP-27**(2), 1979, 165–171.
47. J. Dijk, E. J. Maanders, and J. P. F. Snickers, "On the Efficiency and Radiation Patterns of Mismatched Shaped Cassegrainian Antenna Systems," *IEEE Trans. Antennas and Propagat.*, **AP-20**, 1972, pp. 653–655.
48. P. D. Potter, "Application of Spherical Wave Theory to Cassegrainian-Fed Paraboloids, *IEEE Trans. Antennas and Propagat.*, **AP-15**(6), 1967, 727–736.
49. V. Galindo, "Design of Dual Reflector Antennas With Arbitrary Phase and Amplitude Distribution," *IEEE Trans. Antennas and Propagat.*, **AP-12**, 1964, 403–408.
50. B. Y. Kimber, "On Two Reflector Antennas," *Radio Eng. Electron Phys.*, **6**, 1962, 914.
51. S. K. Buchmeyer, "An Electrically Small Cassegrain Antenna with Optically Shaped Reflectors," *IEEE Trans. Antennas and Propagat.*, **AP-25**, 1977, 346–351.
52. V. Galindo-Israel and R. Mittra, "Synthesis of Offset Dual Shaped Reflector with Arbitrary Control of Phase and Amplitude," in *Proc. URSI Symp. on EM Theory*, 1977, pp. 7–10.
53. V. Galindo-Israel, R. Mittra, and A. Cha, "Aperture Amplitude and Phase Control of Offset Dual Reflectors," in *Proc. IEEE APS and USNC/URSI Symp.*, 1978, p. 239.
54. R. Mittra and F. Hyjazie, "A Method for Synthesizing Offset Dual Reflector Antenna," *Proc. IEEE APS and USNC/URSI Symp.*, 1978, p. 243.
55. O. M. Bucci, G. DiMassa, and C. Savarese, "Control of Reflector Antenna Performance by Rim Loading," *IEEE Trans.* **AP-29**(5), 1981, 773–779.
56. A. Prata Jr., M. R. Barclay, W. V. T. Rusch, and D. C. Jenn, "Dual Reflector Antenna Sidelobe Control through a Dielectric Loaded Subreflector," *IEEE Intl. Symp. Digest*, Syracuse, NY, June 6–10, 1989, 859–862.
57. A. DeBonitatibus, G. DiMassa, and C. Savarese, "Theoretical and Experimental Scattering by Hyperbolic Loaded Dishes," *Electromagnetics*, **5**, 1985, 1–16.
58. H. Yokoi and H. Fukumaru, "Low Sidelobes of Parabolic Antennas with Microwave Absorbers," *Electron Commun. Japan*, **54B**(11), 1971, 34–39.
59. T. B. A. Senior, "Impedance Boundary Conditions for Imperfectly Conducting Surfaces," *Appld. Sci. Res., B*, **8**, 1975, 3326–3332.
60. N. G. Alexopoulos and G. A. Tadler, "Accuracy of the Leontovich Boundary Conditions for Continuous and Discontinuous Surface Impedances," *J. Appld Phys.*, **46**(8), 1975, 3326–3332.
61. O. M. Bucci and G. Franceschetti, "Rim Loaded Reflector Antennas," *IEEE Trans. Antennas and Propagat.*, **AP-28**(3), 1980, 297–305.
62. A. C. Ludwig, "The Definition of Cross Polarization," *IEEE Trans. Antennas and Propagat.*, **AP-21**, 1973, 116–119.
63. E. U. Condon, *Theory of Radiation from Paraboloidal Reflectors, Westinghouse Rep. 15*, Sept. 24, 1941.
64. S. Silver, *Microwave Antenna Theory and Design*, McGraw-Hill, New York, 1949. (Refer to the new book.)

65. C. C. Cutler, "Parabolic Antenna Design for Microwave," *Proc. IRE*, **35**, 1947, 1284–1294.
66. E. M. T. Jones, "Paraboloid Reflector and Hyperbolic Lens Antenna," *IRE Trans. Antennas and Propagat.*, **2**, 1954, 119–127.
67. I. Koffman, "Feed Polarization for Parallel Currents in Reflectors Generated by Conic Sections," *IRE Trans. Antennas and Propagat.*, **AP-14**, 1966, 37–40.
68. M. Afifi, "Scattered Radiation from Microwave Antenna and the Design of a Parabolic Plane Reflector Antenna," Ph.D dissertation, Delft Univ. of Technology, 1966.
69. P. D. Potter, *The Aperture Efficiency of Large Paraboloidal Antennas as a Function of Their Feed System Radiation Characteristics*, JPL, Pasadena, CA, Tech. Rep., Sept. 1961, pp. 32–149.
70. P. D. Potter, "Aperture Illumination and Gain of a Cassegrain System," *IEEE Trans. Antennas and Propagat.*, **AP-11**, 1963, 373–375.
71. M. Safak and P. P. Delogne, "Cross Polarization in Cassegrainian and Front-Fed Paraboloidal Antennas," *IEEE Trans. Antennas and Propagat.*, **AP-24**, 1976, 497–501.
72. S. I. Ghobrial, "Off-Axis Cross-Polarization and Polarization Efficiencies of ReflectorAntennas," *IEEE Trans. Antennas and Propagat.*, **AP-27**(4), 1979, 460–466.
73. B. S. Westcott, F. Brickell, and I. C. Wolton, "Crosspolar Control in Far-Field Synthesis of Dual Offset Reflectors," *IEE Proc., Part H*, **137**(1), 1990, 31–38.
74. K. Miyata, "Center-Fed Parabolic Reflector Antenna Cross Polarization due to Slightly Right-Left Asymmetric Field Patterns," *IEE Proc.* **AP-28**(2), 1980, 203–209.
75. T.-S. Chu, "Polarization Properties of Offset Dual-Reflector Antennas," *IEE Proc.* **AP-39**(12), 1991, 1753–1756.
76. E. M. T. Jones, "Paraboloid Reflector and Hyperbolid Lens Antennas," *IRE Trans. Antennas and Propagat.*, **AP-2**, 1954, 115–127.
77. C. E. Baum, "From Electromagnetic Pulse to High Power Electromagnetics," *Proc. IEEE*, **80**(6), 1992, 789–817.
78. E. G. Farr and C. E. Baum, *Prepulse Associated with the TEM Feed of an Impulse Radiating Antenna*, Sensor and Simulation Note 337, Mar. 1992.
79. H. E. King and J. L. Wong, "Measured RF Characteristics of a TEM-Wire Fed Reflector," *IEEE Antennas and Propagat. Symp.*, 1990, pp. 266–269.
80. Y. Rahmat-Samii, *Analysis of Blockage Effects on TEM-Fed Paraboloidal Reflector Antennas*, Sensor and Simulation Note 347, Oct. 25, 1992.
81. Y. Rahmat-Smaii and D. V. Giri, *Analysis of Blockage Effects on TEM-Fed Paraboloidal Reflector Antennas (Part II: TEM Horn Illumination)*, Sensor and Simulation Note 349, Nov. 3, 1992.
82. C. E. Baum, *Radiation of Impulse-Like Transient Fields*, Sensor and Simulation Note 321, Nov. 25, 1989.
83. C. E. Baum, *Aperture Efficiencies for IRAs*, Sensor and Simulation Note 328, June 24, 1991 and *URSI Symposium, Boulder, CO*, 1992.
84. O. M. Bucci, C. Gennarelli, and L. Palumbo, "Flanged Parabilic Antennas," *IEEE Trans. Antennas and Propagat.*, **AP-30**(6), 1982, 1081–1085.

85. M. S. Narsimhan, K. Raghavan, and P. Ramanujam, "GTD Analysis of the Radiation Patterns of a Prime Focus Paraboloid with Shroud," *IEEE Trans. Antennas and Propagat.*, **AP-31**(3), 1983, 792–794.
86. O. M. Bucci, C. Gennarelli, and L. Palumbo, "Parabolic Antennas with a Loaded Flange," *IEEE Trans. Antennas and Propagat.*, **AP-33**(7), 1985, 755–762.
87. M. S. Narasimhan, P. Ramanujam, and K. Raghavan, "GTD Analysis of a Hyperboloidal Subreflector with Conical Flange Attachment," *IEEE Trans. Antennas and Propagat.*, **AP-29**(6), 1981, 865–871.
88. O. M. Bucci, P. Corona, G. di Massa, G. Franceschetti, G. Pinto, and C. Savarese, *Polarization Properties of Edge Loaded High Gain Reflector Antennas*, Final Rep., ESTEC Contract 2893/76/NL/AK, 1978.

3.13 ADDITIONAL SOURCES

Y. C. Chang and R. C. Rudduck, *Numerical Electromagnetic Code-Reflector Antenna Code, NEC-REF (Version 2) Part II: Code Manual*, Dept. Elec. Eng., Ohio state Univ., Rep. 712242-17. Prepared under contract N00123-79-C-1469 for Naval Regional Procurement Office, Dec. 1982.

J. Chen, P. J. I. de Maagt, and M. H. A. J. Herben, *Wide-Angle Radiation Pattern Calculations of Paraboloidal Reflector Antennas: A Comparative Study*, EUT Rep. 91-E-252, June 1991, Dept. of Elec. Eng., Eindhoven Inst. Tech., Netherlands.

T.-S. Chu, "Polarization Properties of Offset Dual-Reflector Antennas," *IEEE Trans. Antennas and Propagat.*, **AP-39**(12), 1991, 1753–1756.

S. V. Hoerner and W.-Y. Wang, "Improved Efficiency with Mechanically Deformable Subreflector," *IEEE Trans. Antennas and Propagat.*, **AP-27**(5), 1975, 720–723.

B. Houshmond, S. W. Lee, Y. Rahmat-Samii, and P. T. Lam, "Analysis of Near-Field Cassegrain Reflector: Plane Wave versus Element-to-Element Approach," *IEEE Trans. Antennas and Propagat.*, **AP-38**(7), 1990, 1010–1017.

J. Huang and Y. Rahmat-Samii, "Fan Beam Generated by a Linear Array Fed Parabolic Reflector," *IEEE Trans. Antennas and Propagat.*, **AP-38**(7), 1990, 1046–1053.

P. G. Ingerson and W. C. Wong, "The Analysis of Depolyable Umbrella Parabolic Reflectors," *IEEE Trans. Antennas and Propagat.*, **AP-20**(4), 1972, 409–414.

D. C. Jenn and W. V. T. Rusch, "Low-Sidelobe Reflector Synthesis and Design using Resistive Surfaces," *IEEE Trans. Antennas and Propagat.*, **39**(9), 1991, 1372–1375.

V. Kerdemelidis, G. L. James, and B. A. Smith, "Some Methods of Reducing the Sidelobe Radiation of Existing Reflector and Horn Antennas," in *Recent Advances In Electromagnetic Theory*, edited by H. N. Kritikos and J. L. Jaggard. Springer-Verlag, 1990.

C. M. Knop, "An Extension of Rusch's Asymtotic PO Diffraction Theory of a Paraboloid Antenna," *IEEE Trans. Antennas and Propagat.*, **AP-27**(5), 1975, 741–745.

P. J. I. de Maagt, J. Chen, and M. H. A. J. Herben, "A Review and Comparison of Some Asymptotic Techniques for Calculating the Wide-Angle Radiation Pattern of Paraboloid Reflector Antennas," *Electromagnetics*, **12**, 1992, 57–75.

G. Morris, "Coupling Between Closely Spaced Back-to-Back," *IEEE APS* Jan. 1980, pp. 60–64.

P. D. Potter, "Application of Spherical Wave Theory to Cassegrainian-Fed Paraboloids," *IEEE Trans. Antennas and Propagat.*, **AP-15**(6), 1967, 727–736.

P. Ramanujam, P. J. B. Clarricoats, and R. C. Brown, "Offset Spherical Reflector with a Low Sidelobes Radiation," *IEE Proc.*, **134**(2), 1987, 199–204.

R. C. Rudduck and Y. C. Chang, *Numerical Electromagnetic Code—Reflector Antenna Code, NEC-REF (Version 2), Part I: User Manual*, Dept. Elec. Eng., Ohio State Univ., Rep. 712242-16. Prepared under contract N00123-79-C-1469 for Naval Regional Procurement Office, Dec. 1982.

C. J. Sletton (ed.), *Reflector and Lens Design: Software Users' Manual and Example Book*, Artech House, Norwood, MA, 1991.

C. J. Sletton (ed.), *Reflector and Lens Design: Analysis and Design Using Personal Computers*, Artech House, Norwood, MA, 1991.

R. Wohlleben and J. Koehler, "A New Minimum Blocking Condition Including Subreflector and Quadrupole Shadow for the Design of Dual Reflector Antenna," *IEEE Trans. Antennas and Propagat.*, **AP-31**(2), 1989, 342–346.

CHAPTER FOUR

Slot Antennas

4.1 INTRODUCTION

Space vehicles, missiles, and aircraft use flush-mounted aperture antennas. The body of the vehicle if conduting acts as a ground plane, particularly if it is planar. The high-frequency techniques described in Chapter 1 can be successfully used to determine the antenna characteristics on practical vehicles in many cases. One geometry of great interest is the elliptic cylinder since it approximates the fuselage of an aircraft. Several papers deal with antenna radiation in the presence of finite and infinite objects. Modal solutions for slots on the surface of conducting cylinders have been presented in [1]; radiation fields for slot-excited antennas have been discussed in an early paper [2]; and radiation of slots on finite ground planes [3] on PEC circular cylinders [4] and PEC elliptical cylinders [5] have been investigated in a series of papers by C. A. Balanis and L. Peters Jr. Pattern distortion due to edge diffraction was estimated in [6] using a combined boundary value method and wedge diffraction for a rectangular waveguide and a circular waveguide opening into a finite ground plane. Elevation plane patterns of a large TEM coaxial aperture and a TM_{01} mode circular aperture on finite circular and square ground planes have been derived [7] as a superposition of an infinite ground plane solution and first-order diffracted fields. The geometrical theory of diffraction can also be used in a limited sense to antenna impedance calculations as discussed in [8]. The analysis [9] of an infinitesimal dipole mounted on the tip of a finite length PEC cone. The pattern is again a superposition of the contribution from the induced current on the cone surface and the field diffracted by the edge of the cone. High-frequency analysis of a slot-excited antenna [10], radiation from conical surfaces used for high-speed aircraft [11] and mutual impedance between slots using radiation slots on a cone [12] are available. Some discussions on the finite ground plane effects using high-frequency techniques are available in a textbook [13]. Evaluation of the edge effects in a slot array

276 SLOT ANTENNAS

using the GTD has been presented in [14]. An early discussion on slots on cylinders is available in [15].

4.2 SLOTS ON A GROUND PLANE

The geometry and coordinate system of a slot antenna radiating in the presence of a ground plane of finite extent and also of finite thickness is shown in Figure 4.1. The different scattering centers $S_1, S_2, S_3, \ldots, S_6$ are shown on the diagram. All the centers have interior wedge angles equal to $90°$, for which the value of n is 1.5. Analysis of the slot antenna can be divided into two parts; one considers the ground plane to be infinite; the other uses a finite ground plane but takes its very finiteness into consideration.

The total diffracted field at the observation point is the superposition of contributions from the two scattering centers on the edges of wedges 1 and 2.

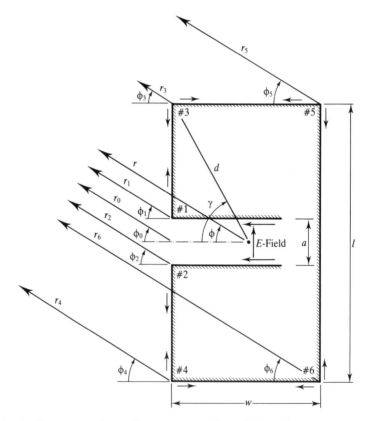

FIGURE 4.1 Geometry and coordinate system of the radiating Slot antenna on a ground plane. (From [6] © 1970 IEEE.)

4.2 SLOTS ON A GROUND PLANE

The expression for the diffracted field is

$$E^d(r_0, \phi_0) = \frac{e^{-j(kr_0 + \pi/4)}}{\sqrt{2\pi k r_0}} D(\phi_0) \quad (4.1)$$

where

$$D(\phi_0) = \exp(j(ka/2)\sin\phi_0)(D_1(\phi_0) + D_2(\phi_0)) \cdot \exp(-jka\sin\phi_0) \quad (4.2)$$

$$D_1(\phi_0) = D_1^{(1)}(\phi_0) + D_1^{(h)}(\phi_0) \quad (4.3a)$$

$$D_2(\phi_0) = D_2^{(1)}(\phi_0) + D_2^{(h)}(\phi_0) \quad (4.3b)$$

where a is the radius of the aperture, the different distances are shown in Figure 4.1. The superscripts in eqn. (4.3a, b) denote the order of diffraction and the subscripts 1 and 2 denote the number of scattering centers.

The primary second-order diffraction will be fields from scattering centers S_1 and S_2 on the aperture and its component towards scattering centers S_3 and S_4 getting diffracted by the 90° wedges at S_3 and S_4. Further, the rays from S_3 and S_4 are again going to be diffracted by the the wedges at S_5 and S_6. To take into account the higher-order diffrations, multiply the incident field by the diffraction coefficient of the wedge, the divergence factor, and the phase factor of the diffracted wavefront each time diffraction takes place at a scattering center.

The singly diffracted ray (SDR) from S_3 is given by

$$E_3^{d,(1)}(r_3, \phi_3) = \frac{e^{-j(kr_3 + \pi/4)}}{\sqrt{2\pi k r_3}} D_3^{d,(1)}(\phi_3) \quad (4.4)$$

where

$$D_{d,(1)}^3(\phi_3) = D_1\left(\frac{\pi}{2}\right)\left[V_B\left(\frac{l-2}{a}, \frac{\pi}{2} + \phi_3, n_3\right)\right] \quad (4.5a)$$

and $V_B(r, \phi, n)$ is the diffraction function defined by

$$V_B(\rho, \phi) = \frac{2e^{j\pi/4} \sin\frac{\pi}{n} \left|\cos\frac{\phi}{2}\right|}{n\sqrt{\pi}\cos\frac{\pi}{n} - \cos\frac{\phi}{n}} \cdot e^{jk\rho\cos\phi} \int_{\sqrt{k\rho(1+\cos\phi)}}^{\infty} e^{j\tau^2} d\tau$$

$$+ \text{ higher-order terms} \quad (4.5b)$$

The diffracted field from S_5 due to illumination from edge S_3 is given by

$$E_5^{d,(1)}(r_5, \phi_5) = \frac{e^{-j(kr_5 + \pi/1)}}{\sqrt{2\pi k r_5}} D_5^{d,(1)}(\phi_5) \quad (4.6)$$

where

$$D_5^{d,(1)}(\phi_s) = D_3^{d,(1)}[V_B(w, \phi_5, n_5)] \tag{4.7}$$

The singly diffracted ray from S_3 diffracted at S_5 ($\phi_3 = 180°$) is given by

$$E^{d,(1)}(r_5, \phi_5) = \frac{e^{-j(kr_5 + \pi/4)}}{\sqrt{2\pi k r_5}} D_5^{d,(1)}(\phi_5) \tag{4.8a}$$

where

$$D_5^{d,(1)}(\phi_5) = D_3^{d,(1)}(\pi)[V_B(w, \phi_5, n_5)] \tag{4.8b}$$

In the same way, one can calculate the first- and second-order fields from many different wedges. Table 4.1 shows the different wedges that contribute to the observation point for the slot on a finite ground plane antenna.

Based on the analytical expression, the radiation patterns of the axially slotted finite ground plane are shown in Figure 4.2. The agreement between theory and experiment is good, with additional lobes appearing for aperture lengths of more than one wavelength. As mentioned earlier, [7] describes the GTD analysis of elevation plane patterns of a large TEM mode coaxial aperture and a TM_{01} mode circular aperture on finite circular and spare ground planes (Fig. 4.3). Once again the solution is a superposition of infinite plane and wedge-diffracted fields. The axial caustic fields are obtained by equivalent edge currents as usual (see Chapter 1 and [127]). The details of the analysis are similar to above and the reader is referred to [7] for analytical details. Figure 4.4 show

TABLE 4.1 Contributions from Different Wedges

Diffracting Wedge	Wedge-Diffracted Field	Angular Region
1, 2	$E^d(\phi) = \exp(kw/2 \cos \phi) D^d(\phi)$	$-\frac{\pi}{2} < \phi < \frac{\pi}{2}$
3	$E_3^{d,(1,2)}(\phi) = \exp(jkd \cos(\gamma - \phi))[D_3^{d,(1)}(\phi) + D_3^{d,(2)}(\phi)]$	$-\frac{\pi}{2} < \phi < \pi$
4	$E_4^{D(1,2)}(\phi) = \exp(-jkd \cos(\pi - \gamma - \phi))$ $\times [D_4^{d,(1)}(\phi) + D_4^{d,(2)}(\phi)]$	$-\pi < \phi < \frac{\pi}{2}$
5	$E_5^{d,(1)}(\phi) = \exp(jkd \cos(\pi - \gamma - \phi))[D_5^{d,(1)}(\phi)]$	$0 < \phi < \frac{3\pi}{2}$
6	$E_6^{d,(1)}(\phi) = \exp(-jkd \cos(\gamma - \phi))[D_6^{d,(1)}(\phi)]$	$-\frac{3\pi}{2} < \phi < 0$

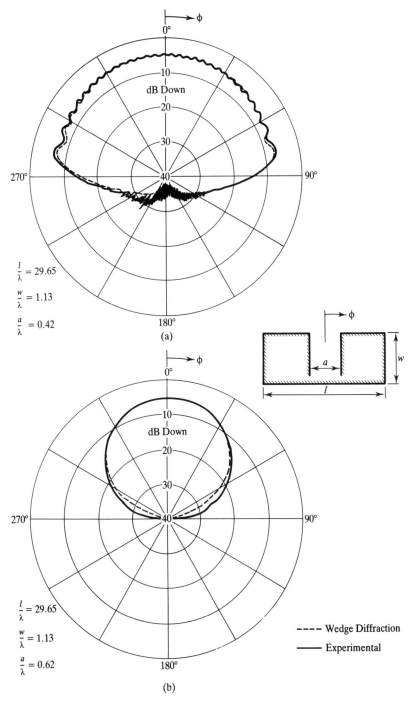

FIGURE 4.2 Radiation patterns of the slot antenna on a ground plane. (From [7] © 1974 IEEE.) (*continued on next page*)

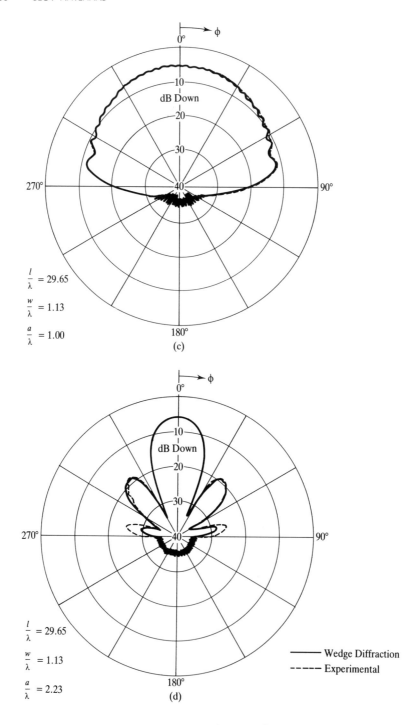

FIGURE 4.2 (*continued*)

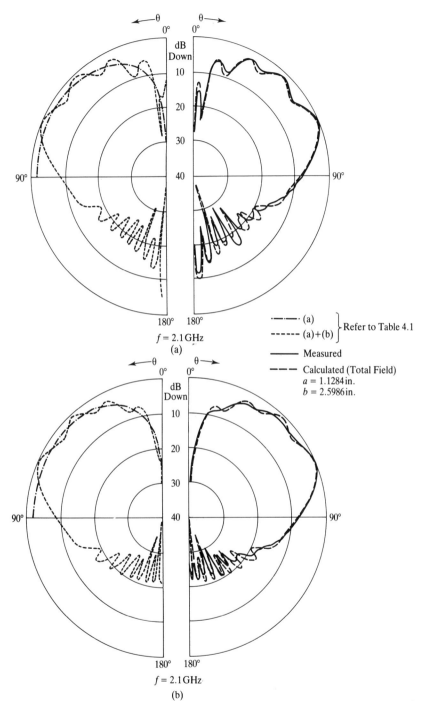

FIGURE 4.3 Elevation plane patterns of a coaxial line opening into a ground plane: (a) circular; (b) square. (From [7] © 1974 IEEE.)

282 SLOT ANTENNAS

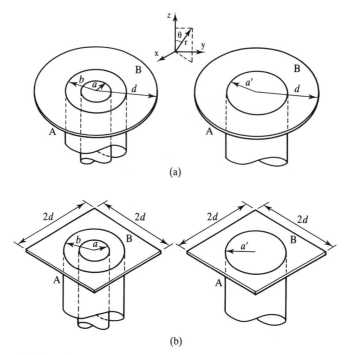

FIGURE 4.4 (a) Coaxial line opening into a ground plane. (b) Circular aperture opening into a ground plane. (From [7] © 1974 IEEE.)

a coaxial line and a cylindrical pipe opening respectively into a 48 in. diameter circular ground plane and a 48 in. square ground plane. Figure 4.5 shows the elevation plane pattern of an 8 in. diameter circular aperture opening into a 48 in. diameter ground plane. The minimum distance between the edge of the aperture and the edge of the ground plane was always greater than 3λ. Figure 4.6 plots the angle of maximum radiation and the maximum relative power intensity versus ground plane size in wavelengths and b/a ratio of the coaxial line. It is interesting that backlobe radiation is primarily affected by choice of ground plane, rectangular or circular. Backlobe radiation is relatively insensitive to the size of the circular ground plane but sensitive to the size of the square ground plane. The angle of maximum radiation increases with ground plane size.

4.3 SLOTS ON CIRCULAR CYLINDERS

Radiation from the axial slots on the circular cylinder has been analyzed in [4] using the geometrical theory of diffraction and creeeping wave theory, since the sidelobes will have contributions from creeping waves. The geometry and the

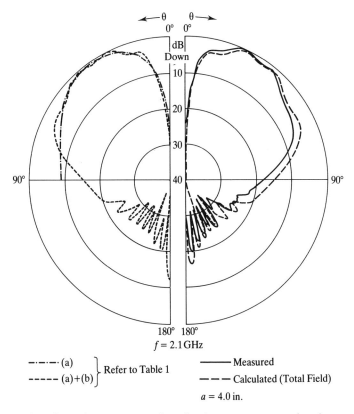

FIGURE 4.5 Elevation plane pattern of a circular aperture opening into a circular ground plane. (From [7] © 1974 IEEE.)

coordinate system of a circular cylinder with a slot is shown in Figure 4.7. The azimuthal radiation field considering only singly diffracted rays is given by

$$E^d = E_1^{d,(1)}(r_1, \phi_1) + E_2^{d,(1)}(r_2, \phi_2)$$
$$= \frac{2\exp[-jkr_1 + \pi/4]}{\sqrt{2\pi kr_1}} D_1^{d,(1)}(\phi_1) + \frac{2\exp[-j(kr_2 + \pi/1)]}{\sqrt{2\pi kr_2}} D_2^{d,(2)}(\phi_2) \quad (4.9)$$

Now one can write a general expression for the total diffracted field as a vector superposition of singly and multiply diffracted rays:

$$Ed(r, \phi) = \frac{2}{\sqrt{2\pi k}} \frac{e^{-jkr}}{\sqrt{r}} \left\{ \exp j \left[k\left(\frac{w}{2}\sin\phi_+ + a\sin\phi\cos\beta - \frac{\pi}{4}\right) \right] D_1 \right.$$
$$\left. + D_2 \exp[-jkw\sin\phi] \right\} \quad (4.10)$$

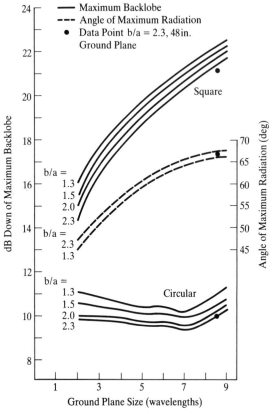

FIGURE 4.6 Angle of maximum radiation and relative power of the maximum backlobe of coaxial apertures in circular and square ground planes. (From [7] © 1974 IEEE.)

where

$$D_1 = D_1^{d,(1)} + D_2^{d,(h)} \quad (4.11a)$$

$$D_2 = D_2^{d,(1)} + D_2^{d,(h)} \quad (4.11b)$$

The wedge-diffracted and creeping wave fields in the different angular sectors are given in Table 4.2 [7]. Figures 4.8 and 4.9 show the radiation patterns of axial slots on a PEC cylinder with the radius of the cylinder and width of the slot as the parameters. The results of single, double, and multiple reflections and boundary value methods have been discussed. Diffraction theory works better for cylinders of large radii. And wedge diffraction theory works very well, since modal analysis can be computationally too expensive because a large number of modes have to be considered.

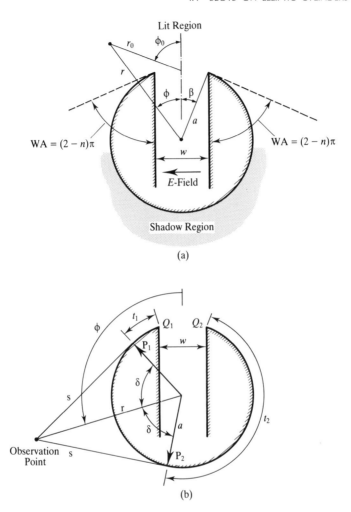

FIGURE 4.7 (a) Geometry and coordinate system of a slot on a circular cylinder. (b) Creeping waves on a circular cylinder. (From [4] © 1969 IEEE.)

4.4 SLOTS ON ELLIPTIC CYLINDERS

The study of slots on an elliptic cylinder is important because an elliptic cylinder represents the fuselage of an aircraft. Figure 4.10(a) shows the geometry and coordinate system of the elliptical cylinder. The creeping wave radiation mechanisms are shown in Figure 4.10(b). The diffracted field is given by

$$E^d(r, \phi) = \frac{1}{\sqrt{2}} \frac{e^{-jkr}}{\sqrt{r}} \{\exp j[k((\omega/2) \sin \phi + r_0 \cos \phi \cos \beta) - \pi/1] D_1$$
$$+ D_2 \exp[-jk\omega \sin \phi]\} \qquad (4.12)$$

TABLE 4.2 Wedge-Diffracted and Creeping Wave Fields in the Different Regions (From [4])

Regions	Wedge-Diffracted Field $E_D(\phi)$	Creeping Wave Field $E_C(\phi)$	t_1	t_2
Region I $0 < \phi \leq \dfrac{\pi}{2} - \beta$	$\dfrac{2}{\sqrt{2\pi k}} \exp\left\{ j\left[k\left(\dfrac{w}{2}\sin\phi + a\cos\phi\cos\beta\right) - \dfrac{\pi}{4} \right] \right\}$ $\times \{R_1 + R_2 \exp[-jkw\sin\phi]\}$	$E_A \exp[-t_1(\alpha_{0h} + jk)]$ $+ E_B \exp[-t_2(\alpha_{0h} + jk)]$	$a\left(\phi + \dfrac{3\pi}{2} - \beta\right)$	$a\left(-\phi + \dfrac{3\pi}{2} - \beta\right)$
Region II $\dfrac{\pi}{2} - \beta < \phi \leq \dfrac{\pi}{2}$	$\dfrac{2}{\sqrt{2\pi k}} \exp\left\{ j\left[k\left(\dfrac{w}{2}\sin\phi + a\cos\phi\cos\beta\right) - \dfrac{\pi}{4} \right] \right\}$ $\times \{R_1 + R_2 \exp[-jkw\sin\phi]\}$	$E_B \exp[-t_2(\alpha_{0h} + jk)]$	—	$a\left(-\phi + \dfrac{3\pi}{2} - \beta\right)$
Region III $\dfrac{\pi}{2} < \phi \leq \dfrac{\pi}{2} + \beta$	$\dfrac{2}{\sqrt{2\pi k}} \exp\left\{ j\left[k\left(\dfrac{w}{2}\sin\phi + a\cos\phi\cos\beta\right) - \dfrac{\pi}{4} \right] \right\}$ $\times \{R_1\}$	$E_B \exp[-t_2(\alpha_{0h} + jk)]$	—	$a\left(-\phi + \dfrac{3\pi}{2} - \beta\right)$
Region IV $\dfrac{\pi}{2} + \beta < \phi \leq \pi$	0	$E_A \exp[-t_1(\alpha_{0h} + jk)]$ $+ E_B \exp[-t_2(\alpha_{0h} + jk)]$	$a\left(\phi - \dfrac{\pi}{2} - \beta\right)$	$a\left(-\phi + \dfrac{3\pi}{2} - \beta\right)$

4.2 SLOTS ON A GROUND PLANE   **287**

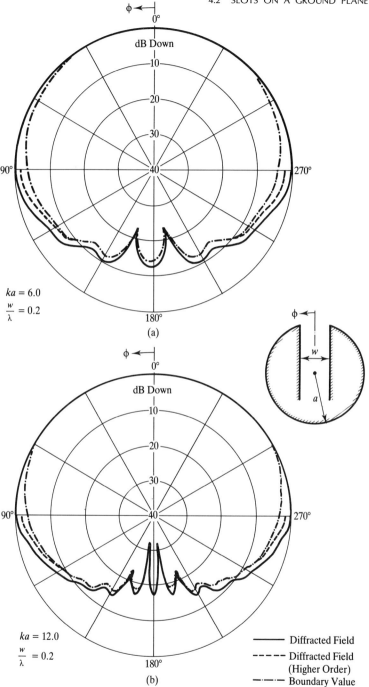

FIGURE 4.8 Radiation patterns of an axial infinite slot on a PEC circular cylinder (TEM mode) with radius and width as the parameters. (From [4] © 1969 IEEE.) (*continued on next page*)

288 SLOT ANTENNAS

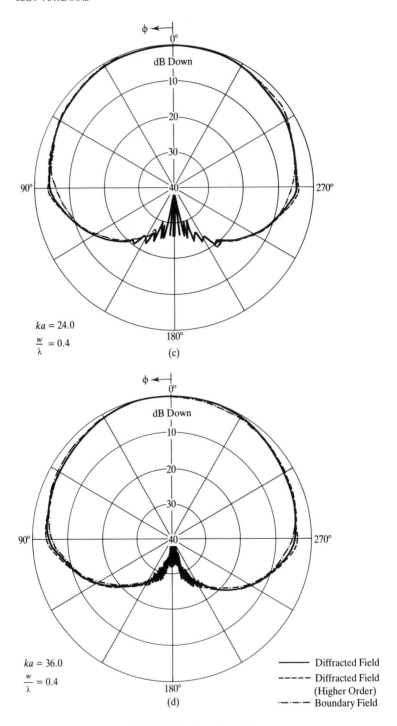

FIGURE 4.8 (*continued*)

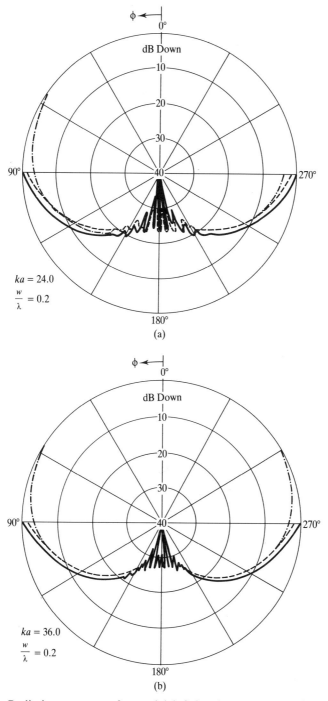

FIGURE 4.9 Radiation patterns of an axial infinite slot on a PEC circular cylinder (TEM mode) with radius and width as the parameters. (From [4] © 1969 IEEE.)

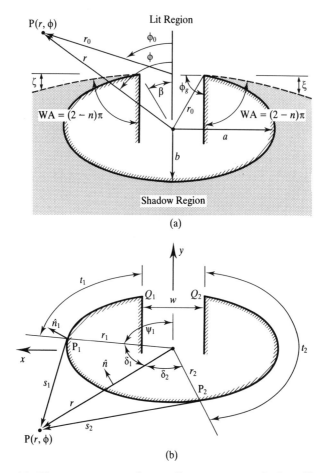

FIGURE 4.10 (a) The geometry and coordinate system of the elliptic cylinder. (b) Creeping wave radiation mechanism. (From [4] © 1969 IEEE.)

where

$$R_0 = (x_0^2 + y_0^2)^{1/2} \tag{4.12a}$$

$$x_0 = \mp \frac{\omega}{2} \tag{4.12b}$$

$$y_0 = \frac{b}{2a}\sqrt{4a^2 - \omega^2} \tag{4.12c}$$

where ω is the width of the slot.

$E^{d,(1)}$ and $E^{d,(2)}$, the total diffracted rays from edges 1 and 2 are super-

4.4 SLOTS ON ELLIPTIC CYLINDERS

positions of first- and higher-order contributions and are given by

$$E^{d,(1)} = E_1^{d,(1)} + E_1^{d,(h)} \qquad (4.12d)$$

$$E^{d,(2)} = E_1^{d,(2)} + E_2^{d,(h)} \qquad (4.12e)$$

where $E_1^{d,(1)}$, $E_1^{d,(2)}$ = singly diffracted rays from edges 1 and 2
$E_1^{d,(h)}$, $E_1^{d,(h)}$ = total higher-order diffracted rays from edges 1 and 2.

The creeping wave contribution (Fig. 4.9(b)) is given by

$$E^C(s, \phi) = E^i(Q_1) \frac{e^{-jk(t_1 + s_1)}}{\sqrt{s_1}} \sum_m D_{mh}(Q_1) D_{mh}(P_1) \cdot \exp\left[-\int_0^{t_1} \alpha_{nh}(\rho) \, dS\right] \qquad (4.12f)$$

where $D_{(m,n)h}(X)$'s are the diffraction coefficients at points X.
For far-field approximations:

For amplitude terms

$$r_1 \approx r_2 \approx r$$

$$\delta_2 \approx \pi - \delta_1$$

For phase terms

$$s_1 \approx r - r \cos \delta_1$$

$$s_2 \approx r + r \cos \delta_1$$

where δ_1 and δ_2 are shown in Figure 4.10(b).
Considering the zeroth-order mode (for details see Section 1.18.2.3) the creeping wave field reduces to

$$E^C(r, \phi) = V_A \frac{e^{-jkr}}{\sqrt{r}} e^{-jkt_1} \cdot \exp\left[-\int_0^{t_1} \alpha_{0h}(\rho) \, dS\right]$$

$$+ V_B \frac{e^{-jkr}}{r} e^{-jkt_2} \cdot \exp\left[-\int_0^{t_2} \alpha_{0h}(\rho) \, dS\right] \qquad (4.13)$$

where

$$V_A = E^A \frac{D_{0h}(P_1)}{D_{0h}(Q_1)} \cdot \exp[jkr \cos \delta_1] \qquad (4.13a)$$

$$V_B = E^B \frac{D_{0h}(P_2)}{D_{0h}(Q_2)} \cdot \exp[-jkr \cos \delta_1] \qquad (4.13b)$$

$$E^A = E^i(Q_1)[D_{0h}(Q_1)]^2 = \frac{\sqrt{r}}{e^{-jkr}} E^d[r, +(\pi/2 + \xi)] \qquad (4.13c)$$

$$E^B = E^i(Q_2)[D_{0h}(Q_2)]^2 = \frac{\sqrt{r}}{e^{-jkr}} E^d[r, +(\pi/2 + \xi)] \qquad (4.13d)$$

From Keller's surface wave theory (Section 1.18.2.3) we have

$$D_{0h}^2(\rho) \approx 1.083 \frac{e^{-\pi/12}}{\sqrt{k}} E^d(k\rho)^{1/3} \tag{4.13e}$$

$$\alpha_{0h}(\rho) \approx \frac{1}{2\rho}(k\rho)^{1/3}\left(\frac{3\pi}{4}\right)^{2/3} e^{j\pi/6} \tag{4.13f}$$

where ρ is the radius of curvature.

The expressions for arc length, complex attenuation factor of the creeping wave, and other parameters for computation of the creeping wave field are available in [5]. Figure 4.11 show the radiation patterns of an axial slot on an elliptical conducting cylinder (TEM mode). A comparison of radiation patterns of smaller elliptical cylinders (TEM mode) is shown in Figure 4.12. Radiation pattern of a thin elliptical cylinder ($ka = 40$, $kb = 8$) and finite ground plane (TEM mode) is shown in Figure 4.13. All the measurements were done at 10 GHz. In general agreement appears to be better in the forward direction. The creeping waves are strongly influenced by the radius of curvature of the surface along the path of the ray.

4.5 SLOTS ON A CONE

GTD has been applied to estimate the radiation patterns of an infinitesimal dipole mounted on the tip of a cone of finite length [10] and from circumferential slots on a conical antenna [11]. Analysis of such antennas is useful in the design of spacecraft antennas.

We briefly present here the analysis of an infinitesimal dipole mounted on the tip of a PEC finite-length cone. The field at an observation point is a superposition of the field from the induced current on an infinite cone and the field diffracted from the cone. The tip of the cone is assumed to be spherical.

The ϕ-independent radiated electric field for an infinitesimal dipole mounted on the tip of an infinite length of the cone is given by [13]

$$E_\theta^s = \sin\theta I_v^{0t}(\cos\theta)\frac{e^{-jkr}}{r} \tag{4.14}$$

with

$$r \approx r_c \quad \text{for amplitude variation} \tag{4.15a}$$

$$r \approx r_c - s\cos(\alpha/2)\cos\theta \quad \text{for phase variations} \tag{4.15b}$$

It turns out the radiated field is given by

$$E_\theta = R(\theta)\exp[jks\cos(\alpha/2)\cos\theta], \quad 0 < \theta < \pi - \frac{\alpha}{2} \tag{4.16a}$$

4.5 SLOTS ON A CONE 293

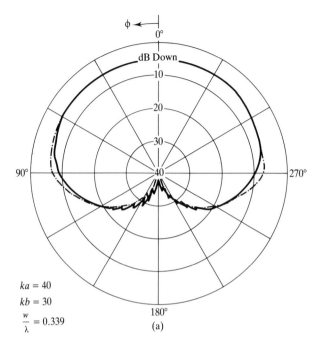

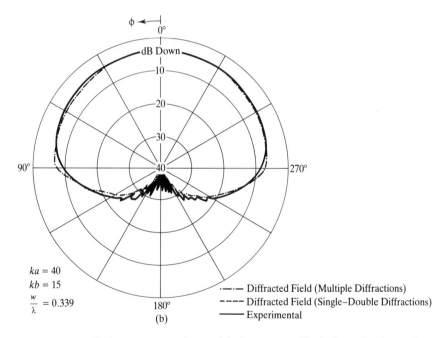

FIGURE 4.11 Radiation patterns of an axial slot on an elliptical conducting cylinder (TEM mode). (From [4] © 1969 IEEE.) (*continued on next page*)

294 SLOT ANTENNAS

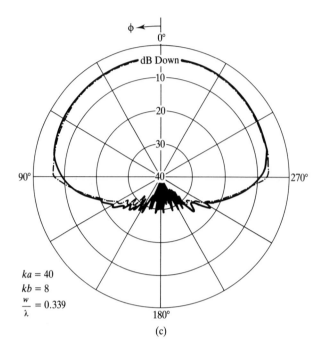

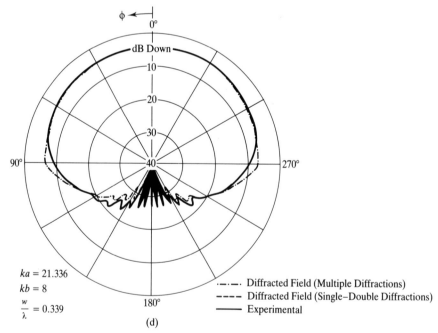

FIGURE 4.11 (*continued*)

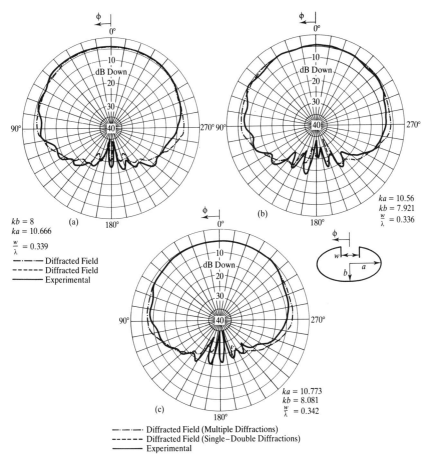

FIGURE 4.12 Comparison of radiation patterns for smaller elliptical cylinders (TEM mode). (From [5] © 1969 IEEE.)

where

$$R(\theta) = \sin\theta I_\nu^{0t}(\cos\theta) \tag{4.16b}$$

The diffracted field once again can be computed using the equivalent edge current method (Chapter 1, [127]) around and away from the caustic direction of the ring source, and is given by

For $\theta = \pi^+$

$$E_\theta = R_\theta\left(\pi - \frac{\alpha}{2}\right)[V_B(s,\psi,n)] \cdot \int_0^{2\pi} \frac{b}{2}\cos\phi \exp\left(j\frac{kb}{2}\sin\theta\cos\phi\right)d\phi \tag{4.17a}$$

$$= j\pi b R_\theta\left(\pi - \frac{\alpha}{2}\right) V_B\left(s, \frac{\alpha}{2}, n\right) J_A\left(\frac{kb}{2}\sin\theta\right) \tag{4.17b}$$

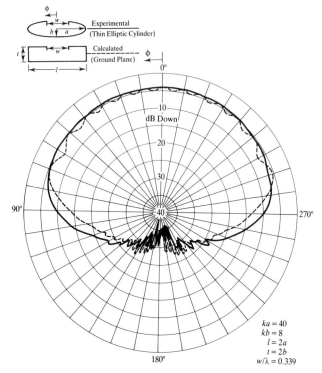

FIGURE 4.13 Radiation pattern of a thin elliptical cylinder and a finite size ground plane (TEM mode). (From [5] © 1969 IEEE.)

where $V_B(.,.,.)$ is given by eqn. (4.5).

$$\psi = \alpha/2 + \theta \approx \alpha/2; \quad b = 2s\sin(\alpha/2); \quad s = 1/2(3 + \alpha/\pi) \qquad (4.18)$$

For $\theta = \pi^-$

$$E_\theta = R_\theta\left(\pi - \frac{\alpha}{2}\right)[V_B(s, \psi, n)] \cdot \int_0^{2\pi} \frac{b}{2}\cos\phi \exp\left(j\frac{kb}{2}\sin\theta\cos\phi\right)d\phi \qquad (4.19a)$$

$$= j\pi b R_\theta\left(\pi - \frac{\alpha}{2}\right)V_B\left(s, \pi + \frac{\alpha}{2}, n\right)J_1\left(\frac{kb}{2}\sin\theta\right) \qquad (4.19b)$$

The diffracted fields away from the caustic can be obtained by superposing the contributions from two scattering centers on the cone edge when the transmitting and receiving points are in the same plane.

The expressions for the field are given by

$$E_\theta = \frac{b}{2} R_\theta\left(\pi - \frac{\alpha}{2}\right) e^{-j\pi/4} \left(\frac{4\pi}{kb \sin \theta}\right)^{1/2} \cdot V_B\left(s + \frac{\alpha}{2} + \theta, \mathbf{n}\right)$$
$$\cdot \exp\left(j\frac{kb}{2} \sin \theta\right) - jV_B\left(s, \frac{\alpha}{2} - \theta, \mathbf{n}\right) \cdot \exp\left(-j\frac{kb}{2} \sin \theta\right) \quad (4.20)$$

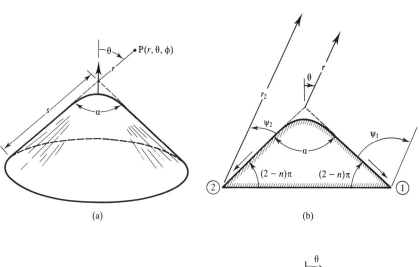

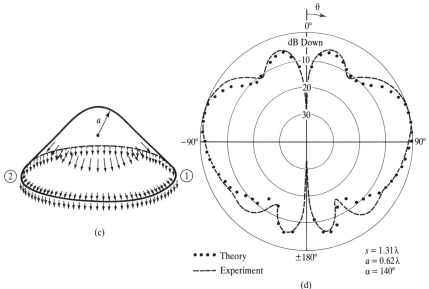

FIGURE 4.14 Radiation pattern of a linear element mounted on the tip of a finite cone [11]: (a) Linear element on the tip of a finite length cone; (b) diffraction from scattering centers on the rim of the cone; (c) rim source around the edge of the cone; (d) radiation pattern.

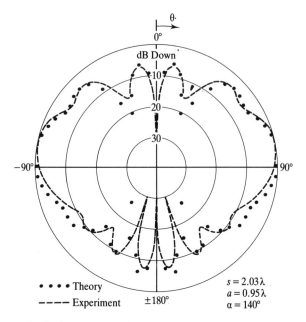

FIGURE 4.15 Radiation pattern of a linear element on a large finite cone [11].

Figures 4.14 and 4.15 show the radiation pattern of linear element mounted on the tip of (a) a finite cone and (b) a large finite cone. The dimensions of the blunted cone are $s = 14.08$ ins. $a = 6.6$ in. and $\alpha = 140°$. There is good agreement between theory and experiment.

Analysis of a slot-excited conical antenna has been presented in [10] using the first-order diffraction coefficient. Avoiding the details [10], the radiated field E_θ due to a narrow (delta source) azimuthal slot at $r = a$ is given by

$$E_\theta = \sqrt{\frac{-\cos\theta_0}{\sin\theta} + 2\frac{-\cos\theta_0^{1/2}}{ka\sin\theta}} |D_m| \cdot \cos[k(R - r - a) - \gamma_m] \cos m\phi \quad (4.21)$$

where

$$D_m(\theta, \theta_0) = |D_m| \exp(j\gamma_m)$$

The radiation pattern of a 10° cone is shown in Figure 4.16.

4.6 SUMMARY

This chapter addressed the application of high-frequency techniques to problems of radiation from slots on finite ground planes: a circular cylinder, an elliptic cylinder, and a cone. The effect of finiteness of a structure (for

4.6 SUMMARY

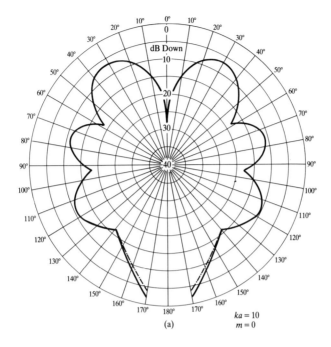

(a)
$ka = 10$
$m = 0$

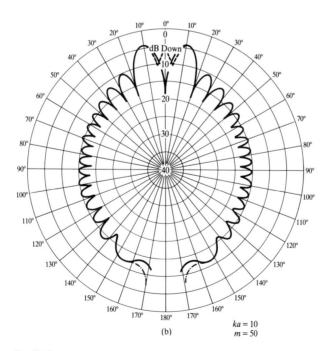

(b)
$ka = 10$
$m = 50$

FIGURE 4.16 Radiation power pattern for a 10° cone using GTD compared with modal analysis [10]. The normalization factor is $2\pi V_0^2/r^2$. (*continued on next page*)

300 SLOT ANTENNAS

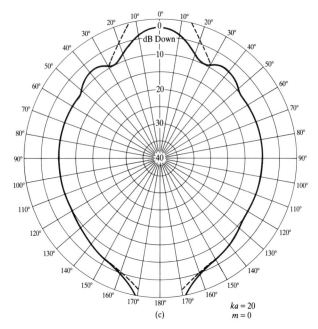

FIGURE 4.16 (continued)

example, a ground plane) can be expressed as a superposition of the field from the infinite ground plane and the wedge/edge-diffracted fields.

4.7 REFERENCES

1. J. R. Wait, *Electromagnetic Radiation from Cylindrical Structures*, Pergamon, New York, 1959, pp. 88–104.

2. D. P.-Brown and G. E. Stewart, "Radiation from Slots Antennas on Cones," *IEEE Trans. Antennas and Propagat.*, **AP-20**(1), 1972, 40–49.

3. C. A. Balanis and L. Peters Jr., "Equatorial Plane Pattern of an Axial Slotted Circular Cylinder using Geometrical Theory of Diffraction, *IEEE Trans. Antennas and Propagat.*, **AP-17**(3), 1969, 351–353.

4. C. A. Balanis and L. Peters Jr., "Analysis of Aperture Radiation from an Axial Slotted Circular Cylinder using Geometrical Theory of Diffraction," *IEEE Trans. Antennas and Propagat.* **AP-17**(1), 1969, 93–97.

5. C. A. Balanis and L. Peters Jr., "Aperture Radiation fron an Axially Slotted Elliptical Conducting Cylinder using Geometrical Theory of Diffraction," *IEEE Trans. Antennas and Propagat.*, **AP-17**(4), 1969, 507–513.

6. C. A. Balanis "Pattern Distortion due to Edge Diffraction," *IEEE Trans. Antennas and Propagat.*, **AP-18**(4), 1970, 551–563.

7. C. R. Cockrell and P. H. Pathak, "Diffraction Theory Techniques Applied to Aperture Antennas on Finite Circular and Square Ground Planes," *IEEE Trans. Antennas and Propagat.*, **AP-22**(3), 1974, 443–448.
8. A. R. Lopez, "The Geometrical Theory of Diffraction Applied to Antenna Pattern Impedance Calculations," *IEEE Trans. Antennas and Propagat.*, **AP-14**(1), 1966, 40–45.
9. C. A. Balanis, "Radiation from Conical Surfaces used for High-Speed Spacecraft," *Radio Science*, **7**(2), 1972, 339–344.
10. D. C. P.-Brown, "Diffraction Coefficients for a Slot-Excited Conical Antenna," *IEEE Trans. Antennas and Propagat.*, **AP-20**(1), 1972, 40–49.
11. C. A. Balanis, "Radiation from Conical Surfaces used for High-Speed Spacecraft," *Radio Science*, **7**(2), 1972, 339–344.
12. S. W. Lee, "Mutual Admittance of Slots on a Cone: Solution by Ray Techniques," *Radio Science*, **AP-26**(6), 1978, 768–773.
13. C. A. Balanis, *Antenna Theory: Analysis and Design*, John Wiley, 1982, Ch. 11.9.
14. G. Mazzareilla and G. Panariello, "Evaluation of Edge Effects in Slot Arrays using the Geometrical Theory of Diffraction," *IEEE Trans. Antennas and Propagat.*, **AP-37**(3), 1989, 392–395.
15. R. F. Harrigton, *Time-Harmonic Electromagnetic Fields*, McGraw-Hill, 1961, p. 316.

CHAPTER FIVE

Radar Cross Sections of Complex Objects

5.1 INTRODUCTION

Many practical targets have complex shapes and need special attention since they can only be handled by approximate solution methods. A complex target may often be considered as a superposition of simple targets. The combination method has already been described in a text [1]. Alternative solutions are integral equation techniques and hybrid methods tabulated in Tables 1.1 through 1.3. In this chapter we describe how to apply high-frequency techniques, to study radiation and scattering from complex targets. Discussions on complex targets are available in many references [1–5]. We concentrate on polygonal and cylindrical plates, plates at or near grazing incidence, the RCS of open-ended cavities with different cross sections the, bistatic RCS of simple and complex objects, and how to control scattering by rim loading and real-time computation of an object's RCS. In my opinion, textbooks have not adequately dealt with these topics up to now.

5.2 RCS OF POLYGONAL PLATES

5.2.1 Using ILDC

Figure 5.1 shows the geometry and coordinate system of scattering from polygonal plates. As mentioned in Section 1.20, ILDC can be used to calculate the scattering from a complex object by computing the incremental scattered field from an element on the edge of the scatterer as a product of the ILDC and the incident field followed by integrating over the whole length of the edge on the scatterer. The analysis of polygonal plates based on this technique is available in [5]. The symbols used are: **S** is the area of the illuminated side of

5.2 RCS OF POLYGONAL PLATES

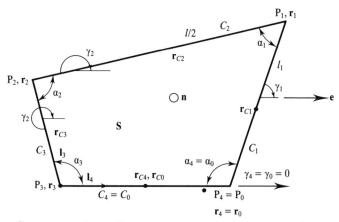

FIGURE 5.1 Geometry and coordinate system of scattering from a polygonal plate [6].

the plate, $\hat{\mathbf{n}}$ is the unit normal on the surface away from the illuminated side, P_n for $n = 0 \cdots N$, is the nth corner of the plate, with n increasing in the anticlockwise direction and $P_N = P_0$, $\mathbf{r}_n$ is the position vector to P_n and is normal to $\hat{\mathbf{n}}$, C_n indicates both edge running from P_{n-1} to P_n and the length of the edge and $C_0 = C_N \mathbf{l}_n$ is the unit tangent to C_n in the direction from P_{n-1} to P_n. The center of C_n is denoted by $\mathbf{r}_{C_n}$ and is given by

$$\mathbf{r}_{C_n} = \tfrac{1}{2}(\mathbf{r}_{n-1} + \mathbf{r}_n) = \mathbf{r}_{n-1} + \tfrac{1}{2}C_n \mathbf{l}_n = \mathbf{r}_n - \tfrac{1}{2}C_n \mathbf{l}_n \qquad (5.1a)$$

where $\hat{\mathbf{e}}$, which is the unit vector along the projection of the sum vector $(\mathbf{e}_r^s + \mathbf{e}_r^i)$ on the surface, and is given by

$$\mathbf{e} = \frac{1}{2\tau}[\hat{\mathbf{n}} \times (\mathbf{e}_r^s + \mathbf{e}_r^i)] \times \hat{\mathbf{n}} \qquad (5.1b)$$

where the parameter τ is given by

$$\tau = \tfrac{1}{2}|\hat{\mathbf{n}} \times (\mathbf{e}_r^s + \mathbf{e}_r^i)| \qquad (5.1c)$$

and the range $1 \geq \tau \geq 0.0$. γ_n is the angle measured anticlockwise from $\hat{\mathbf{e}}$ to $\hat{\mathbf{l}}_n$ and their relationship is

$$\cos \gamma_n = \hat{\mathbf{e}} \cdot \hat{\mathbf{l}}_n, \quad \sin \gamma_n = \hat{\mathbf{n}} \cdot (\hat{\mathbf{e}} \times \hat{\mathbf{l}}_n) \qquad (5.1d)$$

The parameter Y_n is given by

$$Y_n = \tau k C_n \cos \gamma_n \tag{5.2}$$

$$\cos \phi_{sn} = \frac{1}{\cos \beta_{sn}} \mathbf{e}_r^s \cdot (\hat{\mathbf{n}} \times \mathbf{l}_n) \tag{5.3}$$

for $\pi > \phi_{sn} > 0$ for $\mathbf{e}_r^s \cdot \mathbf{r} > 0$

and $2\pi > \phi_{sn} > \pi$ for $\mathbf{e}_r^s \cdot \mathbf{n} < 0$

The current on the polygonal plate consists of two components, the PO component and the fringe wave component. The fringe wave component can be formulated using Ufimtsev's ILDC. The fringe wave contribution can be presented as a combination of either N edges or N corner contributions.

The fringe wave coefficient is given by

$$D^{fw} = \frac{e^{-j\pi/4}}{\sqrt{2\pi}} \sum_{n=0}^{N-1} \exp(-j2\tau k \mathbf{e} \cdot \mathbf{r}_{C_n}) C_n \frac{\sin Y_n}{Y_n d_n^{fw}} \tag{5.4}$$

where d_n^{fw}, the knife-edge Ufimtsev ILDC for $C_n d_n^{fw}$ has the form

$$d_n^{fw} = d_{\perp n}^{fw} e_{\perp n}^s e_{\perp n}^i + d_{Xn}^{fw} e_{\parallel n}^s e_{\perp n}^i + d_{\parallel n}^{fw} e_{\parallel n}^s e_{\parallel n}^i \tag{5.5}$$

where the coefficients $d_{\perp n}^{fw}$, d_{Xn}^{fw} and $d_{\parallel n}^{fw}$ are functions of the parameters k, β_{in}, ϕ_{in}, β_{sn}, and ϕ_{sn}.

We can also obtain eqn. (5.4) in terms of a sum of N corner diffraction coefficients in the form

$$D_n^{fw} = \sum_{n=0}^{N-1} \exp(-j2k\tau \mathbf{e} \cdot \mathbf{r}_n) c_n^{fw} (\mathbf{e}_r^s + \mathbf{e}_r^i) \cdot \mathbf{l}_n \neq 0 \quad \text{for all } n \tag{5.6}$$

where the flat plate corner diffraction coefficients are given by

$$c_n^{fw} = -\frac{\exp(j\pi/4)}{2\sqrt{2\pi\tau}} \left\{ \frac{1}{\cos \gamma_{n+1}} d_{n+1}^{fw} - \frac{1}{\cos \gamma_n} d_n^{fw} \right\} \tag{5.7}$$

Figures 5.2 through 5.6 compare [6] the theoretical results of scattering from polygonal plates using the ILDC formulation and experiment. It appears the agreement is good at and around normal incidence but is poor at large angles and in particular around the nulls.

5.2.2 Using GTD

The RCS of polygonal plates can be obtained by using GTD but there are some difficulties. A very good discussion [6] addresses these difficulties.

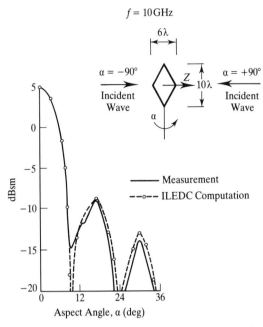

FIGURE 5.2 RCS of a diamond-shaped plate in the plane normal to a 10λ diagonal (VV polarization) [6].

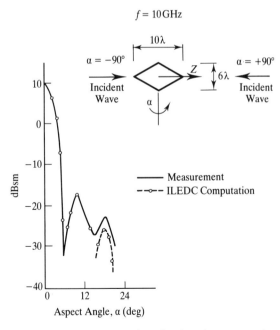

FIGURE 5.3 RCS of a diamond-shaped plate in the plane normal to a 6λ diagonal (HH polarization) [6].

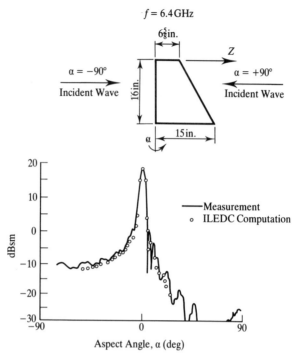

FIGURE 5.4 RCS of a trapezoidal plate in the plane normal to the 16 in. side (VV polarization) [6].

5.3 RCS OF STRIPS AND PLATES AT GRAZING INCIDENCE

Scattering from a wedge, half plane, or any other structure at grazing incidence needs to be specially treated since the incident and reflected shadow boundaries at one surface of the wedge all merge in one plane. An accurate prediction of grazing incidence scattered field and hence the RCS is, in my opinion still an open problem. Modeling the front edge by a radiating wire and then finding the incident field on the rear edge is not adequate. Classical GTD gives singular fields at graving incidence as observed in Ross's work [6] on flat plate scattering. This is an interesting problem in the sense it provides a good test for the capability of different techniques to predict scattered fields in a wide range of aspect angles, including cases of grazing incidence. Cases of grazing incidence for wedges and flat structures have been treated in some references [6–15], including at least one textbook [14].

5.3.1 A Uniform High-Frequency Solution for Strip Scattering at Grazing Incidence

The strip scattering problem at grazing incidence gets complicated by the fact that one edge is in the shadow boundary of the other edge at grazing

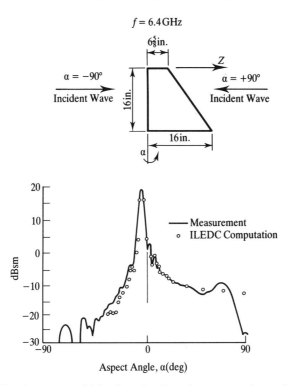

FIGURE 5.5 RCS of a trapezoidal plate in the plane normal to the 16 in. side (HH polarization) [6].

incidence hence the field illuminating the second edge from the first edge is not a ray-optical field. Such a nonray-optical field can be represented as a superposition of inhomogeneous waves as discussed by Tiberio and Kouyoumjian [10, 14]. Lee and Boersma [11] also presented a decomposition of a nonray-optical field in a different way.

The analysis of the grazing incidence has been presented [10, 12] for plane, cylindrical, and spherical incident wavefronts. The high-frequency scattering at grazing incidence for a strip is shown in Figure 5.7. The scattered field can be obtained as a sum of the first- and third-order diffracted fields from the first edge and the second-order diffracted field from the second edge, which is in the shadow boundary of the first edge. While the first- and third-order contributions can be found in a rather straightforward application of UTD, with slope diffraction when the width of the strip is not very large, estimation of the second-order contribution is more involved and has been detailed in [10].

Figure 5.7 shows the geometry and coordinate system of grazing incidence scattering from an infinitely long PEC strip of vanishingly small thickness and width d. The strip is in the xy-plane. Only one is of interest, since for grazing incidence in the TE case there is no diffraction from the edge. Hence, the TM

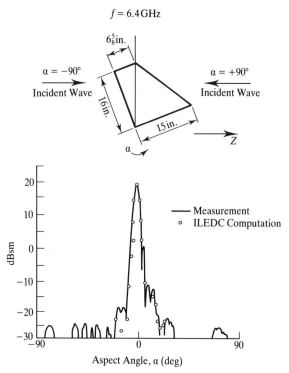

FIGURE 5.6 RCS of a trapezoidal plate in the plane normal to the short diagonal (HH polarization) [6].

case need only be considered. From the scattering center approach, the total field at the observation point has several contributions: (1) the single diffracted ray from Q_1; (2) a doubly diffracted ray from Q_2; and (3) higher-order contributions generated by interactions between Q_1 and Q_2.

The Singly Diffracted Ray. The asymptotic expression for the first-order diffracted field using UTD is given by

$$u_1^d(P) \sim u^i(Q_1) \frac{\exp(-j\pi/4)}{\sqrt{2\pi}} \frac{F\left(2kL \sin\left(\frac{\phi_1}{2}\right)\right)}{\sin\left(\frac{\phi_i}{2}\right)} \cdot f(x) \tag{5.8}$$

where the incident field $u^i(Q_1) = u_0 \cdot f(s)$, u_0 is the amplitude and $f(s)$ is the divergence plus phase factor of the incident field.

5.3 RCS OF STRIPS AND PLATES AT GRAZING INCIDENCE

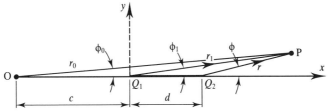

FIGURE 5.7 Scattering from a strip at grazing incidence.

L = the distance parameter = $\dfrac{sr_1}{s+r_1}$

k = the propagation constant

S = the distance of the source from Q_1.

The angles ϕ_0, ϕ_1, and ϕ are defined in the Figure 5.4 and $f(x)$ is given by

$$f(x) = \begin{cases} \exp(-jkx) & \text{for plane waves} \\ \dfrac{\exp(-kx)}{kx} & \text{for spherical wave incidence} \\ \dfrac{\exp(-jkx)}{\sqrt{kx}} & \text{for cylindrical wave incidence} \end{cases}$$

The Doubly Diffracted Ray. The doubly diffracted ray analysis becomes involved since edge Q_2 is in the shadow boundary of Q_1 for grazing incidence illuminating Q_1 first. However, the diffracted ray from Q_2 can be obtained in the following way: (1) the transition region field is expressed as a superposition of inhomogeneous and slowly varying plane, cylindrical or spherical waves; (2) each inhomogeneous wave is multipled by a proper diffraction coefficient, which is an extension of UTD; and (3) all the contributions are summed to give the total contribution. This approach resembles the spectral theory of diffraction described in Section 1.10.

For details of analysis, including evaluation of the diffraction integral using steepest descent path, the reader is referred to [9]. The diffracted field from Q_2 due to incidence from Q_1 is given by

$$u_{12}^{dt}(P) \sim u_0 \frac{2}{\pi} \cos(\phi/2)\left(1+\frac{a}{4}\right)\tan^{-1}\sqrt{\sigma} - j2\sqrt{\sigma}\cdot\int_0^{\sqrt{k_2 a}} \exp(-j\sigma x^2)$$

$$\times \int_x^\infty \exp(-j\tau^2)\,d\tau\,dx \cdot \exp(jk_3 a) f(S+d+r)$$

$$+ ju_0 \frac{\cos(\phi/2)}{4\pi\sqrt{kd}}\sqrt{\frac{s}{s+d}}\cdot f(s+d)f(r) \qquad (5.9)$$

where

$$k_2 = k\frac{dr}{d+r} \tag{5.9a}$$

$$k_3 = k\frac{r(s+d)}{s+r+d} \tag{5.9b}$$

$$\sigma = \frac{k''}{k'} = \frac{cr}{d(c+d+r)} \tag{5.9c}$$

$$a = 2\sin^2(\phi/2) \tag{5.9d}$$

When the observation is also along the tangent to the plane of the strip, that is with $\phi = 0$, the expression for $u_{12}^d(P)$ is given by

$$u_{12}^d(P) \sim u_0 \frac{2}{\pi}\tan^{-1}\sqrt{\sigma} + j\frac{1}{kd\sqrt{\sigma}}\left(\frac{s}{s+d}\right)^2 \cdot f(s+d+r) \tag{5.10}$$

The expression for a triply diffracted ray is given by

$$u_{121}^d(P) \sim u^i(Q_1) \cdot j\frac{\exp(-j\pi/4)}{\sqrt{2\pi}}\frac{\exp(-j2kd)}{2\pi kd}\sin(\phi_1/2)$$

$$\cdot (1-F)2k\frac{dr_1}{d+r_1}\cos^2(\phi_1/2) \cdot \frac{s}{s+d}\frac{r_1}{d+r_1}f(r_1) \tag{5.11}$$

This higher-order contribution becomes important in the backscattered direction for narrow strips.

Figure 5.8 shows the scattered power with observation angles for plane wave incidence for medium and large values of kd, where k is the propagation constant and d is the width of the strip. The results using the present theory have been compared with results obtained using the method of moments.

The backscattering cross sections of a PEC plate for the two orthogonal polarizations with a stress on grazing incidence have been computed [8] using PO, Keller's GTD, and Ufimtsev's PTD. The results are compared in Figure 5.9. Nonuniform cylindrical waves were used to compute higher-order diffractions. They show poor agreement using nonuniform cylindrical waves beyond 80° from normal, since at or near grazing the assumption of nonuniformity is not valid in the presence of the plate surface. PTD with second-order diffraction predicts low RCS at grazing incidence. The PTD agrees much better with experiment than any other technique near incidence. Backscattering from a flat plate with edge-on incidence and the characteristics of surface currents have also been discussed in [9] with the flat plate having center-fed slots milled on it.

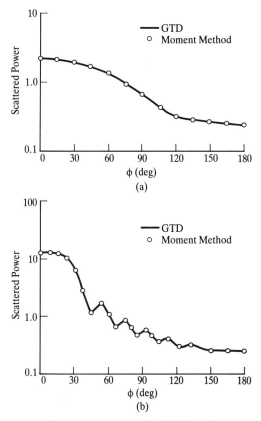

FIGURE 5.8 Scattered far-field power pattern for incident plane wave grazing incidence: (a) $kd = 20$; (b) $kd = 3.0$ [9].

5.4 RCS OF CYLINDRICAL PLATES

The RCS of cylindrical plates is an interesting problem because of their curvature. This problem has been treated in the literature using various high-frequency techniques: physical optics [16], geometric and physical theory of diffraction [17], and uniform asymptotic theory (UAT) [18] of diffraction.

5.4.1 Physical Optics Current on Curved Smooth Conducting Surfaces

The current on a curved surface induced by a source can be found by a physical optics approximation. If the curved plate terminates in an edge, like in a finite curved plate, then a correction current can be found using the physical theory of diffraction. The structure and its geometry are shown in Figure 5.10.

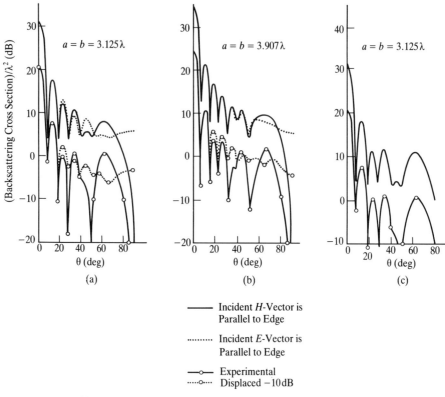

FIGURE 5.9 RCS computed using nonuniform cylindrical waves in higher-order diffraction, for different plate size and orientations of the field vectors with reference to the edges. (From [10] © 1971 IEEE.)

The Maue integral equation for the unknown current **J(r)** is given by [17]

$$\mathbf{j}(\mathbf{r}) = 2\hat{\mathbf{n}}(r) \times \mathbf{H}^i(\mathbf{r}) + 2\hat{\mathbf{n}}(\mathbf{r}) \times \iint_S \mathbf{j}(\mathbf{r}) \times \iint_S \mathbf{J}(\mathbf{r}) \times \nabla' G(\mathbf{r}, \mathbf{r}') \, dS(\mathbf{r}'), \quad \mathbf{r}, \quad \mathbf{r}' \in S \quad (5.12)$$

The Green's function G is given by

$$G(\mathbf{r}, \mathbf{r}') = \frac{\exp(-jk|\mathbf{r} - \mathbf{r}'|)}{4\pi|\mathbf{r} - \mathbf{r}'|} \quad (5.13)$$

The incident field is given by

$$\mathbf{H}^i(\mathbf{r}) = \mathbf{H}_0^i(\mathbf{r}) \exp(-jks(\mathbf{r})) \quad (5.14)$$

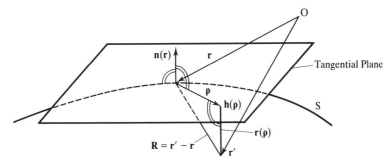

FIGURE 5.10 Geometry and coordinate system of a curved surface. (From [18] © 1978 IEEE.)

The total induced current is given by

$$\mathbf{J} = \mathbf{J}_0 + \mathbf{J}_1 \tag{5.15a}$$

where,

$$\mathbf{J}_0, \text{ the classical PO current} = 2(\hat{\mathbf{n}} \times \mathbf{H}^i(\mathbf{r}')) \tag{5.15b}$$

$$\mathbf{J}_1 = L\{2\hat{\mathbf{n}} \times \mathbf{H}^i\} \tag{5.15c}$$

where L is an operator and J_1 turns out to be

$$J_1 = \frac{1}{\pi}\hat{n}(\mathbf{r}') \times \iint_S \exp(-jkg(\hat{\rho}))\frac{1 + jk\sqrt{\rho^2 + h^2}}{\rho}\mathbf{A}(\rho)\,d\mathbf{S}(\rho)\,d\mathbf{S}(\rho) \tag{5.15d}$$

and the phase function $g(\rho)$ is given as

$$g(\rho) = \sqrt{\rho^2 + h^2(\rho)} + s(\hat{r}(\rho)) \tag{5.16}$$

$$\mathbf{A}(\rho) = \frac{\rho w(\rho)}{(\rho^2 + h^2)^{3/2}} \cdot \left\{[(\rho - h\hat{n}(\mathbf{r}'))\cdot \mathbf{H}_{10}]\,\text{grad}\,h + [h - \rho\cdot\text{grad}\,h]\mathbf{H}_1\right\} \tag{5.17}$$

5.5 RCS OF OPEN-ENDED CAVITIES

Scattering from and coupling into open-ended cavities are of great interest to engineers since an open-ended cavity simulates jet intakes of aircraft. Analysis of a simple open-ended circular cylinder, albeit a different shape, does help to explain scattering mechanisms of actual jet intakes. A good recent reference is the proceedings [20] of a workshop on high-frequency electromagnetic jet engine cavity modeling and a dissertation and technical report [21]. There are many different approaches to estimate scattering from and coupling into an open-ended cavity. Figure 5.11 shows different types of open-ended cavities, which may or may not be loaded or terminated. A cavity may be loaded, for

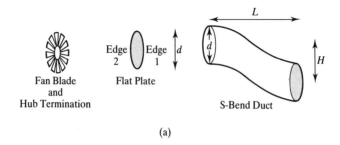

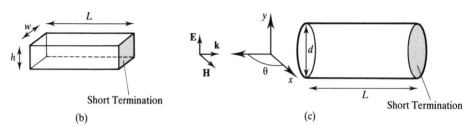

FIGURE 5.11 Different configurations of open-ended cavities: (a) a general cavity inlet with a termination; (b) a rectangular cavity; (c) a cylindrical cavity.

example, by coating the inner surfaces fully or partially with a radar-absorbing material. The different techniques [20–24] can be classified as: aperture integration (AI) using a Kirchhoff-type (PO) approximation, and the GO shooting and bouncing Ray (SBR) method; the Gaussian beam (GB) method; and hybrid techniques.

5.5.1 Aperture Integration (AI) and the GO Shooting and Bouncing Ray (SBR) Method

In this method the incident wave at the aperture at the open end is expressed as a superposition of a sufficiently large number of narrow ray tubes. It assumes the effect of the internal walls is small enough for the wave inside the cavity to be approximated by the GO rays. As described in [22, 23], the expression at the second reflection point of the pth ray tube field reflected from the first reflection point is given by

$$\mathbf{E}^r_{1,p}(Q_2) \sim \mathbf{E}^i(Q_1) \cdot \bar{\mathbf{R}}_{1,p} \sqrt{\frac{\delta S_0}{\delta S_1}} e^{-jkr_1} \qquad (5.18)$$

Generalizing for the pth ray tube and nth reflection, the field at the $(n + 1)$th

5.5 RCS OF OPEN-ENDED CAVITIES

point of reflection is given by

$$\mathbf{E}_{n,p}^r(Q_{n+1}) \sim \mathbf{E}_{n-1,p}^r(Q_n) \cdot \bar{\mathbf{R}}_{n,p} \sqrt{\frac{\delta S_{n-1}}{\delta S_n}} e^{-jkt_n} \qquad (5.19)$$

where

$$l_n = |Q_n \bar{Q}_{n+1}|$$

Assume the reflecting surfaces are smooth and there is no cross-polarization associated with the reflection process, the reflection coefficient dyadic is given by

$$[R_{np}] = \begin{bmatrix} R_{n,p}^{\parallel} & 0 \\ 0 & R_{n,p}^{\perp} \end{bmatrix} \qquad (5.20)$$

where $R_{n,p}^{\parallel}$ and $R_{n,p}^{\perp}$ are the plane wave Fresnel reflection coefficients with the tangent plane approximation

$$\begin{bmatrix} E_{n,p}^{\parallel t}(Q_n+1) \\ E_{n,p}^{\perp r}(Q_{n+1}) \end{bmatrix} \sim [R_{np}][T_{np}] \begin{bmatrix} E_{n-1,p}^{\parallel t}(Q_n) \\ E_{n-1,p}^{\perp r}(Q_n) \end{bmatrix} \cdot \sqrt{\frac{\Delta S_{n-1}}{\Delta S_n}} e^{-jkt_n} \qquad (5.21)$$

$$E_{n,p}^{\parallel r} = \hat{e}_{\parallel}^r \cdot \mathbf{E}_{n,p}^r(Q_{n+1}) \qquad (5.22\text{a})$$

$$E_{n,p}^{\perp r}(Q_{n+1}) = \hat{e}_{\perp} \cdot \mathbf{E}_{n,p}^r(Q_{n+1}) \qquad (5.22\text{b})$$

and

$$E_{n-1,p}^{\parallel r}(Q_n) = \hat{e}_{\parallel}^r \cdot \mathbf{E}_{n-1,p}^r(Q_n) = \hat{e}_{\perp} \cdot \mathbf{E}_{n-1,p}^r(Q_n) \qquad (5.22\text{c})$$

$$E_{n-1,p}^{\perp r}(Q_n) = \hat{e}_{\perp} \cdot \mathbf{E}_{n-1,p}^r(Q_n) \qquad (5.22\text{d})$$

It turns out that the reflected field column vector can be expressed as

$$\begin{bmatrix} E_{n,p}^{\parallel t}(Q_{n+1}) \\ E_{n,p}^{\perp r}(Q_{n+1}) \end{bmatrix} \sim \prod_{q=1}^{n} \left| [R_{qp}][T_{qp}] \cdot [E_{\parallel}^i(Q_1) E_{\perp}^i(Q_1)] \right|$$

$$\cdot \sqrt{\frac{\Delta S_0}{\Delta S_n}} e^{-jk} \sum_{q=1}^{n} t_q e^{-jk(m/2)\pi} \qquad (5.23\text{a})$$

$$\mathbf{E}_{n-1,p}^r(Q_n) = \mathbf{E}^i(Q_1) \quad \text{if} \quad n=1 \qquad (5.23\text{b})$$

and

$$E_{\parallel}^i(Q_1) = \hat{e}_{\parallel}^i \cdot \mathbf{E}^i(Q_1) \qquad (5.23\text{c})$$

$$\mathbf{E}_{\perp}^i = \hat{e}_{\perp}^i \cdot \mathbf{E}^i(Q_1) \qquad (5.23\text{d})$$

The electric and magnetic sources $d\mathbf{p}_{ep}$ and $d\mathbf{p}_{mp}$ which radiate the field $\mathbf{E}_c^s$ outside the cavity are given by

$$d\mathbf{p}_{ep}(\mathbf{r}') = [-\hat{n} \times \mathbf{H}_p^r(\mathbf{r}')]\Delta A_N \qquad (5.23\text{e})$$

$$d\mathbf{p}_{mp}(\mathbf{r}') = [-\mathbf{E}_p^r(\mathbf{r}') \times \hat{n}]\Delta A_N \qquad (5.23\text{f})$$

Using aperture integration, the GO/AI result for the scattered field is

$$\mathbf{E}_c^s \sim \frac{jkZ_0}{4\pi} \frac{e^{-jkr}}{r} \sum_p \iint_{\Delta A_N} [\hat{r} \times \hat{r} \times d\mathbf{p}_{ep} + Y_0 \hat{r} \times d\mathbf{p}_{mp}] e^{jk\hat{r}\cdot\mathbf{r}'} \, ds' \quad (5.24)$$

The integration in eqn. (5.24) can be evaluated in closed form in some cases. Some results using this technique are discussed in Section 5.6.1.

5.5.2 The Gaussian Beam (GB) Method

In this section, we discuss the GB method to estimate scattering from an open-ended cavity. The problem of coupling into the open end is also an important problem and will be discussed in Chapter 7. In this method, a part of the EM field which is incident on the aperture is expressed as a superposition of a number of well-focussed (or, spectrally narrow) GBs. Advantage of the GB technique in the frequency domain are [22–23]

1. The GB technique retains the simplicity and generality of the classic geometrical optics (GO) approach.
2. If the density of the GBs is increased to $100\lambda^2$ in the entry aperture at the open end of the cavity or any structure, GBs become independent of the incident angle over sufficiently large frequency bands. So the work need not be repeated for each incident angle of the external excitation.
3. It is possible to preselect the most strongly excited GBs and discard the weak ones. This offers a potential saving of computer time in ray tracing, particularly inside a complex cavity.

A summary of the GB technique is available [21, 22].

The Gaussian Beam. The basic GB expression is given by

$$\Psi = u \exp(-jk_0 z) \quad (5.25)$$

where

$$u = A \frac{W_0}{W} \exp(-r^2/W^2) \exp(-j\pi r^2/\lambda R + j \tan^{-1}(z/z_R)) \quad (5.26)$$

where, A, the complex amplitude, and W_0, the width of the beam, are constants. The beam width at $z = z$ is given by

$$W(z) = W_0[1 + (z/z_R)]^{1/2} \quad (5.27a)$$

$$R(z) = z + z_R^2/z \quad (5.27b)$$

$$z_R = \pi W_0^2/\lambda \quad (5.27c)$$

5.5 RCS OF OPEN-ENDED CAVITIES

At $z = 0$, the width W of the beam is minimum and is equal to W_0, termed the waist beam radius. $R(z)$ is the radius of curvature of the surfaces of constant phase.

Theory of Coupling into the Cavity. The frequency domain field at any point inside the cavity is given by

$$U(P_c) = \sum_{m=-M}^{M} \sum_{n=-N}^{N} A_m(\theta_0, \theta_n) B_{mn}(P_c) \tag{5.28}$$

where $A_m(\theta_0, \theta_n)$ = the frequency domain expansion coefficient for the mnth GB. This depends on the angle of incidence θ_i and the angle of observation (scattering) θ_0.

$B_{mn}(P_e)$ = the amplitude at the point of observation due to the GB launched from subaperture m at an Angle θ_n, which has been traced inside the cavity.

GB as an FD Basis Function. Figure 5.12 shows a number of rays launched from the subaperture. These rays get reflected inside the cavity and ultimately return to the entry aperture from the end of the termination after reflections. The beam suffers a broadening after reflection, as shown in Figure 5.13.

The amplitude of the beam, also known as the beam basis function, in the transverse direction is a GB centered on the z-axis and is given by

$$B(x, z) = \sqrt{\frac{jb}{z + jb}} e^{-jkz} \left(1 + \frac{1}{2}\frac{x^2}{s^2 + b^2}\right) \cdot e^{-(1/2)(hb/2) \left(\frac{x^2}{z^2 + b^2}\right)} |z + jb|^2 \gg |x|^2 \tag{5.29}$$

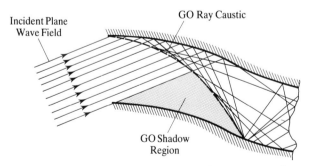

FIGURE 5.12 A number of rays launched from the subaperture, getting reflected inside the cavity and from the termination, and ultimately coming back to the entry aperture from the termination end after some reflections. For clarity GBs are represented by the axial ray.

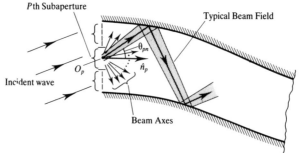

FIGURE 5.13 GB reflection from the wall; broadening of the beam.

The far field from the beam basis function in the frequency domain is

$$B(\rho, \theta) = \sqrt{\frac{jb}{\rho}}\, e^{-jk\rho} e^{-1/2 k b \theta^2} \qquad (5.30)$$

In the frequency domain, the beam waist is proportional to the square root of b, and the width is inversely proportional to the square root of b.

Subaperture Expansion Using GBs. The field due to the launched GB is expressed as

$$U_{GB} = \sum_{m=-M}^{M} \sum_{n=-N}^{N} A_m(\theta_i, \theta_n) B(x_{mn}, z_{mn}) \qquad (5.31)$$

where $2N + 1 =$ the number of beams from each aperture
$2M + 1 =$ the number of subapertures
$\theta_n =$ the angle between the nth beam and the z-axis, which is the axis of the cylinder
$x_{mn} = (x - m\delta) \cos \theta_n + z \sin \theta_n$
$z_{mn} = (x - m\delta) \sin \theta_n + z \cos \theta_n$
$\delta = \dfrac{d}{2M + 1}$
$\theta_n = n\Delta\theta$
$\Delta\theta =$ the angular separation between adjacent launched beams

The far-field pattern of the mth subaperture illuminated by a plane wave can be easily obtained using the PO method. This method is accurate until the off-axis angle of launch is more than 40°. It turns out the PO expression for the subaperture field at any point is given by

$$E_\phi = \frac{-j\beta e^{jkr}}{4\pi r} \sum_{m=1}^{M} \sum_{n=1}^{N} \int_{\rho_{1m}}^{\rho_{2m}} \int_{\phi_{1m}}^{\phi_{2m}} H_{0y} e^{j\beta\rho' \sin\theta \cos(\phi-\phi')} \rho' \, d\rho' \, d\phi'$$
$$- \frac{j\beta e^{-j\beta r}}{4\pi r} \sum_{m=1}^{M} \sum_{n=1}^{N} \int_{\rho_{1m}}^{\rho_{2m}} \int_{\phi_{1m}}^{\phi_{2m}} H_{0x} \sin 2\phi' e^{j\beta \sin\theta \cos(\phi-\phi')} \rho' \, d\rho' \, d\phi' \quad (5.32)$$

where, ρ' varies from a/Mm to $a/M(m+1)$, ϕ' varies from $2\pi/N$ to $2\pi/N(n+1)\rho_m = \rho_{1m} + \rho_{2m}/2$.

The result is that the expansion coefficient $A(\theta_i, \theta_n)$ and the field $B(\rho_m, \theta_m)$ due to GB are transforms of the following expressions:

$$A(\theta_1, \theta_n) = \sqrt{\frac{\rho_m}{jb}} e^{jk\rho_m} \frac{1}{\log 4} U_{PO}(\rho_m, \theta_m) \quad (5.33)$$

$$B(\rho_m, \theta_m) = \sqrt{\frac{jb}{\rho_m}} e^{-jk\rho_m} e^{(-1/2kb)(\theta_m - n\Delta\theta)^2} \quad (5.34)$$

Ray Tracing. Given each ray at the time of launch from the subaperture, the ray should be traced from the open end to the termination along the whole length of the cavity, reflected from the flat termination at $z = -2L$, then retraced back to the open end. Whenever there is a reflection, a reflection coefficient and a divergence factor must be associated. For the reflection we might assume the GBs are sufficiently narrow to consider the inside surface of the cylinder as locally flat, so the plane surface Fresnel reflection coefficients can be used for the two polarizations. This assumption can lead to error particularly when the reflecting surface is curved. The divergence factor after each reflection at the curved surface is a function of the principal radii of curvature of the incident wavefront and the reflecting surface, though the divergence factors are independent of frequency and hence are not a function of time.

The TD field of a ray at any instant is given by

$$u(s) = A(f) \cdot R \cdot \text{phase} \quad (5.35)$$

where

$$A(f) = -j(\hat{\theta}A_1 + \hat{\phi}A_2) \left\{ \frac{\exp(j\omega_2 f)}{f_2} - \frac{\exp(j\omega_1 f)}{f_1} \right\} \quad (5.36a)$$

with a frequency domain expression given by

$$A(f) = \begin{cases} A & \text{for } f_1 < f \le f_2 \\ 0 & \text{otherwise} \end{cases} \quad (5.36b)$$

where $A(f)$ is the reference amplitude, which could be a specified waveform in FD or, might be the sampled numerical values of a numerically specified waveform. R is the reflection coefficient, equal to ± 1 for parallel/perpendicular polarization. Phase is the phase factor associated with the process.

The divergence factor is given by

$$\Gamma = \sqrt{\frac{\rho_1^r \rho_2^r}{(\rho_1^r + s)(\rho_2^r + s)}} \qquad (5.36c)$$

where

$$\frac{1}{\rho_1^r} = \frac{1}{2}\left(\frac{1}{\rho_1^i} + \frac{1}{\rho_2^i}\right) + \frac{1}{f_1} \qquad (5.36d)$$

$$\frac{1}{\rho_2^r} = \frac{1}{2}\left(\frac{1}{\rho_1^i} + \frac{1}{\rho_2^i}\right) + \frac{1}{f_2} \qquad (5.36e)$$

where s = is the distance the ray travels from the reference ($x = 0, y = 0, z = 0$), the origin of the coordinate system
ρ_1^r, ρ_2^r = principal radii of curvature of the reflected wavefront
ρ_1^i, ρ_2^i = principal radii of curvature of the incident wavefront
f_1, f_2 = the focal lengths

It can be proved that in this case

$$f_1 = \frac{a \cos \theta_n}{2 \sin^2 \theta_n} \quad \text{and} \quad f_2 = \infty \qquad (5.36f)$$

where θ_n is the angle the GB makes with the z-axis of the cylinder.

It turns out that [22]

$$\rho_1^r = \frac{2a \sec \theta_n}{1 + 4 \sin^2 \theta_n} \qquad (5.36g)$$

$$\rho_2^r = 2a \sec \theta_n \qquad (5.36h)$$

where a is the radius of the cylinder. Γ can be found using eqn. (5.36f,g).

Once the launching angle of a particular beam is known, it is relatively simple in the case of the cavity with circular cross-section and plane termination to trace the ray until it is back at the aperture. This does not require a general algorithm to search for reflection points. Whenever there is a reflection, a reflection coefficient of ± 1 is used, depending upon the polarization.

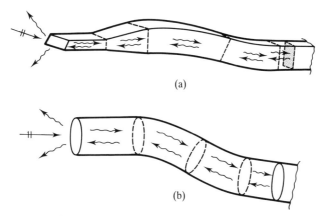

FIGURE 5.14 Open-ended cavities made up of piecewise uniform waveguide sections: (a) varying rectangular cross section; (b) varying circular cross section.

Hybrid Techniques. Regular cavities are amenable to modal solutions since the EM modes are in exact forms. In other cases, a hybrid method conbining the modal and ray techniques may be used. In this approach [21], the open-ended cavity (Fig. 5.14) is considered as a series combination of waveguide sections where adjacent sections are almost the same. An analysis of the first piece can be extended [26, 27] to include other pieces, leading to the complete structure.

The total field scattered from the cavity is a vector combination of the field scattered by the rim of the aperture, the field scattered from the interior of the cavity, and the field scattered by the external constituents of the cavity. The total field $\mathbf{E}^{t,s}$ is given by

$$\mathbf{E}^{t,s} = \mathbf{E}_R^s + \mathbf{E}_I^s + \mathbf{E}_{ext}^s \tag{5.37}$$

The incident field may be of different waveforms, like a step function, narrow pulse in time domain or, a pulse in the frequency domain.

The scattered fields are given by

$$\mathbf{E}_R^s \sim (A_\theta \hat{\theta} + A_\phi \hat{\phi}) \frac{e^{-jkr}}{r} \tag{5.38a}$$

$$\mathbf{E}_I^s \sim (B_\theta \hat{\theta} + B_\phi \hat{\phi}) \frac{e^{-jkr}}{r} \tag{5.38b}$$

The electromagnetic field in the modes coupled to the cavity propagates through the cavity and gets reflected from the termination of the cavity.

In matrix notation

$$\begin{bmatrix} A_\theta \\ A_\phi \end{bmatrix} = [S_{11}] \begin{bmatrix} A_\theta^i \\ A_\phi^i \end{bmatrix} \cdot \frac{e^{-jkr}}{r} \qquad (5.39)$$

$$\begin{bmatrix} B_\theta \\ B_\phi \end{bmatrix} = [S_{12}][C_n^+] \frac{e^{-jkr}}{r} \qquad (5.40)$$

$$[C_n^-] = [S_{22}](A)_\theta^i A_\phi^i \qquad (5.41)$$

$$[C_n^+] = [S_\Gamma][C_n^-] \qquad (5.42)$$

With considerable mathematical exercise [21] the expression for $(\mathbf{E}_R^s + \mathbf{E}_I^s)$ is given by

$$\left(\begin{bmatrix} \hat{\theta} \cdot (\mathbf{E}_R^s + \mathbf{E}_I^s) \\ \hat{\phi} \cdot (\mathbf{E}_R^s + \mathbf{E}_I^s) \end{bmatrix} \right) = \left\{ [S_{11}] + [S_{12}][P][S_\Gamma][P] \cdot [[I] - [S_{22}][P]][S_\Gamma][P] \right\}^{-1}$$

$$\times [S_{21}] \left[\begin{pmatrix} \hat{\theta}^i \cdot E^i \\ \hat{\phi}^i \cdot E^i \end{pmatrix} \right] \frac{e^{-jkr}}{r} \qquad (5.43)$$

This approach can be extended to multiple junctions.

5.5.3 Results and Discussion on Scattering from Open-Ended Cavities

In this section, we present results on scattering from an open-ended cavity obtained by different techniques. Numerical results from modal and SBR approach have been presented in [25]. In modal analysis only the propagating modes were considered to avoid underflow in the computation. This assumption is justified as long as the depth of the cavity is at least nine wavelengths, for shorter cavities the evanescent modes do not attenuate enough. For SBR calculations, the linear density was 5 rays per wavelength. Figure 5.15 gives RCS versus θ plots for an open cavity of length 30λ with a $10\lambda \times 10\lambda$ square cross section in the xz-plane. The agreement between the two methods is very good up to $40°$ from normal incidence including only the interior scattering mechanisms and without backscattering from external features of the cavity. Figure 5.16 shows the RCS of an open rectangular cavity with dimensions $a = b = 10\lambda$; $L = 30\lambda$ and a plane PEC termination with $\phi = 45°$. Figure 5.17 presents the RCS of an open rectangular cavity with dimensions $a = b = 10\lambda$, $L = 30\lambda$ and terminated by a PEC termination coated with RAM. The coating layer has parameters $\varepsilon_r = 2.5 - j1.25$, $\mu_r = 1.6 - j0.8$ and thickness $\tau = 0.15\lambda$ $\phi = 0°$. Figure 5.18 shows the RCS of an open circular cylinder with dimensions $2a = 10\lambda$, $L = 30\lambda$, with a PEC termination (a) $\phi - \phi$ polarization (b) $\theta - \theta$

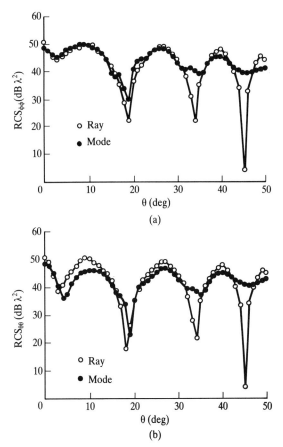

FIGURE 5.15 RCS of an open rectangular cavity with dimensions $a = b = 10\lambda$, $L = 30\lambda$ and terminated by a PEC termination, $\phi = 0°$ (xz-plane): (a) $\phi\phi$ polarization (b) $\theta\theta$ polarization [25].

polarization. The RCS characteristics of an offset rectangular cylinder using SBR methods are shown in Figure 5.19. Figure 5.19(a) shows the structure, Figure 5.19(b) shows the RCS variations for $\phi = 0°$ (xz-plane), and Figure 5.19(c) shows the RCS variations for $\phi = 90°/180°$ (yz-plane). It may be noted that the size of the ray remains the same in a rectangular cavity it changes in a circular cylindrical cavity because the beam diverges.

A discussion on the relative merits and demerits of the different approaches is in order. Referring to Figure 5.19(b) and (c) the agreement between the SBR and modal results are good through the SBR results are not so smooth, as the ray is incident at large angles. It is observed that a significant amount of RCS reduction occurs with the rounding of the end plates, as shown in Figure 5.20.

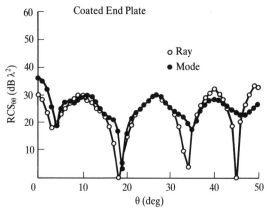

FIGURE 5.16 RCS of an open rectangular cavity with dimensions $a = b = 10\lambda$, $L = 30\lambda$ and terminated by a PEC termination, $\phi = 45°$ [25].

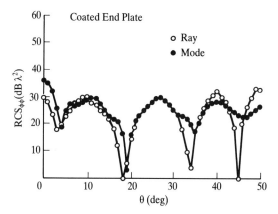

FIGURE 5.17 RCS of an open rectangular cavity with dimensions $a = b = 10\lambda$, $L = 30\lambda$ and terminated by a PEC termination coated with a RAM. The coating layer has parameters $\varepsilon_r = 2.5 - j1.25$, $\mu_r = 1.6 - j0.8$, and thickness $\tau = 0.15\lambda$; $\phi = 0°$ [25].

To make SBR a practical tool for RCS prediction, it is believed a linkup with a geometrical modeling package will be necessary.

The RCS of a coated circular cylindrical cavity with a PEC termination has been investigated in [24]. The rim diffraction contributions and interior contributions were formulated respectively by GTD and PO approximations. The theory has been detailed in [24] and we choose not to repeat here and discuss important results. The experimental results of RCS measurements of hollow pipes, cylinders, and cylindrical cavities were reported by Brooks and Crispin Jr. [28]. Figures 5.21 through 5.24 compare the theoretical results of RCS versus θ of a PEC terminated waveguide and of an open-ended waveguide with the experiment in [28]. The conclusions made in [24] are: (1) the

5.5 RCS OF OPEN-ENDED CAVITIES 325

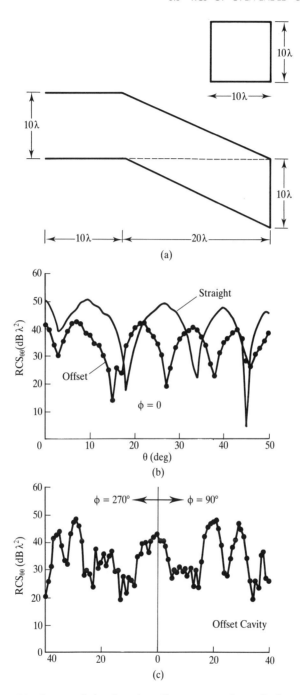

FIGURE 5.18 RCS characteristics for the offset rectangular cylinder using the SBR method: (a) offset rectangular cylinder; (b) $\phi = 0°$ (xz-plane); (c) $\phi = 90°/270°$ (yz-plane) [25].

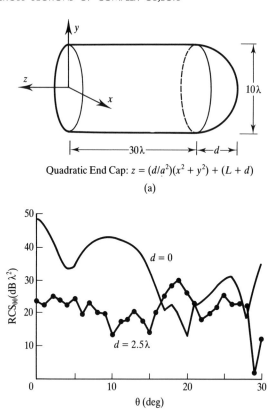

FIGURE 5.19 RCS results using the SBR approach with and without rounding of end plate [25]: (a) open cylinder with rounded plates; (b) open cylinder without rounding of plates.

contributions to the RCS from the rim become more and more insignificant as a/λ increases, improving agreement between experiment and a theory which does not consider the rim diffraction effects; (2) the interior contributions are much more than the rim contributions to the RCS, particularly at high frequencies (more than 10 dB for a/λ). Figure 5.25 RCS of a PEC terminated and open-ended circular guides at axial incidence as a function of a/λ, the broken line indicates the RCS of a circular plate. Figure 5.26 shows the RCS of a PEC terminated circular waveguide coated with a lossy material (Crowloy BX 113, $\varepsilon_r = 12 - j0.144$, $\mu_r = 1.74 - j3.306$) with a coating thickness $\tau = 0.025$ cm ($a = 3.95$ cm, $f = 9.2$ GHz, $a/\lambda = 1.2$, $L = 26.46$ cm, $L_1 = 1$ cm, VV polarization). Figures 5.29 and 5.30 show the low-frequency RCS versus incident angle θ with the experimental and theoretical results (with and without rim diffraction) for VV and HH polarizations. For more detailed discussion on

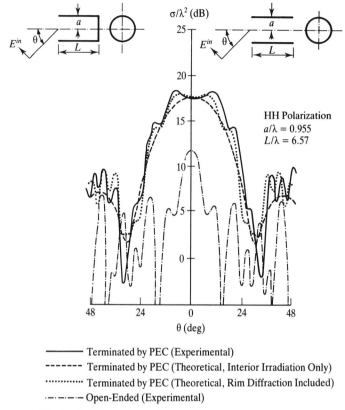

- —— Terminated by PEC (Experimental)
- ------ Terminated by PEC (Theoretical, Interior Irradiation Only)
- ············ Terminated by PEC (Theoretical, Rim Diffraction Included)
- —·—·— Open-Ended (Experimental)

FIGURE 5.20 RCS of a PEC terminated waveguide and an open-ended waveguide as a function of the incident angle (HH polarization, $a = 3.137$ cm, $f = 9.13$ GHz, $a/\lambda = 0.955$, length $= 21.59$ cm).

the characteristics of RCS for the RAM-coated circular cylindrical cavity, the reader is referred to [24, 25]. The mode conversion problems have not been discussed in the research reported in [24, 25] hence the theory does not correctly predict high-frequency RCS reduction with the coating.

Apart from the open-ended cavity with and without RAM coating with flat PEC termination, more complicated structures with real terminations and real geometries are of great interest. The hybrid method outlined above has been used in [22] to calculate the RCS of an open-ended cavity with terminations other than flat. Figures 5.27 and 5.28 respectively show the comparison of RCS measured and theoretically calculated using the hybrid method versus aspect angle and frequency. Figure 5.29 shows the theoretical backscatter patterns of a 2D S-shaped waveguide cavity with a perfectly conducting plane termination and absorber-lined inner walls at 35 GHz for perpendicular polarization and

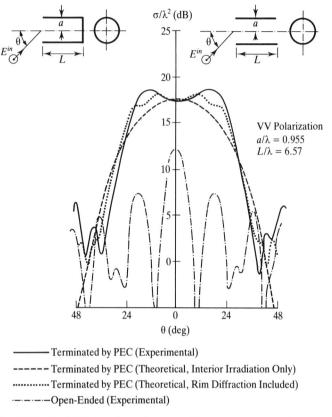

—————— Terminated by PEC (Experimental)
- - - - - - Terminated by PEC (Theoretical, Interior Irradiation Only)
············· Terminated by PEC (Theoretical, Rim Diffraction Included)
—·—·—· Open-Ended (Experimental)

FIGURE 5.21 RCS of a PEC terminated waveguide and an open-ended waveguide as a function of the incident angle (VV polarization, $a = 3.137$ cm, $f = 9.13$ GHz, $a/\lambda = 0.955$, length $= 21.59$ cm).

(a) using hybrid asymptotic/modal methods or (b) using the GO/AI method. We consider three cases: without absorber lining, with 1 dB/reflection, and with 5 dB/reflection. Figure 5.30 shows the calculated backscatter patterns of an open-ended parallel-plate waveguide with planar short-circuit termination. The fields generated inside the structures are of great interest since they determine the hot and cold spots inside a given structure for a given incident field, particularly for high-power microwave (HPM). Figure 5.31 plots the interior field [22] in dB at a cross section of an open-ended semi-infinite parallel-plate waveguide due to an incident plane wave for perpendicular and parallel polarizations. Figure 5.32 shows the backscattering from a 2D open-ended semi-infinite S-shaped waveguide with an incident plane wave using hybrid modal and paraxial GB techniques with 25 beams/subaperture and beam waist $b = 77.5\lambda$.

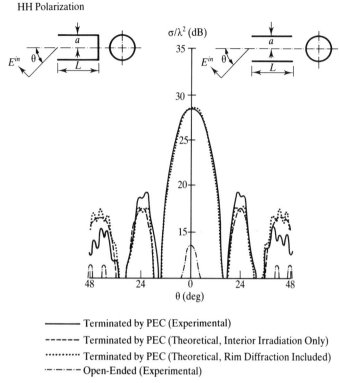

———— Terminated by PEC (Experimental)
------ Terminated by PEC (Theoretical, Interior Irradiation Only)
·········· Terminated by PEC (Theoretical, Rim Diffraction Included)
—·—·— Open-Ended (Experimental)

FIGURE 5.22 RCS of a PEC terminated waveguide and an open-ended waveguide as a function of the incident angle (HH polarization, $a = 3.137$ cm; $f = 15.20$ GHz, $a/\lambda = 1.589$, length = 21.59 cm).

5.6 BISTATIC RCS OF RADAR TARGETS

The radar cross section and EM scattering literature contains more information on monostatic cases than on bistatic scattering. With the availability of large high-speed computers and advanced measurement capabilities, computation and measurement of bistatic scattering has advanced over the past decade. Some of the early modeling work is available in [29–31]. The problem of bistatic scattering of chaff dipoles with applications to communications was treated in [29]. PO has been used to find approximate solutions to the bistatic scattering cross section of a finite-length PEC cylinder [30]. The bistatic scattering cross section of a randomly oriented dipole was discussed in [31] and for a finite PEC cylinder in [32]. A simplified approach to bistatic radar reflectivity computations was presented in [33] using Stokes parameters. A review paper [34] on the bistatic scattering of complex objects describes the history and recent advances in instrumentation and analytical modeling of complex objects.

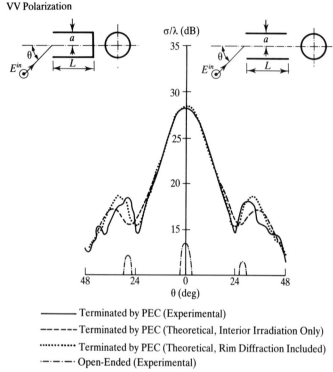

FIGURE 5.23 RCS of a PEC terminated waveguide and an open-ended waveguide as a function of the incident angle (VV polarization, $a = 3.317$ cm, $f = 15.20$ GHz, $a/\lambda = 1.589$, length = 21.59 cm).

A recent article [35] describes the application of PO for near-field monostatic and bistatic calculations.

The bistatic cross section of a dipole (Fig. 5.33) in terms of the orientation angles of the transmitter and receiver can be cast in the form of an integral equation and a closed-form expression of the cross section can be obtained with certain simplified assumptions. For a thin half-wave dipole and making a zeroth-order approximation, the cross section is given by

$$\frac{\sigma(\theta, \theta', \psi')}{\lambda^2} = 0.85 \cos^2(\psi) \cos^2(\psi') \cdot \frac{\cos^2[(\pi/2) \cos \theta] \cos^2[(\pi/2) \cos \theta']}{\sin^2 \theta \sin^2 \theta'} \quad (5.44)$$

where the angles are shown in Figure 5.33. The theoretical values [29] of average bistatic scattering cross section versus receiver elevation angle for different cases are shown in Figure 5.34. Three types of distributions are considered: (1) uniform orientation, which means all orientations are assumed to be equally likely; (2) horizontal distribution, where all orientations in a

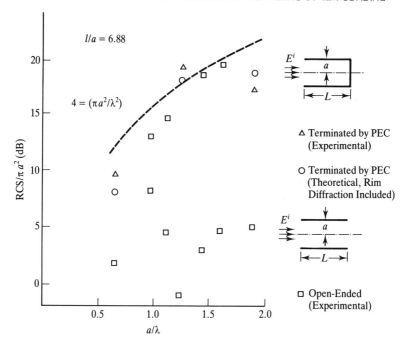

FIGURE 5.24 RCS of PEC terminated and open-ended circular guides at axial incidence as a function of a/λ; (—) approximate RCS of a circular plate. (From [26] © 1989 IEEE.)

horizontal plane are assumed to be equally likely; and (3) vertical distribution, which means all dipoles are assumed to fall vertically. Figure 5.35 compares theoretical and measured average bistatic cross sections. The bistatic cross sections for a PEC finite length cylinder have been computed [30] using a physical optics integral. The bistatic radar cross section of a PEC cylinder 23.2λ long and 2.71λ in. in diameter is shown in Figure 5.36. Figure 5.37 shows the bistatic cross section of a PEC Starship Enterprise in the elevation plane calculated by SBR using a BRL-CAD package developed at the Army Ballistic Research Laboratory (Army BRL). The incident field is from the direction $EL = 45°$, and $AZ = 45°$ (which is exactly the viewing angle of the top picture). The rays for this example have been launched using BRL-CAD at a density of 10 rays/λ. The results of the bistatic far-zone scattering by a square cylinder (each side 3λ) with $\beta_0 = 60°$, $\phi_0 = 45°$ are shown in Figure 5.38.

Uniform and nonuniform PO currents have been applied to the illumination of a smooth surface and to bistatic specular reflection [16].

5.7 CONTROL OF SCATTERING BY RIM LOADING

The modification of the radiation pattern of antennas and the scattering characteristics of an object by uniform and nonuniform loading using radar-

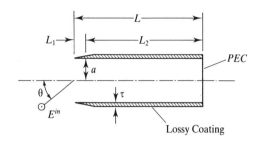

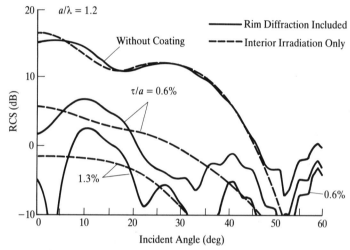

FIGURE 5.25 RCS of a PEC terminated circular waveguide coated with a lossy material (Crowloy BX 113, coating thickness $\tau = 0.025$ cm (0.6 percent coating), 0.05 cm (1.3ε_r = 12.0 − j0.144, μ_r = 1.74 − j3.306) as a function of the incident angle (a = 3.95 cm, f = 9.2 GHz, a/λ = 1.2, L = 26.46 cm, L_1 = 1 cm, VV polarization).

absorbing materials RAMs has been of interest to radar engineers for quite some time. Bucci and Fransceschetti [36] and Bucci, Di Massa and Savarese [37] analyzed the effect of rim-loading reflector antennas using the diffraction coefficient due to Maliuzhinets [38]. Their results lead to the conclusions that an effective control of radiation characteristics can be achieved by rim loading. Yokoi and Fukumuro [39] studied the effect of using absorbers on the sidelobes of paraboloidal reflectors. Hurst and Mittra [40, 41] discussed the effect of resistive edge loading and corner rounding on the scattering center description of targets. Their conclusion is that resistive edge loading, when used properly, can have a dramatic effect on the magnitude of the field scattered from an edge and can lead to a significant reduction in RCS. Reduction of the RCS of a flat plate with resistive and RAM loading has been described [42–45]. Since diffraction is a local phenomenon, in line with the scattering center approach, a given target may not need to be fully coated. It should be enough to coat

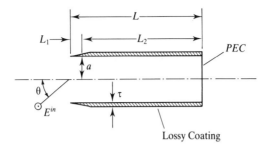

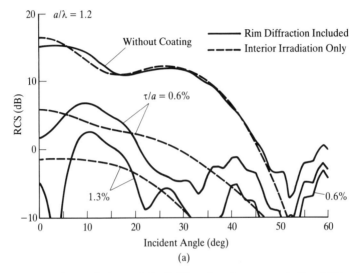

FIGURE 5.26 Experimental data of the RCS at low frequency ($a/\lambda = 0.98$) of a PEC terminated waveguide coated with a lossy magnetic material ($\varepsilon_r = 24 - j1.7$, $\mu_r = 2.9 - j2.3$), compared with theoretical values: $a = 10.16$ cm, $f = 2.9$ GHz, $L = 63.5$ cm, $L_1 = 1.27$ cm, $\tau/a = 1.18$ (a) VV (b) HH. (*continued on next page*)

around the scattering centers. For a flat plate, the next question is What should be the extent of the coating around its edges and what is the minimum width to achieve an RCS reduction about the same as for fully flat plate. An answer to this question may lead to considerable RAM saving in quantity of material and in the cost of the coating process. This problem was addressed in [46]; an attempt was made to estimate the width of the loading layer for a reduction in scattered field equal to the fully loaded plate, and a percent reduction in the cost of the material has been predicted.

The geometry and coordinate system of the rim-loaded plate are shown in Figure 5.39. The dimensions of the flat plate, which has coating around its perimeter, are ($2A \times 2B$) meters. In the yz-plane, the dominant scattering centers are A, A_1, B, and B_1. The width of the coatings on the edges A and B

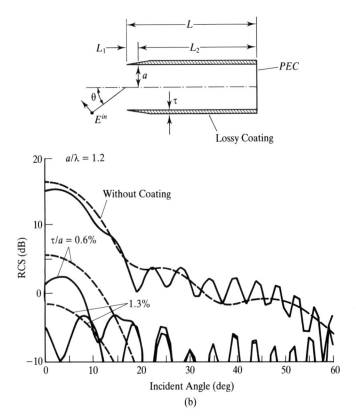

FIGURE 5.26 (*continued*).

and $AA_1 = d_1$ and $BB_1 = d_2$. The diffracted field due to scattering by a wedge of exterior wedge angle $n\pi$ and with two face impedances is

$$u^d \sim u^i D(\phi, \phi, \theta, \theta, k\rho) \frac{\exp(-jk\rho)}{\sqrt{\rho}} \exp(-j\phi) \exp(j\omega t) \quad (5.44)$$

where u^i is the incident plane wave field at the edge and ϕ_N is the phase of the Nth scattering center. The diffraction coefficient $D(\cdot)$ is expressed as

$$\begin{aligned}
D(\phi, \phi, \theta, k\rho) = {} & M(\phi, \phi, \theta, \theta) \cdot C_n(\phi, \phi_0, \theta_0, \theta - n) \\
& \times [D^-(\beta^-, k\rho) + \Gamma_N(\theta, \theta_0) D^-(\beta^+, k\rho)] \\
& + C_n(n\pi - \phi, n\pi - \phi_0, \theta_0, \theta_n) \\
& \cdot [D^+(\beta^-, k\rho) + \Gamma_n(\phi_0, \phi_n) \cdot D^+(\beta^+, k\rho)]
\end{aligned} \quad (5.45)$$

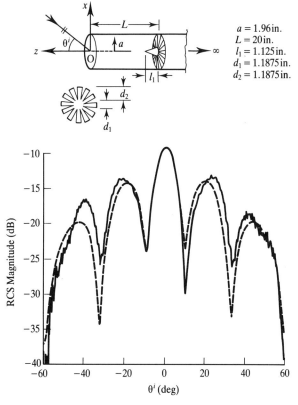

FIGURE 5.27 Calculated and measured RCS patterns of an open-ended circular waveguide with a hub and blade termination at 10 GHz for vertical polarization: (solid line) measured; (dashed line) theoretical RCS using the hybrid asymptotic and modal expansion method.

where

$$M_n(\phi, \phi_0, \theta_0, \theta_n) = \exp[(M_n(\phi, \phi_0, \theta_0) + M_n(n\pi - \phi, n\pi - \phi_0, \theta_n)] \quad (5.46)$$

and

$$M_n(\alpha, |\alpha_0, \theta) = \int_0^{\pi + (\pi - \alpha')} [f_n(t + n\pi - \alpha - \pi/2 + \theta) - f_n(t + n\pi - \alpha - \pi/2 - \theta)] \, dt \quad (5.47)$$

with

$$f_1(X) = \frac{(2X - \pi \sin X)}{4\pi \cos X} \quad (5.48a)$$

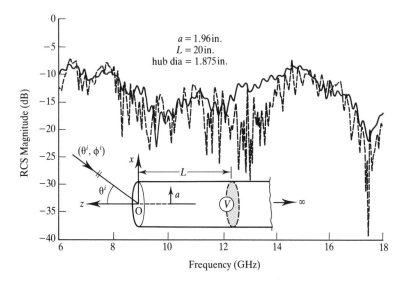

FIGURE 5.28 Calculated and measured RCS versus frequency of an open-ended circular waveguide cavity with a disk and hemispherical hub termination, horizontal polarization, angle of incidence $= 15°$: (dashed line) measured RCS; (solid line) theoretical RCS using the hybrid asymptotic and modal expansion method.

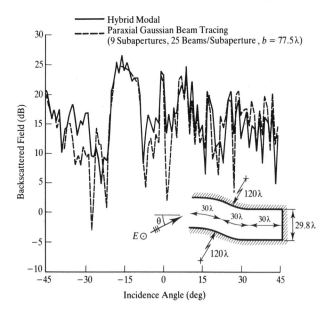

FIGURE 5.29 Backscattering from a two-dimensional, open-ended, semi-infinite, S-shaped waveguide due to an incident plane wave.

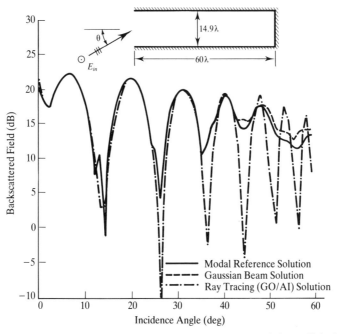

FIGURE 5.30 Theoretical backscattering patterns of an open-ended parallel-plate waveguide cavity with a planar short-circuit termination using three methods.

$$f_2(X) = -\frac{(\pi \sin X - 2\sqrt{2\pi} \sin(X/2) + 2X)}{8\pi \cos X} \quad (5.48b)$$

$$\beta^{\mp} = \phi \pm \phi_0$$

θ and θ_n are the Brewsters angles for faces 0 and n and are obtained from

$$\sin \theta^e_{0,n} = \eta_0/Z_{0,n} \quad \text{for TM polarization} \quad (5.49a)$$

$$\sin \theta^m_{0,n} = Z_{0,n}/\eta_0 \quad \text{for TE polarization} \quad (5.49b)$$

$Z_{0,n}$ are the surface impedances of the 0 and N faces, η_0 is the intrinsic impedance of free space.

The expressions for $c_n(\alpha, A_\theta, \theta - 1, \theta_2)$, $s_n(x, x_1, x_2)$, $\Gamma_n(\phi_0, \theta)$, and $D^\pm(\beta, K)$ are given in [45].

The evaluation of the function $M_n(\phi, \phi_0, \theta_0, \theta_n)$ is described in Appendix D.

The round-trip total phase factors associated with four different scattering centers are given by

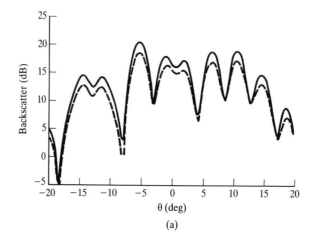
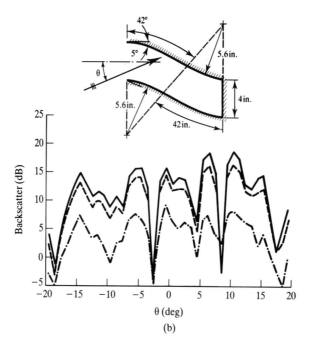

FIGURE 5.31 Calculated backscatter patterns of a 2D S-shaped waveguide cavity with a perfectly conducting plane termination and absorber-lined inner walls at 35 GHz, perpendicular polarization: (a) calculations using the hybrid asymptotic/modal method; (b) calculations using the GO/AI method [22]. (———) no absorber lining; (- - -) 1 dB loss/reflection absorber lining; (—·—·) 5 dB loss/reflection absorber lining.

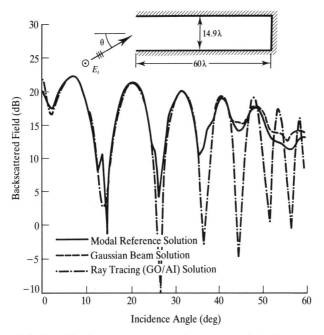

FIGURE 5.32 Calculated backscatter patterns of an open-ended plate waveguide cavity with a planar short-circuit termination [22].

$$\text{At A,} \quad (\rho - A(\cos\phi + \cos\phi - 0))$$
$$\text{At A}_1, \quad (\rho - (A - d_1)(\cos\phi + \cos\phi_0))$$
$$\text{At B,} \quad (\rho + A(\cos\phi + \cos\phi_0))$$
$$\text{At B}_1, \quad (\rho + (A - d_2)(\cos\phi + \cos\phi_0))$$

The coordinate system reference is the point O, d_1 and d_2 are the widths of the coating around edges A and B. The total scattered field at an observation point is the superposition of scattered fields from all the dominant scattering centers seen by the receiver. The 2D field is converted to a 3D field using Siegel's formula $E_{3D} = L \exp(-j\pi/4)(E_{2D}/\sqrt{\lambda\rho})$ where L is the length of the edge. Some results have been presented in [46]. Figure 5.40 shows the variation of the maximum value of the scattered field for TE polarization in the broadside direction versus different values of $d_1 = d_2 = d$ in wavelengths with $Z_c = (200.0, 0.0)$ ohms and $Z_c = (496.8, 242.5)$ ohms; $Z_p = (0.0, 0.0)$ ohms where Z_c is the surface impedance of the shaded portion in Figure 5.39. The maximum field oscillates in a triangular fashion about a mean value equal to the maximum field generated when the illuminating surface has a uniform surface impedance. This happens for all values of lit surface impedance. The coating width recommended for $Z_c = (200.0, 0.0)$ ohms is 0.37λ as shown by the dotted line.

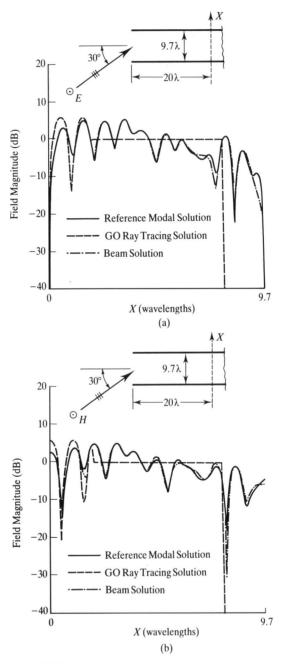

FIGURE 5.33 Interior fields at a cross section of an open-ended semi-infinite parallel plate waveguide due to an incident plane wave: (a) perpendicular polarization; (b) parallel polarization [22].

5.7 CONTROL OF SCATTERING BY RIM LOADING

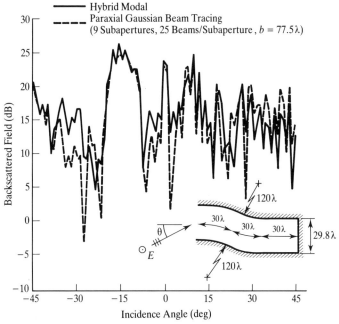

FIGURE 5.34 Backscattering from a two-dimensional open-ended, semi-infinite s-shaped waveguide due to an incident plane wave [22].

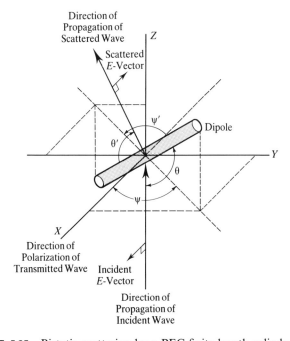

FIGURE 5.35 Bistatic scattering by a PEC finite length cylinder [32].

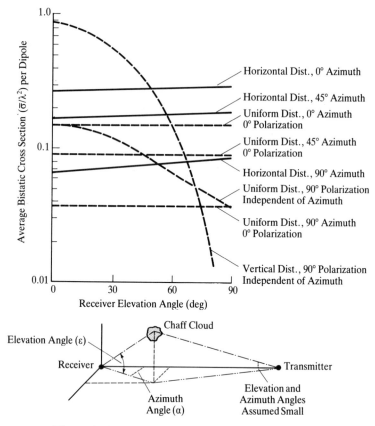

FIGURE 5.36 Theoretical values of average bistatic scattering cross section [32].

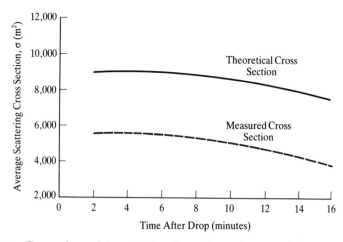

FIGURE 5.37 Comparison of theoretical and experimental average bistatic cross section [32].

5.7 CONTROL OF SCATTERING BY RIM LOADING

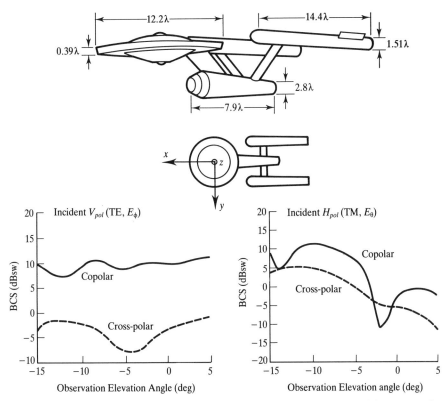

FIGURE 5.38 Bistatic RCS in the elevation plane of a PEC Starship Enterprise calculated by SBR. The incident field is from the direction EL = 5°, and AZ = 45°. Rays are launched with the aid of BRL-CAD at a density of 10 rays/λ [35].

Figure 5.40 compares the scattered field patterns for three cases: (1) fully coated plate, graph raised by 5 dB; (2) rim-loaded plate with $d = \lambda/4$; and (3) rim-loaded plate with $d = \lambda/2$, graph lowered by 5 dB. Figure 5.41 compares field patterns for a fully coated plate and $d = 2.5\lambda$. There is little change in the shape of the pattern with changes in the value of d. There is of course a change in the maximum scattered field. Figure 5.42 shows the relative contributions due to edges A and A_1, B and B_1 to the scattered field. It is found from Figure 5.42 that contributions from A_1 and B_1 to the total field are much less than from edges A and B. This causes hardly any difference in the shape of the patterns. Figure 5.43 shows a comparison of RCS patterns with two values of Z_c equal to 200 ohms and (496.7, 242.5) ohms. With epoxy resin, the RCS is reduced by 10 dB and there is practically no change in the shape of the patterns with d for a given Z_c and Z_p. Variation of RCS about the mean is ± 0.275 dB; the mean value is the RCS of the fully covered plate with the same value of Z_c and the variation is not large. Figure 5.44 shows a comparison of the relative

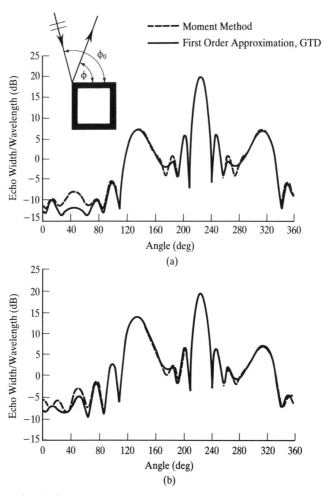

FIGURE 5.39 Bistatic far-zone scattering echo width per wavelength by a square cylinder (each side 3λ) with $\beta_0 = 60°$, $\phi_0 = 45°$: (a) $\eta = 0.5$ for all sides; (b) top and bottom faces perfectly conducting and side faces having $\eta = 0.5$.

strength of contribution from edges A and B combined and edges A_1 and B_1 individually. An estimation of RAM saving is in order. The minimum value d for which the maximum value of field corresponding to the fully coated plate is found to be 0.37λ. The percentage of RAM saving is given by

$$\frac{(2A \cdot 2B - 0.37\lambda \cdot 2B)}{(2A \cdot 2B)} \times 100$$

With $2A = 2B = 6$ in.; $\lambda = 3.061$ cm ($f = 9.8$ GHz), the RAM saving by rim loading is 78.386%; the saving is frequency sensitive. For a wider bandwidth,

5.8 REAL-TIME PREDICTION OF RCS OF COMPLEX OBJECTS

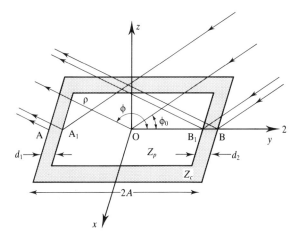

FIGURE 5.40 Geometry and coordinate system of the rim-loaded plate. (From [47] © 1989 IEEE.)

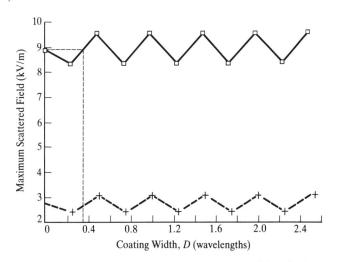

FIGURE 5.41 Backscattered field of the plate versus width of the coating; $f = 9.8$ GHz; $d_1 = d_2 = d$; $Z_0 = (0.0, 0.0)$ ohms; $(\square) - jZ_c = (200.0, 0.0)$ Ohms; $(+) - jZ_c = (496.8, 242.5)$ ohms.

RAM layers of unequal thicknesses may be of help but this reduces the material saving to some extent.

5.8 REAL-TIME PREDICTION OF RCS OF COMPLEX OBJECTS USING HIGH-FREQUENCY TECHNIQUES

This section briefly describes a new approach in a forthcoming paper [48] to compute the high-frequency radar cross section (RCS) of complex radar targets

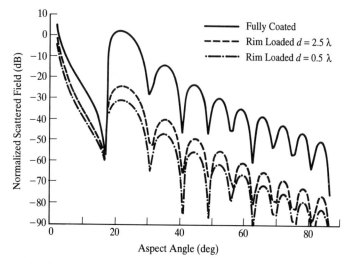

FIGURE 5.42 Backscattered field versus aspect angle, $Z_c = 50$ ohms, $f = 9.8$ GHz, backscatter.

using PO, MEC, PTD, and IBC in real-time with a 3D graphics workstation. The aircraft is modeled using a parametric software approach with I-DEAS solid modeling software. This approach is also called graphic electromagnetic computing (GRECO). The visible surfaces of the radar target are indentifed by graphical processing of an image of the target on the workstation screen. Through knowledge of the unit normal on the illuminated surfaces of a target the high-frequency approximation to the RCS prediction is accomplished. The image is obtained in real time from the I-DEAS geometric model using the 3D graphics hardware accelerator. The time-consuming geometric modeling is done by the graphics package and the CPU time is used to do the electromagnetic part.

5.8.1 Graphical Electromagnetic Computing (GRECO)

GRECO finds RCS using the following steps: (1) geometric modeling of the aircraft with a CAD package; (2) real-time imaging of the target on the workstation using the graphics hardware accelerator; (3) graphical processing to obtain the coordinates (x, y, z, n_x, n_y, n_z) of each point of the illumination; (4) computing of reflection from the PEC surfaces by GO and PO, reflection at the coated surfaces by PO and IBC approximations, diffraction at the edges by MEC and PTD ILDC.

In this analysis, the computation of a unit normal to the surface was done using the Phong local illumination model [48], the color of each pixel depends only on the normal to the surface element associated with this pixel and the location of the observer and light sources. Details of the analysis are given in

5.8 REAL-TIME PREDICTION OF RCS OF COMPLEX OBJECTS

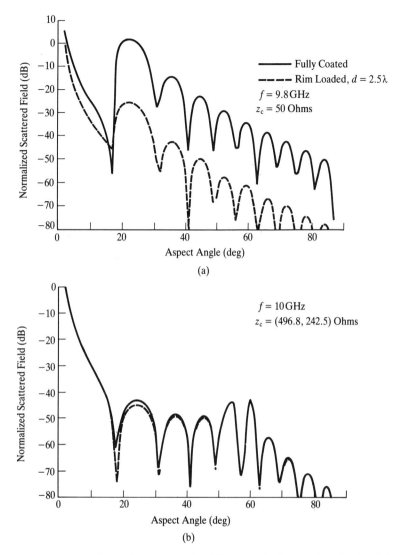

FIGURE 5.43 Comparison of patterns of a fully coated plate and a rim-loaded plate; $d = 2.5\lambda$; plate size = (6 in. × 6 in.); $f = 9.8$ GHz; $Z_p = 0.0$ ohms: (a) $Z_c = 50$ ohms and (b) $Z_c = (496.8, 242.5)$.

348 RADAR CROSS SECTIONS OF COMPLEX OBJECTS

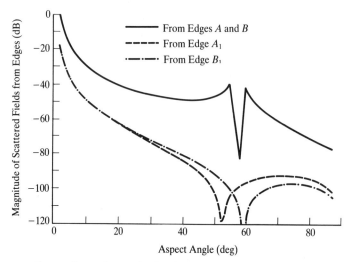

FIGURE 5.44 Comparison of magnitude of scattered field from individual contributions from A, A_1, B, B_1; $f = 9.8$ GHz; $d_1 = d_2 = d = \lambda/4$; $Z_c = (496.8, 242.5)$ ohms. (From [47] © 1989 IEEE.)

[49–51]. Figure 5.45 shows the 2D airfoil section and its RCW plots. Some details of the computer program are presented in Chapter 10.

5.9 SUMMARY

This chapter dealt with the RCS of selected complex targets. It covered rectangular plates at grazing incidence, polygonal cylindrical plates and open-ended cavities, bistatic scattering of radar targets, the role of rim loading and rim-loaded flat plates. The chapter ended by discussing real-time RCS prediction for complex targets using high-frequency techniques.

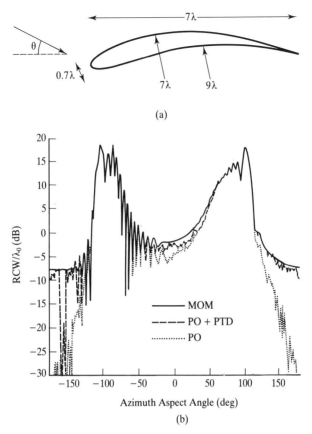

FIGURE 5.45 (a) The two-dimensional airfoil section. Radar cross section width (RCW) normalized with respect to wavelength λ for (b) TM and (c) TE polarization. Results for a perfectly conducting airfoil surface are compared to a rigorous solution presented by Aerospatiale [49]. RCW normalized by λ for (d) TM and (e) TE polarizations. Results for a coated surface are compared to a rigorous solution presented by Aerospatiale [49]. The coating has a relative dielectric permittivity $\varepsilon_r = 7.4 - j1.11$, relative magnetic permeability $\mu_r = 1.4 - j0.672$ and thickness $d = 0.06\lambda_0$, (the wavelength in free space). (*continued on next page*)

350 RADAR CROSS SECTIONS OF COMPLEX OBJECTS

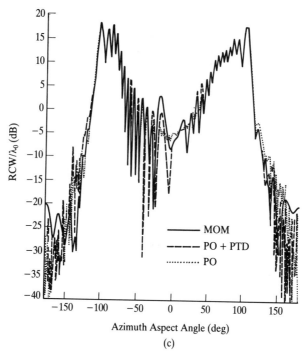

(c)

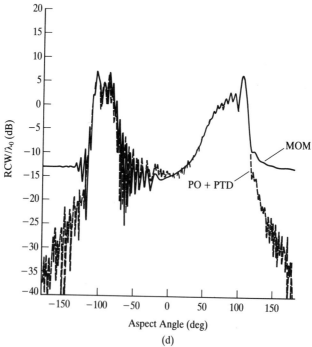

(d)

FIGURE 5.45 (*continued*).

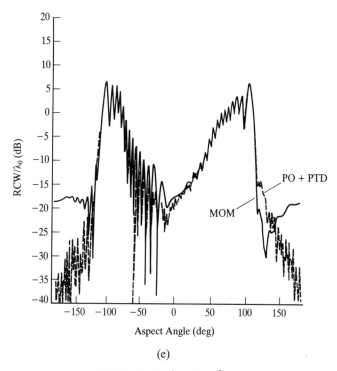

(e)

FIGURE 5.45 (*continued*).

5.10 REFERENCES

1. G. T. Ruck (ed.), *Radar Cross Section Handbook, Vol.* 2, Plenum Press, New York, 1970, Ch. 8.
2. A. K. Bhattacharyya and D. L. Sengupta, *Radar Cross Section Analysis and Control*, Artech House, Norwood, MA, 1991, Section 2.10.3.
3. W. R. Stone (ed.), Special issue on radar cross section of complex objects, *IEEE Trans. Antennas and Propagat.*, **AP-37**(5), 1989, 525–676.
4. W. R. Stone (ed.), Special issue on radar cross section of complex objects, *Proc. IEEE*, **77**, 1989.
5. Section 1.20 in this book.
6. K. M. Mitzner, *Incremental Length Diffraction Coefficients*, Tech. Rep. AFAL-TR-73-296, Aircraft Division, Northrop Corporation, Apr. 1974.
7. G. Pelosi, R. Tiberio, S. Puccini, and S. Maci, "Applying GTD to Calculate the RCS of Polygonal Plates," *IEEE Trans. Antennas and Propagat.*, **AP-38**(8), 1990, 1294–1298.
8. R. A. Ross, "Radar Cross Section of Rectangular Flat Plates as a Function of Aspect Angle," *IEEE Trans. Antennas and Propagat.*, **AP-14**(3), 1966, 329–335.
9. J. S. Yu, "Radar Cross Section of a Thin Plate Near Grazing Incidence, *IEEE Trans. Antennas and Propagat.*, **AP-18**(5), 1970, 711–713.

10. E. F. Knott, V. V. Liepa, and T. B. A. Senior, "Plates and Edges," *IEEE Trans. Antennas and Propagat.*, **AP-19**(6), 1971, 788–789.

11. R. Tiberio and R. G. Kouyoumjian, "A Uniform GTD Solution for the Diffraction by Strips Illuminated at Grazing Incidence," *Radio Science*, **14**, 1979, 933–941.

12. S. W. Lee and J. Boersma, Ray-Optical Analysis of Fields on Shadow Boundaries of Two Parallel Plates," *J. Math. Phys.*, **16**, 1975, 1756–1764.

13. R. Tiberio and R. G. Kouyoumjian, "An Analysis of Diffraction at Edges Illuminated by Transition Region Field," *Radio Science*, **17**, 1982, 323–336.

14. D. A. McNamara, C. W. I. Pistorious, and J. A. G. Malherbe, *Introduction to the Uniform Geometrical Theory of Diffraction*, Artech House, Norwood, MA, 1990, Section 4.4.5.

15. R. Tiberio and R. G. Kouyoumjian, "Calculation of the High Frequency Diffraction by Two Nearly Edges Illuminated at Grazing Incidence," *IEEE Trans. Antennas and Propagat.*, **AP-32**(11), 1984, 1186–1196.

16. R. Tiberio and R. G. Kouyoumjian, "Application of the Uniform GTD to the Diffraction at the Edges Illuminated by Transition Region Fields," presented at the *USNC/URSI Annual Meeting, Stanford University, CA.*, 1977.

17. H. Chaloupka and H.-J. Meckelburg, "Improved High-Frequency Current Approximation For Curved Conducting Surfaces," *AEU*, Band 39, Heft 4, 1985.

18. M. A. Plonus, R. Williams, and S. C. H. Wang, "Radar Cross Section of Curved Plates Using Geometrical and Physical Diffraction Techniques," *IEEE Trans. Antennas and Propagat.*, **AP-26**(3), 1978, 8488–8493.

19. S. Sanyal and A. K. Bhattacharyya, "Electromagnetic Scattering by a Curved Plate—Solution by Uniform Theory of Diffraction," *IEEE Trans. Antennas and Propagat.*, **AP-32**(2), 1984, 187–189.

20. M. Ideman, "Diffraction of an Obliquely Incident High-Frequency Wave by a Cylindrically Curved Sheet," *IEEE Trans. Antennas and Propagat.*, **AP-34**(2), 1986, 181–187.

21. *High Frequency Electromagnetic Jet Engine Cavity Modeling Workshop*, August 1–2, 1992, WL/AARA and AFIT/ENG, Wright-Patterson AFB, Dayton, OH.

22. R. J. Burkeholder and P. H. Pathak, *High-Frequency Asymptotic Methods for Analyzing the EM Scattering by Open-Ended Cavities*, Ohio State Univ. Electroscience Lab., Tech. Rep. 719630-3, Sept. 1989, prepared for NASA Lewis Research Center, Cleveland, OH. (This is also the PhD thesis of Burkeholder, Ohio State Univ., Columbus, OH.)

23. P. H. Pathak and R. J. Burkeholder, "Modal, Ray, and Beam Techniques for Analyzing the EM Scattering by Open-Ended Cavities," *IEEE Trans. Antennas and Propagat.*, **AP-37**(5), 1989, 635–647.

24. C. S. Lee and S. W. Lee, "Radar Cross Section of an Open-Ended Circular Waveguide: Calculation of Second-Order Diffraction Terms," *Radio Science*, **22**(1), 1987, 2–12.

25. C. S. Lee and S. W. Lee, "RCS of a Coated Circular Waveguide Terminated by a Perfect Conductor," *IEEE Trans. Antennas and Progagat.*, **AP-38**(4), 1987, 391–398.

26. H. Ling, S. W. Lee and R.-C. Chou, "High-Frequency RCS of Open Cavities With Rectangular and Circular Cross Sections," *IEEE Trans. Antennas and Propagat.*, **AP-37**(5), 1989, 648–654.

27. A. Altintas, P. H. Pathak, and W. D. Burnside, *Electromagnetic Scattering From a Class of Open-Ended Waveguide Discontinuities*, Ohio State Univ. Electroscience Lab. Tech. Rep. 716148-9. Prepared under grant NSG-1613, for NASA/Langley Research Center, Hampton, VA, Mar. 1986.

28. R. J. Burkeholder, C. W. Chuang, and P. H. Pathak, *Electromagnetic Fields Backscattered From an S-Shaped Inlet Cavity with an Absorber Coating on its Inner Walls*, Ohio State Univ. Electroscience Lab., Final Rep. 715723-2. Prepared under grant NAG 3-476 for NASA/Lewis Research Center, Cleveland, OH, July 30, 1987.

29. H. A. Brooks and J. W. Crispins Jr., *Comments On the RCS Characteristics of Cylinders, Hollow Pipes and Cylindrical Cavities*, Conductron Corp. Rep. 1801-2-T(0043-147), Ann Arbor, MI, August 1966.

30. J. W. Crispin Jr. and A. L. Maffett, "Radar Cross Section Estimates for Simple Shapes," *Proc. IEEE*, **53**, 1965, 833–848.

31. C. J. Palermo and L. H. Bauer, "Bistatic Scattering Cross Section of Chaff Dipoles with Application to Communications," *Proc. IEEE* 1965, 1119–1121.

32. V. J. DiCaudo and W. W. Martin, "Approximate Solution to Bistatic Radar Cross Section of Finite Length, Infinitely Conducting Cylinder," *IEEE Trans. Antennas and Propagat.*, **AP-14**(6), 1966, pp. 668–669.

33. S. L. Borrison, "Bistatic Scattering Cross Section of a Randomly-Oriented Dipole," *IEEE Trans. Antennas and Propagat.*, **AP-7**(2), 1967, 320–321.

34. D. Tarducci and S. Tao, "A Simplified Approach to Bistatic Reflectivity Computations," *Radio Science*, **23**(3), 1988, 443–449.

35. J. I. Glaser, "Some Results on the Bistatic Radar Cross Section (RCS) of Complex Objects," *Proc. IEEE*, **77**(5), 1989, 639–649.

36. P. Pouliguen and L. Desclos, "An Application of Physical Optics for Near-Field Mono- or Bistatic Calculations," *Micro. Opt. Tech. Lett.*, **5**(12), 1992, 653–657.

37. O. M. Bucci and G. Franceschetti, "Rim-Loaded Reflector Antennas," *IEEE Trans. Antennas and Propagat.*, **AP-28**(3), 1980, 297–305.

38. O, M. Bucci, G. Di Massa, and C. Savarese, "Control of Reflector Antenna Performances by Rim-Loading," *IEEE Trans. Antennas and Propagat.*, **AP-29**(5), 1981, 773–779.

39. G. D. Maliuzhinets, "Excitation, Reflection and Emission of Surface Waves from a Wedge with Given Face Impedances," *Sov. Phys. Dokl.*, **3**, 1958, 752–755.

40. G. Fransceschetti, *Polarization Properties of Edge Loaded High Gain Reflector Antennas*, ESTEC Contract 2893/76/NL/AK, Final Rep., May 1978.

41. H. Yokoi and H. Fukumuro, "Low Sidelobe Paraboloidal Antenna with Microwave Absorbers," *Elec. Commun. Japan*, **54B**(11), 1971, 34–39.

42. M. Hurst and R. Mittra, *Scattering Center Analysis for Radar Cross Section Modifications*, EM Commun. Lab., Univ. of Illinois, Urbana, IL, July 1984, Ch. 4.

43. M. Hurst and R. Mittra, "The Effect of Resistive Loading and Corner Rounding on Scattering Center Locations," Presented at the *IEEE Antennas and Propagation Symposium and URSI Meeting, Boston, MA*, 1984.

44. C. M. Knopp and G. I. Cohn, "On the RCS of a Coated Plate," *IEEE Trans. Antennas and Propagation*, **AP-13**, 1963, 719–721. (Correction **AP-14**, p. 516.).
45. A. K. Bhattacharyya, "Radar Cross Section Reduction of a Flat Plate By RAM Coating," *Micro. Opt. Tech. Lett.*, **3**(4), 1990, 324–327.
46. A. K. Bhattacharyya and S. K. Tandon, "Electromagnetic Scattering from a Flat Plate Structure Coated With a Lossy Dielectric," *IEEE Trans. Antennas and Propagat.*, **AP-32**(11), 1984, 1003–1007.
47. A. K. Bhattacharyya, "Electromagnetic Scattering from a Flat Plate with Rim Loading and RAM Saving," *IEEE Trans. Antennas and Propagat.*, **AP-37**(5), 1989, 659–663.
48. J. M. Rius, M. Ferrando, and L. Jofre, "High-Frequency RCS of Complex Radar Targets In Real-Time," *IEEE Trans. Antennas and Propagat.*, **AP-41**(9), 1993, 1308–1319.
49. B. T. Phong, "Illumination for Computer Generated Images," PhD dissertation, Univ. of Utah, 1973.
50. J. M. Rius Casals, "Section Recta de Blancos Radar Complejos en Tiempo Real," PhD thesis, Universitat Politecnica de Catalunya, Barcelona, July 1991 (in Spanish).
51. *Workshop on RCS of PEC or Coated Bodies, La Turbie, Nice, France*, organized by Dassault Aviation, Societe Mothesim and CNET-PAB, Nov. 16, 1990.

5.11 ADDITIONAL SOURCES

K. Barkeshi and J. L. Volakis, "Scattering from Narrow Rectangular Filled Grooves," *IEEE Trans. Antennas and Propagat.*, **39**(6), 1991, 804–810.

J.-M. Lin and J. L. Volakis, "Electromagnetic Scattering by and Transmission through a Three-Dimensional Slot in a Thick Conducting Plane," *IEEE Trans. Antennas and Propagat.*, **39**(4), 1991.

T. B. A. Senior and J. L. Volakis, "Scattering from Gaps and Cracks," *IEEE Trans. Antennas and Propagat.*, **37**(6), 1989, 744–750.

CHAPTER SIX

High-Frequency Treatment of Antennas in Complex Environments

6.1 INTRODUCTION

Any practical antenna has to work not in an infinitely extending media but in an environment. This environment may consist of a variety of structures, such as the ground, aircraft, ships, buildings, and satellites. The aircraft surface might be an advanced composite material. The presence of such structures may cause interference in various ways: (1) blockage of the antenna beam; (2) multipath propagation and consequent fading and mutual coupling between the antenna and the surroundings. These may lead to a change in antenna characteristics namely, near- and far-field patterns and input impedance. The antenna characteristics need to be reevaluated taking into account the surroundings. This chapter looks at high-frequency techniques for antennas on aircraft and other finite bodies. Starting from the monopole on a finite PEC and composite ground plane and a microstrip antenna on a finite ground plane, we discuss antennas on cylinders, cones, rocket-shaped bodies, and aircraft. A book chapter [1], and a review article [2] summarize high-frequency techniques for antennas on complex platforms. We describe the analytical formulations and discuss results in this chapter and, if appropriate, discuss a typical computer code in Chapter 10.

6.2 MONOPOLE ON A FINITE PEC AND COMPOSITE GROUND PLANES

The characteristics of a monopole on a finite circular PEC plate are of interest for qualifying the antenna range, establishing antenna standards with which to measure test antennas and modeling candidate antennas. The communication, navigation, and identification (CNI) blade antennas are mounted on finite composite ground planes. In this section, we study the monopole on a finite circular PEC plate. A monopole on a ground plane [3, 4] and an aperture on ground plane [5] have been discussed. Wedge diffraction theory has been applied to a slotted ground plane [5]. The total field at any point is a superposition of the incident, reflected, and diffracted fields from the two edges and the surface waves, if any, generated from the two edges. No surface waves are excited at normal incidence. All the field components do not exist in all angular sectors. In the discussion to follow the surface wave contributions have been ignored. The monopole on a finite PEC circular composite ground plane is shown in Figure 6.1(a) and the different regions are shown in Figure 6.1(b).

A. Incident and Reflected Fields With the phase reference at the origin (Fig. 6.1(a)), the GO incident and reflected fields associated with the edges 1 and 2 are given by

$$E_1^i = I_1 |\sin\theta| \frac{e^{-jkr_1}}{r_1} \sim I_1 |\sin\theta| e^{jkh_1 \cos\theta} \frac{e^{-jkr}}{r} \tag{6.1a}$$

$$E_1^i = I_2 |\sin\theta| \frac{e^{-jkr_2}}{r_2} \sim I_2 |\sin\theta| e^{jkh_2 \cos\theta} \frac{e^{-jkr}}{r} \tag{6.1b}$$

$$E_1^r \sim RI_1 |\sin\theta| e^{-jkh_1 \cos\theta} \frac{e^{-jkr}}{r} \tag{6.2a}$$

$$E_2^r \sim RI_2 |\sin\theta| e^{-jkh_2 \cos\theta} \frac{e^{-jkr}}{r} \tag{6.2b}$$

where, I_1 and I_2 are the currents in the two segments of the dipole and h_1 and h_2 are the heights of the midpoints of the two segments of the current elements from the ground plane. The reflection coefficient R as a function of the surface impedance Z_s is given by

$$R = \frac{\sin(\pi/2 + \theta) - Z_s/\eta_0}{\sin(\pi/2 - \theta) + Z_s/\eta_0} \tag{6.3a}$$

with

$$Z_s = \sqrt{\frac{j\omega\mu}{\sigma + j\omega\varepsilon}} \tag{6.3b}$$

A PEC surface is obtained by simply equating the surface impedance Z_s to zero.

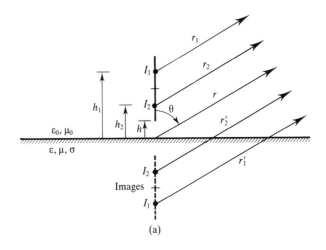

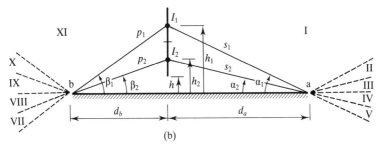

FIGURE 6.1 (a) Monopole on finite PEC/composite ground plane (b) two-segment linear dipole radiation sectors around the finite ground plane [4].

B. Diffracted Fields The expressions for diffracted fields can be obtained by using the diffraction coefficient (eqn. (1.122)) for the two-face impedance wedge [6] with interior wedge angle equal to zero to reduce the wedge to an impedance half plane.

The diffracted fields from the edges a and b are given by

$$E^d_{a1} = E^i_1(r_1 = s_1, \theta = \pi/2 + \alpha_1) D_B(s_1, \Psi_{a1}, \alpha_1, n = 2) \cdot A_{a1} e^{jkd_a \sin\theta} e^{-jkr} \quad (6.4)$$

$$\Psi_{a1} = \begin{cases} \pi/2 + \theta, & 0° \leq \theta \leq 270° \\ \theta - 3\pi/2, & 270° \leq \theta \leq 360° \end{cases} \quad (6.5)$$

The divergence factor

$$A_{a1} = \frac{\sqrt{s_1}}{r} \quad (6.8)$$

$$E^d_{a2} = +E^i_2(r_2 = s_2, \theta = \pi/2 + \alpha_2) \cdot D_B(s_2, \Psi_{a2}, \alpha_2, n = 2) \cdot A_{a2} e^{jkd_a \sin\theta} e^{-jkr} \quad (6.6)$$

$$\Psi_{a2} = \begin{cases} \pi/2 + \theta, & 0° \leq \theta \leq 270° \\ \theta - 3\pi/2, & 270° \leq \theta \leq 360° \end{cases} \quad (6.7)$$

The divergence factor

$$A_{a2} = \frac{\sqrt{s_2}}{r}$$

$$E_{b1}^d = -E_1^i(r_1 = p_1, \theta = \pi/2 + \beta_1) \cdot D_B(p_1, \Psi_{b1}, \beta_1, n=2) \cdot A_{b1} e^{-jkd_b \sin\theta} e^{-jkr} \quad (6.9)$$

$$\Psi_{b1} = \begin{cases} \pi/2 - \theta, & 0° \leq \theta \leq 90° \\ 5\pi/2 - \theta, & 90° \leq \theta \leq 360° \end{cases} \quad (6.10)$$

The divergence factor

$$A_{b1} = \frac{\sqrt{p_1}}{r}$$

$$E_{b2}^d = -E_2^i(r_2 = p_2, \theta = \pi/2 + \beta_2) \cdot D_B(p_2, \Psi_{b2}, \beta_2, n=2) \cdot A_{b2} e^{-jkd_b \sin\theta} e^{-jkr} \quad (6.11)$$

$$\Psi_{b2} = \begin{cases} \pi/2 - \theta, & 0° \leq \theta \leq 90° \\ 5\pi/2 - \theta, & 90° \leq \theta \leq 360° \end{cases} \quad (6.12)$$

The divergence factor

$$A_{b2} = \frac{\sqrt{p_2}}{r} \quad (6.13)$$

The diffraction coefficient $D_B(\ldots)$ has been discussed in Chapter 1 and is given by substituting the appropriate angles in eqn. (1.122).

Figure 6.2 shows the elevation plane pattern of a $\lambda/4$ CNI blade antenna above a square PEC ground plane. The dimensions of the square plate are (1.22 m × 1.22 m) and the frequency $f = 1$ GHz. The same patterns are shown in Figure 6.3 with graphite composite material of conductivity 10^4 S/m replacing PEC material. A more generalized treatment on impedance wedges and half-planes can be found in [6] and can be used to refine what is described above.

6.3 MICROSTRIP PATCH ANTENNA ON A FINITE PEC GROUND PLANE

6.3.1 Introduction

The effect of a finite ground plane on the performance of a microstrip antenna can be studied using high-frequency techniques and a reasonable size of ground

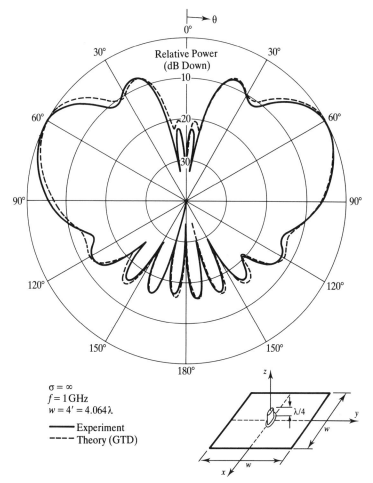

FIGURE 6.2 Elevation plane radiation patterns of $\lambda/4$ communication, navigation, and identification (CNI) blade antenna above a square PEC ground plate [4].

plane can be predicted in a given practical situation. The microstrip rectangular patch antenna is shown in Figure 6.4. Two acceptable methods of microstrip patch modeling are: (1) slot theory and (2) the modal expansion method. Microstrip patch antennas with finite ground planes have been treated in [7–10] and analytical and experimental results compared.

6.3.2 Radiation Pattern Calculations Using Slot Theory and GTD

In the slot model configuration (Fig. 6.4(b)), the radiation from the rectangular microstrip patch is equivalent to two parallel slots as shown. The GO fields

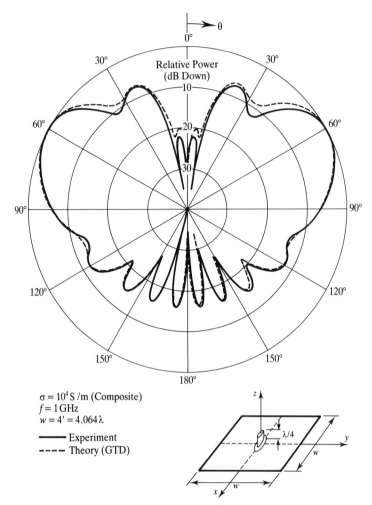

FIGURE 6.3 Elevation plane radiation patterns of $\lambda/4$ communication, navigation, and identification (CNI) blade antenna above a square PEC ground plate [4].

from each of these slots are of the form $\sin(X)/X$ and are given by

$$E_{GO} = \hat{\mu} \frac{\sin(\pi W \sqrt{\varepsilon_r} \cos \mu)}{\pi W \sqrt{\varepsilon_r} \cos \mu} \cdot \frac{e^{-jkS}}{\sqrt{S}} \qquad (6.14)$$

where $\hat{\mu}$ = orientation of the incident electric field
W = slot width in terms of wavelength
ε_r = relative permittivity of the material of the substrate
s = distance from the center of the slot to the observation point.

6.3 MICROSTRIP PATCH ANTENNA ON A FINITE PEC GROUND PLANE

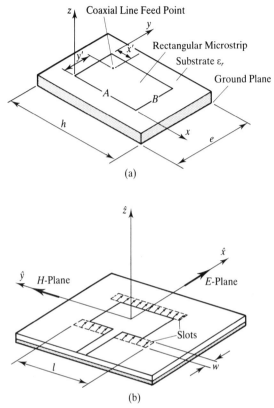

FIGURE 6.4 (a) Microstrip antenna configuration; (b) the slot model of the microstrip patch.

The diffraction mechanisms for E- and H-planes are shown in Figure 6.5. The E-plane edge-diffracted field due to a slot from each of the edges is given by

$$E^d = \hat{\mu} E_i D_h \frac{e^{-jkS_i}}{\sqrt{S_i}}$$

$$= \hat{\mu} \frac{\sin(\pi W \sqrt{\varepsilon_r})}{\pi W \sqrt{\varepsilon_r}} \frac{e^{-jkd_i}}{\sqrt{d_i}} D_h \frac{e^{-jkS_i}}{\sqrt{S_i}} \quad (6.15)$$

where D_h is a hard diffraction coefficient assuming a PEC case only, which neglects the effect of the presence of the dielectric. The effect of the dielectric can be incorporated by considering the edge diffraction as from the half plane with two face impedances, one side is PEC and the other side has a dielectric surface impedance.

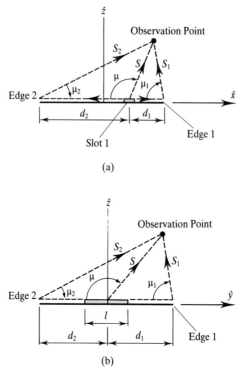

FIGURE 6.5 Diffraction mechanisms: (a) in the E-plane and (b) in the H-plane for the microstrip patch.

For the H-plane, slope diffraction correction will be necessary. The direct GO field from the slot is given by

$$E_{GO} = \hat{x} \frac{\sin(\pi/\cos\mu)}{\pi/\cos\mu} \sin\mu \cdot \frac{e^{-jkS}}{\sqrt{S}} \qquad (6.16)$$

The first-order diffracted field from each edge is zero since it is a case of grazing incidence and the tangential component of the E-field is zero on the PEC surface. The second-order field is given by

$$E_{\text{slope}} = \hat{x}\frac{1}{jk}\frac{1}{d_i}\frac{\delta E_{GO}}{\delta\mu}\bigg|_{\mu=0°\text{ or }180°} D_{sp} \frac{e^{-jkS_i}}{\sqrt{S_i}} \qquad (6.17)$$

where D_{sp} is the slope-diffracted field as given in Appendix A.1, and $\mu = 0°$ and $180°$ are the two shadow boundaries.

6.3.3 Modal Expansion and GTD

For the rectangular microstrip, the modes in the cavities can be expanded in the form

$$E_{m,n}(x, y) = \sum_{m=0}^{\infty} \sum_{n=0}^{\infty} C_{mn} \phi_{mn}(x, y) \phi_{mn}(x', y') \qquad (6.18)$$

where, C_{mn} are the excitation coefficients depending on the mode numbers m and n, the patch dimensions A and B, ε_r, and feed size. The primes indicate the location of the feed and the coordinates (x, y) any point on the patch. The modal function $\phi_{mn}(x, y)$ is given by

$$\phi_{mn}(x, y) = \cos(m\pi x/A) \cos(n\pi y/B) \qquad (6.19)$$

The details of the field configurations are discussed in [7].

Some computations were made and compared with experiment in [6] for a microstrip patch antenna and they are shown in Figures 6.6 through 6.8. Figure 6.6 shows the radiation patterns in the E- and H-plane of a rectangular microstrip antenna with antenna dimensions $A = 2.126$ in., $B = 1.488$ in., $e = 1.488$ in., $h = 14.0$ in., substrate thickness $= 0.125$ in., $\varepsilon_r = 2.55$, and $f = 2.295$ GHz. One question of special interest is the effect of finiteness of the substrate on the radiation pattern of the microstrip patch. This is depicted in Figure 6.7 where the calculated E-plane patterns with finite and infinite ground planes are compared. It is observed that the far-angle radiation is significantly affected by the finiteness of the ground plane. Figure 6.8 shows the E- and H-plane patterns of a square microwave antenna with the theoretical formulation done by combining modal expansion and GTD. The antenna dimensions are $A = B = 1.8$ in., $e = h = 38.7$ in., substrate thickness $= 0.125$ in., $\varepsilon_r = 2.17$, and $f = 2.115$ GHz. The cross-polarization levels are also shown in Figure 6.8; as expected they are low at or near broadside and increase in the direction of large angles. The effect of a finite ground plane on the input conductance of a microstrip antenna array has been analyzed in [8] using GTD. Diffraction from the end edges of the ground plane changes the input conductance and causes impedance matching problems. The calculated and measured radiation patterns of circularly polarized circular disk microstrip antennas are shown in Figure 6.9. The frequency used in these calculations is 2.295 GHz and a 38 cm square ground plane was used. The first one was constructed from a honeycomb substrate with relative dielectric constant 1.2, a substrate thickness of 1.27 cm, and a patch diameter of 11.18 cm. The second one had a thickness of 0.32 cm and a patch diameter of 14.8 cm. The theoretical results were obtained by using multimode cavity theory [8] together with aperture integration. GTD was used to calculate the edge contributions particularly for (b) since the main beam is at a low angle. In [9], ground plane edge diffraction has been taken into account to study the effect of the ground plane on input impedance and resonant frequency.

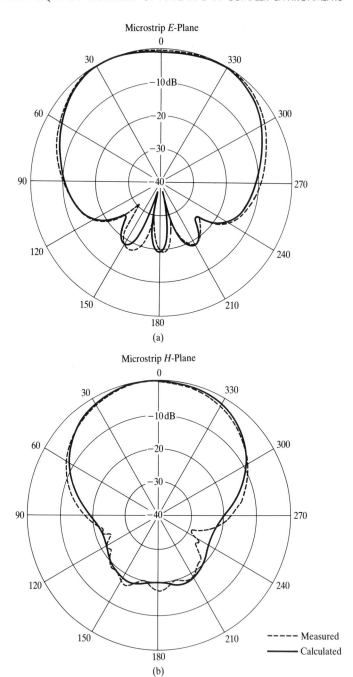

FIGURE 6.6 Radiation patterns of a rectangular microstrip patch antenna of dimensions: $A = 2.126$ in.; $B = 1.488$ in.; $e = 10.5$ in.; $h = 14.0$ in.; substrate thickness $= 0.125$ in.; $\varepsilon_r = 2.55$; $f = 2.295$ GHz. (a) E-plane; (b) H-plane [7].

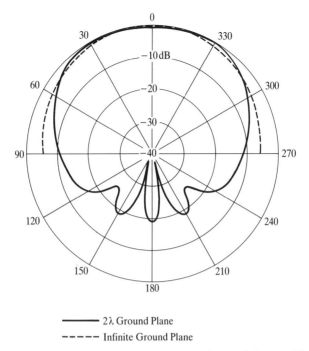

FIGURE 6.7 Comparison of calculated patterns when the patch is on a 2λ square PEC plate and on an infinite plate [7].

6.4 MONOPOLE ON A FINITE CYLINDER

Analysis of monopoles on a finite cylindrical structure is important for modeling antennas mounted on the fuselage or the wings of aircraft. A finite cylinder simulates the fuselage and thin wedges simulate the wings. This modeling significantly helps in saving the cost and complexity of an in situ measurement, for example, the radiation pattern of an antenna mounted on the aircraft when it is flying. There are several key papers [10–14]. An analysis [10] using image theory and GTD was used to calculate elevation plane patterns of radial infinitesimal current elements mounted near or on PEC cylinders of finite length; the results were compared with experiment in [7]. Wedge diffraction (Section 1.9) and equivalent edge currents (Section 1.19.1) were applied in [11] to compute the patterns of radial monopoles and circumferential slot antennas on finite PEC circular cylinders with conical or disk endcaps. In [12], an integral equation formulation was numerically solved to compute the induced currents and scattered field by the finite PEC cylinder. The same problem was treated in [13, 14] to estimate radiation from quarter-wavelength monopoles on finite cylindrical, conical, and rocket-shaped conducting bodies. Though the material in [12–13] does not fall within

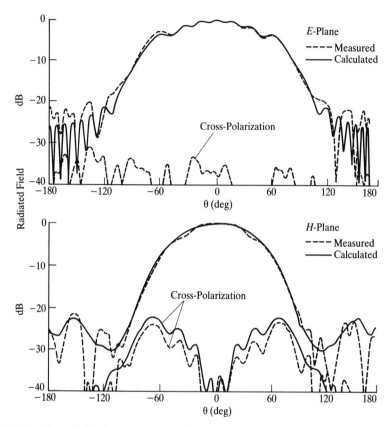

FIGURE 6.8 E- and H-plane patterns of a square microstrip antenna using modal expansion and GTD. Dimensions: $A = B = 1.8$ in.; $e = h = 38.7$ in.; substrate thickness $= 0.125$ in.; $\varepsilon_r = 2.17$; $f = 2.115$ GHz [7].

high-frequency techniques, it is useful to compare results obtained by different methods.

The problem of the monopole near a finite cylinder can be considered as a superposition of two parts: (1) the current element near an infinitely extending flat PEC surface and (2) the diffraction of the fields due to the current element by the edges of the finite cylinder. The first part can be taken care of by image theory and the second part by UGTD. The tangent plane approximation associated with the image concept assumes an electrically large cylinder radius. This can be corrected by multiplying the reflection coefficient by the divergence factor.

The geometry of the current element–cylinder combination is shown in Figure 6.10. The following gives the key expressions [10] to calculate the elevation plane patterns.

6.4 MONOPOLE ON A FINITE CYLINDER   367

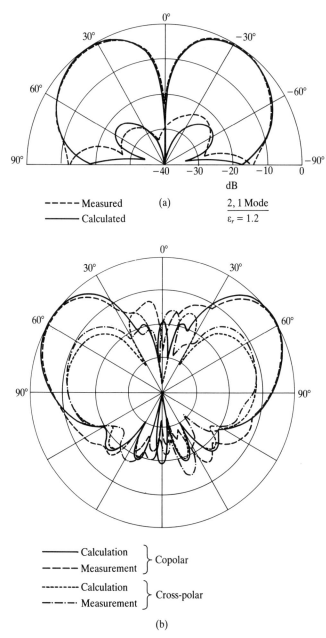

FIGURE 6.9 Calculated and measured radiation patterns of circularly polarized circular microstrip antennas: (a) TM_{21} mode; (b) TM_{41} [8].

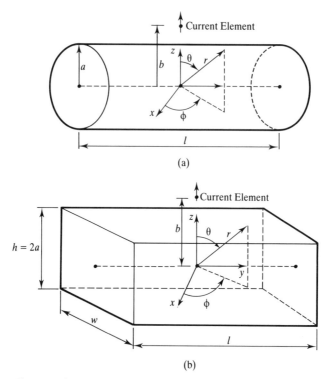

FIGURE 6.10 Current element near finite structures: (a) a finite circular PEC cylinder and (b) a finite rectangular PEC cylinder [11].

A. Pattern Function of the Current Element The normalized pattern function is given by (Fig. 6.10)

$$E_p(\theta) = \cos[k_0(b-a)\cos\theta]\sin\theta \cdot \exp(jk_0 a \cos\theta) \quad \text{for} \quad -\pi/2 < \theta < \pi/2 \quad (6.20)$$

where a and b are respectively the radius of the cylinder and the height of the center of the current element from the axis of the cylinder.

B. The Diffracted Field from Edge 1 The diffracted field from edge 1 is given by

$$E_{D1}^{(1)}(\theta) = R_{D1}^{(1)}(\theta) \cdot \exp -j[-k_0 d \cos(\gamma - \theta) + \pi/4] \quad -\pi/2 < \theta < \pi \quad (6.21)$$

where

$$R_{D1}^{(1)}(\theta) = R_D(\pi/2)[V_B(l/2, \pi/2 + \theta, n)] \quad (6.22a)$$

$$d = \sqrt{[(l/2)^2 + a^2]} \quad (6.22b)$$

$$\gamma = \tan^{-1}[l/2a] \quad (6.22c)$$

l is the length and a is the radius of the cylinder.

The contributions from edges 2, 3, and 4 can be estimated in the same way.

The above theory has been validated in [10] where the current elements were simulated by extending the inner conductor of the coaxial line. The frequency range was 1.25 to 3.75 GHz. Typical elevation plane patterns are shown in Figure 6.11. The effect on the standing wave patterns of changing the cylinder length, keeping a constant cylinder length-to-radius ratio, is described in [14] along with the effects on the tilt of the patterns and the change in the maximum field. Some near-zone scattering calculations were done in [13] for finite PEC cylinders. Figure 6.12 compares measured and theoretical results for vertical polarization. The cylinder has a length of 24 in. and a diameter of 2.45 in., the frequency is 9.4 GHz, and the radar-to-target separation is 8.0 in. The vertical polarization results have an offset of -9 dB with respect to the horizontal polarization. There is a significant amount of disagreement in the cone pattern for low-intensity scattered fields. The radiation patterns of axial slots on an elliptic cylinder [15] are shown in Figure 6.13(b) and (c); Figure 6.13(a) gives the elliptic cylinder geometry.

6.5 MONOPOLE ON A CONE

In this section and the following sections, we discuss the patterns of antennas on conical structures without repeating their mathematics. Figure 6.14 shows a frustum of a cone. Comparison of theoretical and experimental radiation patterns for a bent monopole on a frustum of a cone are shown in Figure 6.15. Some near-zone calculations using GTD were done and compared with experiment for some antenna pattern functions on a finite cone in Figure 6.16. The dimensions of the cone are height 8 in. and diameter 2.06 in. with $f = 9.4$ GHz. The polarization is vertical and the target-to-radar separation is 8.0 in.

6.6 MONOPOLE ON A ROCKET-SHAPED BODY

The geometry and coordinate system of the monopole on a rocket-shaped structure is shown in Figure 6.17. The distribution of current densities on the rocket-shaped structure is shown in Figure 6.18 [16] and the normalized radiation patterns for a bent monopole on rocket-shaped bodies is shown in Figure 6.19 [16].

6.7 ANTENNAS ON AIRCRAFT

Analysis of antennas is important for aircraft operations since the antenna has to work in the presence of the aircraft structure. A single modern aircraft may

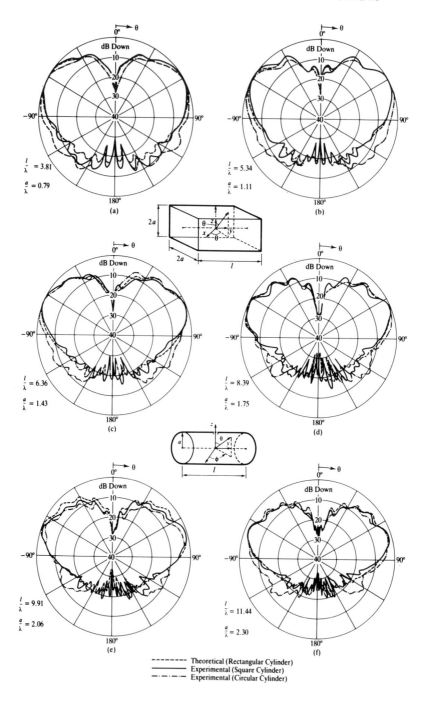

FIGURE 6.11 Elevation plane patterns ($\phi = \pm\pi/2$) of radial current elements on finite rectangular and circular cylinders. The parameters are on the diagram [11].

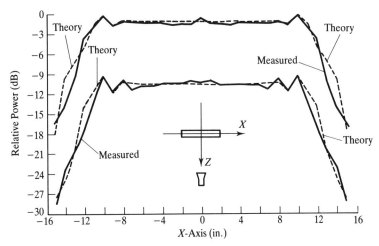

FIGURE 6.12 Comparison of measured and theoretical results of relative scattered power for a finite cylinder with antenna weighting function. Cylinder length = 24 in.; cylinder diameter = 2.45 in.; $f = 9.4$ GHz; radar to target separation = 8.0 in. Vertical polarization results offset -9 dB from horizontal. (From [11] © 1972 IEEE.)

have as many as 200 antennas mounted on its body and therefore an efficient algorithm is needed to decide the positions of the antennas for the point of electromagnetic compatibility and interference. Usually, the fuselage is modeled by a circular or elliptic cylinder and the wings by thin wedges or polygonal flat plates. A considerable amount of literature is available on the modeling of aircraft antennas. They are described in [2, 16–20]. An overview of antennas in a complex environment is available in [2]. The roll-plane analysis of antennas on a simply modeled aircraft was done in [16]. A study of KC-135 aircraft antenna patterns with a monopole, slots mounted on the wings was done in [17]. The theoretical results agreed very well with experiments done on a scaled-down model of the KC-135 aircraft. In [17], an efficient technique for volumetric antenna pattern calculations blended the patterns for roll and elevation; the results are compared with measurements on scale models. The near-field patterns of an aperture or monopole antenna mounted on the fuselage has been formulated in [18]. High-frequency radiation patterns on a curved surface have been predicted [19].

6.7.1 Near-Field Patterns

The importance of the near-field patterns in antenna and scattering analysis is well known since the near-field pattern with a suitable transformation gives a far-field pattern. The subject of curved surface diffraction has been discussed

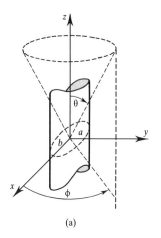
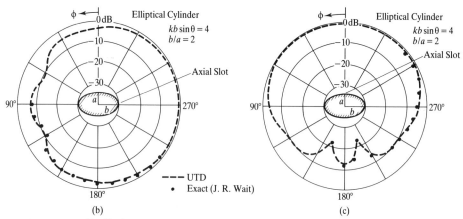

FIGURE 6.13 Radiation patterns of small antennas on PEC elliptic cylinder: (a) elliptic cylinder geometry; (b) E_θ; and (c) E_ϕ radiation patterns in elliptic cylinder [15].

in Section 1.18. The near-field patterns of two types of antennas are discussed here; one is the flush-mounted slot, the other is a monopole on or near the curved surface of a cylinder the approximation to the fuselage. A more accurate fuselage model is a PEC composite ellipsoid. When high accuracy is necessary, the fuselage should be modeled very accurately, particularly near the antenna. The dominant mode is assumed on the exciting aperture of size ($A \times B$). It is cosine in the A-direction and uniform in the B-direction. For the monopole, a sinusoidal distribution with unity amplitude is assumed on the monopole. The analysis is detailed in [16, 17]. The geometry and coordinate system for lit region calculations are shown in Fig. 6.20. The expressions for the fields in the lit region are for the slot and the monopole.

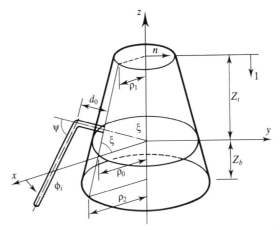

FIGURE 6.14 The geometry and coordinate system of a bent monopole on the frustum PEC cone [16].

For the Slot

Lit region. The field components in the lit region are given by [17, 18]

$$\mathbf{E}^a_{\phi\phi'} \sim \frac{-j}{2} P^a_{lit} \sin\theta_i G(\xi_{lit}) \frac{e^{-jks}}{s} \hat{\phi} \tag{6.23a}$$

$$\mathbf{E}^c_{\phi\phi'} \sim \frac{-j}{2} P^a_{lit} \sin\theta_i (\hat{e}_1 \cdot \hat{t}) G(\xi_{lit}) \frac{e^{-jks}}{2} \hat{\phi} \tag{6.23b}$$

$$\mathbf{E}^c_{\theta\hat{\theta}'} \sim \frac{-j}{2} P^c_{lit} \cdot -j \left(\frac{2}{k\rho_r(Q')\sin\theta_i}\right)^{1/3} G(\xi_{lit}) \frac{e^{-jks}}{s} \tag{6.23c}$$

where

$$P^a_{lit} = \frac{kB}{\pi^2} \frac{\cos\left(\frac{kB\cos\theta'}{2}\right)}{1 - \left(\frac{kB\cos\theta'}{\pi}\right)^2} \cdot \frac{\sin\left(\frac{kA}{2}\sin\theta'\right)\hat{e}'(\hat{e}_1 \cdot \hat{e})}{\frac{kA}{2}\sin\theta'(\hat{e}_1 \cdot \hat{t})} \tag{6.23d}$$

$$P^c_{lit} = \frac{kB}{\pi^2} \frac{\cos\left(\frac{kB}{2}\sin\theta'\right)(\hat{e}_1 \cdot \hat{t})}{1 - \left(\frac{kB\sin\theta'(\hat{e}_1 \cdot \hat{t})}{\pi}\right)^2} \cdot \frac{\sin\left(\frac{kA}{2}\cos\theta'\right)}{\frac{kA}{2}\cos\theta'} \tag{6.23e}$$

Shadow Region. The field components in the shadow region (Fig. 6.18) are given by [16, 17]

$$\mathbf{E}^a_n \approx \text{AMP} \cdot \cos\omega_i g(\xi_i) \cdot \text{DF} \cdot \text{EXP} \cdot \hat{n}_i \tag{6.24a}$$

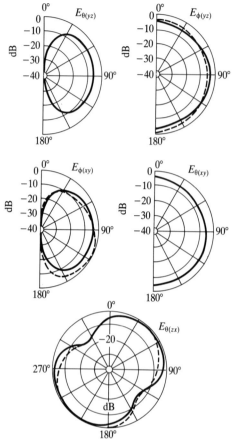

FIGURE 6.15 Computed and measured radiation patterns for a bent monopole on a frustum of a cone. ρ_0/λ, $\xi = 85°$, $\rho_1/\lambda = 0.11333$, $\rho_2/\lambda = 0.15113$, $d_0/\lambda = 0.1$, $Z_1/\lambda = 0.15008$. (From [14] © 1979 IEEE.)

where
$$\text{EXP} = \frac{e^{-jk(l_i + s_i)}}{\sqrt{(l_i + s_i)s_i}}$$

DF, the divergence factor $= \left(\dfrac{\rho_g(Q_i)}{\rho_g(Q_i)}\right)^{1/6}$

and AMP, the complex amplitude $= \dfrac{-j}{2} P^c_{dark}(-1)^{i-1}$

$$\mathbf{E}^c_n \approx \text{AMP} \cdot \cos w_i g(\xi_i) \cdot \text{DF} \cdot \text{EXP} \cdot \hat{n}_i \qquad (6.24\text{b})$$

$$\mathbf{E}^c_n \approx \text{AMP} \cdot -j\left(\frac{2}{k\rho_r(Q') \cos w_i}\right)^{1/3} g(\xi_i) \cdot \text{DF} \cdot \text{EXP} \cdot \hat{b}_i \qquad (6.24\text{c})$$

6.7 ANTENNAS ON AIRCRAFT   375

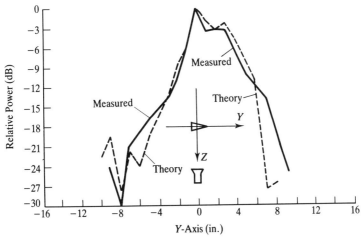

FIGURE 6.16 Comparison of measured and theoretical results of relative near-zone scattered power for a finite cone with antenna weighting function. Cone length = 8 in.; diameter of the base = 2.06 in.; $f = 9.4$ GHz; radar-to-target separation = 8 in. (From [14] © 1979 IEEE.)

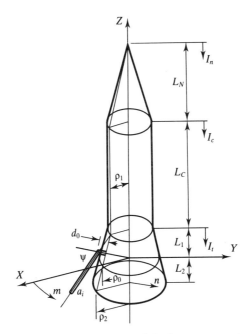

FIGURE 6.17 The geometry and coordinate of the bent monopole on a rocket-shaped structure. (From [14] © 1979 IEEE.)

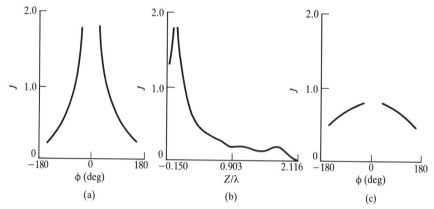

FIGURE 6.18 Distribution of current densities: (a) circumferential distribution for side cells with centers at $z/\lambda = 0.012$; (b) longitudinal distribution for side cells with centers at $\phi = 30°$; (c) current on bottom cells with centers at $\rho/\lambda = 0.075$. The parameters are $\rho_0/\lambda = 0.138$, $\zeta = 85°$, $L_1/\lambda = 0.28193$, $L_2/\lambda = 0.1515008$, $\rho_1/\lambda = 0.11333$, $\rho_2/\lambda = 60.15113$, $L_c/\lambda = 1.07634$, $L_N/\lambda = 0.75834$, $\zeta = 85°$, $d_0/\lambda = 0.1$, and $\psi = 78°$.

where

$$P_{dark}^{a} = \frac{kB}{\pi^2} \frac{\cos\left(\frac{kB}{2}\sin w_i\right)}{\left(1 - \frac{kB\sin w_i}{\pi}\right)^2} \frac{\sin\left(\frac{kA}{2}\cos w_i\right)}{\frac{kA}{2}\cos w_i} \quad (6.24\text{d})$$

$$P_{dark}^{c} = \frac{kB}{\pi^2} \frac{\cos\left(\frac{kB}{2}\sin w_i\right)}{\left(1 - \frac{kB\sin w_i}{\pi}\right)^2} \frac{\sin\left(\frac{kA}{2}\sin w_i\right)}{\frac{kA}{2}\sin w_i} \quad (6.24\text{e})$$

$$\xi_i, \text{ the detour parameter} = \int_{Q'}^{Q} \left(\frac{k}{2\rho_g^2}\right)^{1/3} dl_i \quad (6.24\text{f})$$

The axial and circumferential slots are represented by superscripts a and c respectively. The vectors $(\hat{t}_1, \hat{n}_1, \hat{b}_1)$ represent the three directions at any point along the trajectory of the ray path $Q'\bar{Q}_1 P$.

Now we can arrive at the expressions for the cases of monopole and slots.

For the Monopole

Lit Region

$$\mathbf{E}_{\phi'}^{m} \approx P_{lit}^{m} \cdot -(\hat{e}_1 \cdot \hat{t}) G(\xi^{lit}) \frac{e^{-jks}}{s} \phi' \quad (6.25\text{a})$$

$$\hat{\mathbf{E}}_{\theta'}^{m} \approx P_{lit}^{m} \cos\theta' \cdot -j\left(\frac{2}{k\rho_r(Q')\sin\theta'}\right)^{1/3} \cdot G(\xi^{lit}) \frac{e^{-jks}}{s} \theta' \quad (6.25\text{b})$$

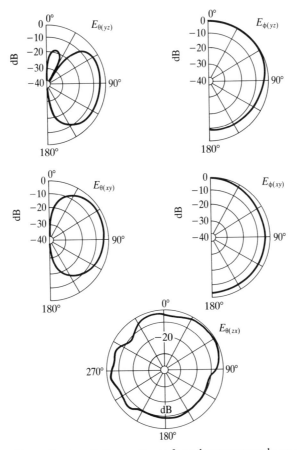

FIGURE 6.19 Normalized radiation patterns for a bent monopole on a rocket.

where

$$P_{lit}^m = \frac{-jz_0 p_e}{2} \quad \text{for a short stub}$$

$$P_{lit}^m = \frac{-jz_0}{4\pi} \frac{\cos(kl(\hat{n}' \cdot \hat{s})) - \cos(kl)}{[1 - (\hat{n}' \cdot \hat{s})^2] \sin(kl)} \quad \text{for a linear monopole}$$

Shadow Region

$$\mathbf{E}_n^m \sim P_{dark}^m \sum_{i=1}^{2} g(\xi_i) \cdot \text{DF} \cdot \text{EXP} \cdot \hat{n}' \qquad (6.26\text{a})$$

$$\mathbf{E}_b^m \sim P_{dark}^m \sum_{i=1}^{2} (-1)^{i-1} \sin w_i \cdot -j\left(\frac{2}{k\rho_r(Q')\cos w_i}\right)^{1/3} \tilde{g}(\xi_i) \cdot \text{DF} \cdot \hat{b}_i \qquad (6.26\text{b})$$

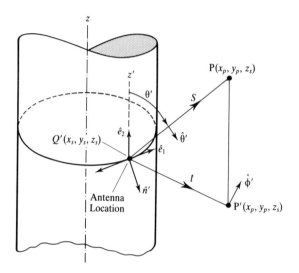

FIGURE 6.20 An elementary antenna located on a PEC cylinder and its coordiant coordinate system.

The expressions for P_{dark}^m are given by

$$P_{dark} = \frac{-jz_0 P_e}{2} \quad \text{for a short stub}$$

$$P_{dark}^m = \frac{-jz_0}{4\pi}\left(\frac{1-\cos(kl)}{\sin(kl)}\right) \quad \text{for a linear monopole}$$

In the above expressions, $\rho_g(Q')$ is the effective radius of the ray path at the point under consideration and is equal to $(a/\sin\theta_i)$ at the point Q'. The functions $g(.)$ and $G(.)$ are complex conjugates of Fock functions and are available in [21].

The total fields in the lit and shadow regions of the near zone of an antenna (slot or, monopole) mounted on the PEC cylinder are given by

For the Slot

Axial

$$\mathbf{E}_t = \mathbf{E}_\phi^a \quad \text{in the lit region} \tag{6.27a}$$

$$= \mathbf{E}_n^a \quad \text{in the shadow region} \tag{6.27b}$$

Circumferential

$$\mathbf{E}_t = \mathbf{E}_{\theta'}^c + \mathbf{E}_\phi^{c'} \tag{6.27c}$$

Solution is obtained by superposition of the following fields:

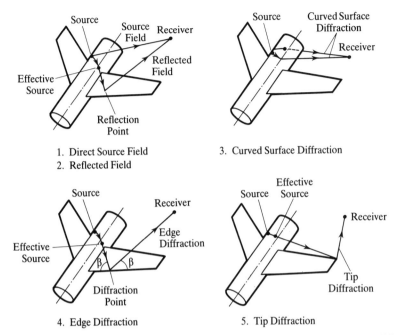

1. Direct Source Field
2. Reflected Field
3. Curved Surface Diffraction
4. Edge Diffraction
5. Tip Diffraction

FIGURE 6.21 A square wing (plate) mounted on a cylinder to illustrate the different contributions in three views: oblique, top, and side.

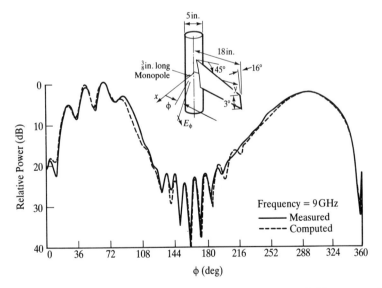

FIGURE 6.22 Comparison of measured and calculated E_ϕ near-field patterns for test geometry composed of a circular cylinder and one plate. The receiver is at a distance of 18 in. from the origin. (From [17] © 1975 IEEE.)

For the Monopole

Lit Region
$$\mathbf{E}_t = \mathbf{E}_{\theta'}^m + \mathbf{E}_{\phi'}^m \qquad (6.27d)$$

Shadow Region
$$\mathbf{E}_t = \mathbf{E}_n^m + \mathbf{E}_b^m \qquad (6.21e)$$

The analysis takes into account contributions from different sources, including corner diffraction (Section 1.9.2). Again, we plan to omit the steps. Essentially, a cylinder–plate model was used to analyze the near-field patterns. The different contributions to the total fields are shown in Figure 6.20. Some of the patterns are shown in Figures 6.21 through 6.23. The comparison of measured and calculated near-field E_ϕ patterns for the test geometry, a circular cylinder and a plate, is shown in Figure 6.21. The same pattern is shown in Figure 6.22. A comparison is made of the theoretical and experimental patterns of a monopole antenna mounted on a Boeing 737 aircraft. The simplified model used is shown below the aircraft. Some of the comparisons of near-field analytical and experimental patterns are shown in Figures 6.23 and 6.24. In conclusion, the method described does calculate the near field accurately and this analysis could be extended to any antenna position on the fuselage.

6.7.2 Far-Field Patterns

Far field analyses of aircraft antennas were done in a number of references [15–17, 20]. The geometry of the monopole and slot on an elliptic cylinder is shown in Figure 6.25. The treatment of antennas on a curved PEC surface is

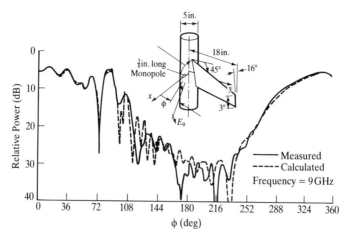

FIGURE 6.23 Comparison of measured and calculated near-field patterns for our test geometry: E_θ. The patterns are 124° conical and the range is 24 in. from the origin. (From [17] © 1975 IEEE.)

6.7 ANTENNAS ON AIRCRAFT 381

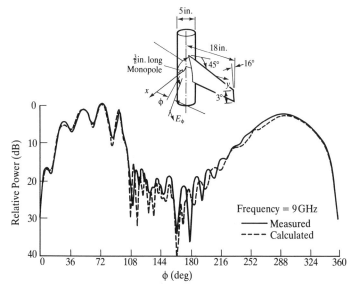

FIGURE 6.24 Comparison of theoretical and experimental elevation plane patterns of a monopole antenna mounted on a Boeing 737 aircraft. The model used in the analysis is shown below the actual aircraft. (From [17] © 1975 IEEE.)

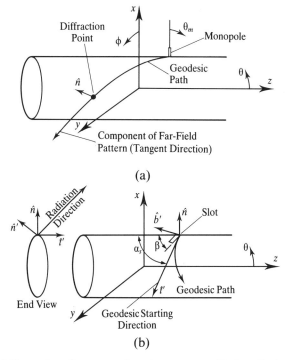

FIGURE 6.25 (a) Geometry of a monopole antenna on an elliptic cylinder; (b) geometry of a slot antenna on an elliptic cylinder.

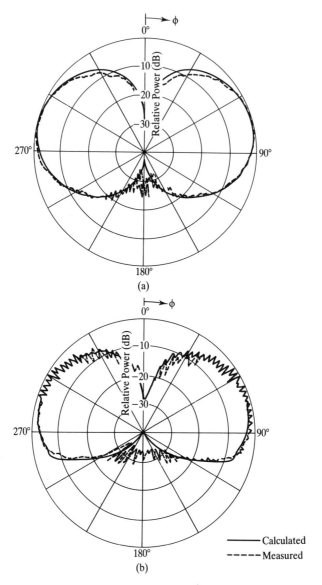

FIGURE 6.26 (a) Roll plane pattern (E_ϕ) for $\frac{1}{25}$ scale model of KC-135 with $\lambda/4$ monopoles on fuselage forward of wings at frequency of 34.92 GHz (model frequency). (b) Roll plane pattern (E_ϕ) for $\lambda/4$ monopole over wings. (From [16] © 1973 IEEE.)

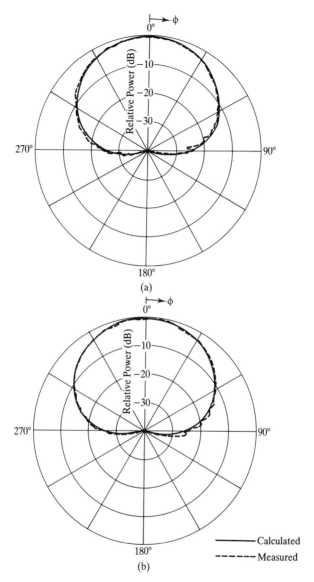

FIGURE 6.27 (a) Roll plane pattern (E_ϕ) for Ka band circumferential waveguide forward of wings. (b) Roll plane pattern (E_θ) for Ka band circumferential waveguide over wings. (From [16] © 1973 IEEE.)

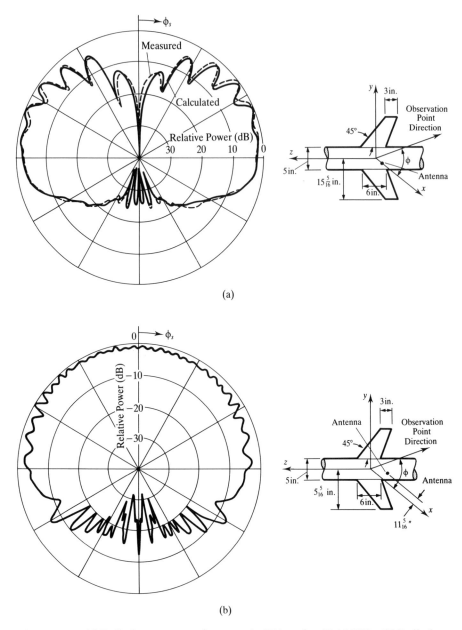

FIGURE 6.28 (a) Roll plane pattern of monopole (E_ϕ) at $f = 11.45$ GHz. (b) Roll plane pattern of axial slot (E_ϕ) at $f = 11.45$ GHz. (c) Roll plane pattern of circumferential slot (E_θ) at $f = 11.45$ GHz [15].

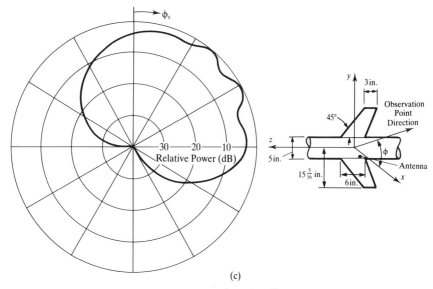

FIGURE 6.28 (*continued*)

discussed in Section 1.18.2.4. For many antennas mounted on aircraft, when part of the body of the aircraft is not directly illuminated, a virtual source has to be postulated. Also, the ray paths to the receiving point have to be traced on the body by starting from the source. This includes torsion when the transmitting and receiving locations are not in the same equatorial plane. Some of the radiation patterns are shown in Figures 6.26 through 6.28.

6.8 SUMMARY

This chapter dealt with high-frequency antennas on finite bodies. It discussed effects of a finite ground plane, both PEC and composite, on a monopole and a microstrip patch. Monopoles on finite cylinders, cones, and rocket-shaped bodies were treated along with a discussion of near- and far-field patterns for antennas on aircraft.

6.9 REFERENCES

1. W. D. Burnside and R. J. Marhefka, "Antennas on Aircraft, Ships or Any Large, Complex Environment," Ch. 20 in *Antenna Handbook: Theory, Applications and Design*, edited by Y. T. Lo and S. W. Lee, Van Nostrand Reinhold, NY, 1980.

2. R. J. Marhefka and W. D. Burnside, "Antennas on Complex Platforms," *Proc. IEEE*, **80**(1), 1992, 204–208.

3. M. M. Weiner, S. P. Cruze, C. C. Li, and W. J. Wilson, "*Monopole Elements on Circular Ground Planes*," Artech House, Norwood, MA, 1987.
4. C. A. Balanis and D. Decarlo, "Monopole Antenna Patterns on Finite Size Composite Ground Planes," *IEEE Trans. Antennas and Propagat.*, **AP-30**(4), 1982, 764–768.
5. C. A. Balanis and L. Peters Jr, "Equatorial Plane Patterns of an Axial-TEM Slot on a Finite Size Ground Plane," *IEEE Trans. Antennas and Propagat.*, **AP-17**, 1969, 507–513.
6. R. Tiberio, G. Pelosi, and G. Manara, "A Uniform GTD Formulation for the Diffraction by a Wedge with Impedance Faces," *IEEE Trans. Antennas and Propagat.*, **AP-33**(8), 1985, 867–873.
7. J. Huang, "The Finite Ground Plane Effect on the Microstrip Antenna Radiation Patterns," *IEEE Trans. Antennas and Propagat.*, **AP-31**(4), 1983, 649–653.
8. J. Huang, "Microstrip Antenna Development at JPL," *IEEE Antennas and Propagat. Magazine*, 33(3), 1991, 33–41.
9. E. Lier and K. R. Jakobsen, "Rectangular Microstrip Patch Antennas with Infinite and Finite Ground Plane Dimensions," *IEEE Trans. Antennas and Propagat.*, **AP-31**(6), 1983, 978–984.
10. C. A. Balanis, "Radiation Characteristics of Current Elements near a Finite-Length Cylinder," *IEEE Trans. Antennas and Propagat.*, **AP-18**(3), 1970, 352–359.
11. C. E. Ryan Jr., "Analysis of Antennas on Finite Circular Cylinders with Conical or Disk End Caps," *IEEE Trans. Antennas and Propagat.*, **AP-20**, 1972, 474–476.
12. E. C. Burdette and C. E. Ryan Jr., "Near-Zone Scattering by Plates, Cylinders, and Cones," *IEEE Trans. Antennas and Propagat.*, **AP-22**, 1974, 823–826.
13. K. M. Chen, D. E. Livesay, and B. S. Guru, "Induced Currents in and Scattered Field from a Finite Cylinder with Arbitrary Conductivity and Permittivity," *IEEE Trans. Antennas and Propagat.*, **AP-24**, 1976, 330–336.
14. M. N. I. Fahmy and A. Z. Botros, "Radiation from Quater-Wavelength Monopole on Finite Cylindrical Conical and Rocket-Shaped Structures," *IEEE Trans. Antennas and Propagat.*, **AP-27**, 1979, 615–623.
15. P. H. Pathak, N. Wang, W. D. Burnside, and R. G. Kouyoumjian, "A Uniform GTD Solution for the Radiation from Sources on a Convex Surface", *IEEE Trans. Antennas and Propagat.*, **AP-29**(4), 1981, 609–622.
16. W. D. Burnside, R. J. Marhefka, and L. Y. Chong, "Roll-Plane Analysis of On-Aircraft Antennas," *IEEE Trans. Antennas and Propagat.*, **AP-21**(6), 1973, 780–786.
17. W. D. Burnside, M. C. Gilbreath, R. J. Marhefka, and L. Y. Chong, "A Study of KC-135 Aircraft Antenna Patterns," *IEEE Trans. Antennas and Propagat.*, **AP-23**, 1975, 309–316.
18. C. L. Yu, W. D. Burnside, and M. C. Gilreath, "Volumetric Pattern Analysis of Airborne Antennas," *IEEE Trans. Antennas and Propagat.*, **AP-26**(5), 1978, 636–641.
19. W. D. Burnside, N. Wang, and P. L. Pelton, *Near Field Pattern Computations for Airborne Antennas*, Ohio State Univ., Electroscience Lab., Dept. Elect. Eng., Rep. 78685-4, June 1978. Prepared under contract N00019-77-C-0299 for Naval Air Systems Command.

20. W. D. Burnside, N. Wang, and E. L. Pelton, "Near-Field Pattern Analysis of Airborne Antennas," *IEEE Trans. Antennas and Propagat.*, **AP-28**(3), 1980, 318–327.
21. J. J. Kim and W. D. Burnside, "Simulation and Analysis of Antennas Radiating in a Complex Environment," *IEEE Trans. Antennas and Propagat.*, **AP-34**(4), 1984, 554–562.
22. W. D. Burnside, R. J. Marhefka, and N. Wang, *Computer Programs, Subroutines and Functions for the Short Course on the Modern Geometrical Theory of Diffraction*, Ohio State Univ. Electroscience Lab. Collumbus, OH.

6.10 ADDITIONAL SOURCES

K.-K. Chan, L. B. Felsen, A. Hessel, and J. Shmoys, "Creeping Waves on a Perfectly Conducting Cone," *IEEE Trans.*, **AP-25**(5), 1977, 661–670.

A. W. Glisson and C. M. Burtler, "Analysis of a Wire Antenna in the Presence of a Body of Revolution," *IEEE Trans. Antennas and Propagat.*, **AP-28**, 1980, 604–609.

Per-Simon Kildal, "A Formula for Efficient Computation of Radiation from a Current Source in Proximity to Cylindrical Scatterers," *IEEE Trans.*, **AP-32**(7), 1984, 754–757.

CHAPTER SEVEN

Estimation of Mutual Coupling Between Antennas on Structures: EMC and EMI Studies

7.1 INTRODUCTION

Specifications for potential radio frequency and electromagnetic interference must be provided to define an electromagnetic environment for the equipment designer and certifier. Compliance with realistic specifications provides a reasonable guarantee that a system will be operational. It is then necessary to develop a definition of the threat. Mutual coupling and electromagnetic compatibility between antennas are very important in many practical applications. Aircraft avionics are affected by radiated and leakage electromagnetic waves. The specifications of interference levels are decided by the Federal Aviation Administration (FAA) and the Electromagnetic Compatibility Analysis Center (ECAC). Some of the information regarding accepted interference levels are available in [1]. In the case of aircraft antennas, for example, compliance with these specifications requires a knowledge of mutual coupling between the elements on planar and conformal arrays and a study of the smooth surface waves that are generated, particularly in the shadow region.

Estimation of surface waves is described in Section 1.18 and among the pertinent sources [1-9] is a book [3]. A number of computer codes [10, 11] are available to generate results on SHF/EHF coupling between antennas: intrasystem electromagnetic compatibility program (IEMCAP), aircraft interantenna propagation with graphics (AAPG), general 3D airborne antenna radiation pattern code (fuselage code), basic scattering code (BSC), general electromagnetic model for the analysis of complex systems (GEMACS), and GTD code. All are discussed briefly in Chapter 10.

7.2 MUTUAL COUPLING BETWEEN ANTENNAS ON A CYLINDER

7.2.1 Using Uniform Theory of Curved Surface Diffraction

The surface field components excited on a smooth convex surface due to infinitesimal electric and magnetic dipole moments are given by eqns. (1.213) through (1.217). These expressions consider torsional effects for nonequatorial excitations, and the surface field components can be directly used to compute the mutual coupling between two antennas on a cylinder. Since thin slots are approximately equivalent to elementary electric and magnetic dipole moments we describe here thin slot coupling on a PEC cylinder. The mutual admittance between two slots at positions Q and Q' (Fig. 7.1) are given by

$$Y_{21} = -\frac{1}{V_{11}V_{22}} \iint_{S_2} \iint_{S_1} d\mathbf{H}_m(Q/Q') \cdot d\mathbf{p}_m(Q) \tag{7.1}$$

where V_{11} and V_{22} are the voltages across the slots. The dominant mode functions $\hat{\mathbf{e}}_1$ and $\hat{\mathbf{e}}_2$ are only different if the slots are different. They are given by

$$\mathbf{E}_1^a = V_{11}\hat{\mathbf{e}}_1 \tag{7.2a}$$

$$\mathbf{E}_2^a = V_{22}\hat{\mathbf{e}}_1 \tag{7.2b}$$

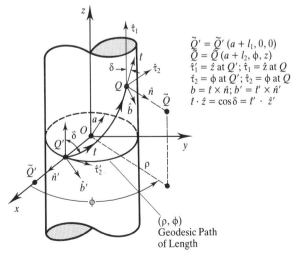

FIGURE 7.1 Positions of transmitter and receiver and the geodesic path on the cylinder [2].

where $\mathbf{E}_1^a$ and $\mathbf{E}_2^a$ are the electric fields in slots 1 and 2. Q and Q' are the locations of the first and second slots with the second slot short-circuited. $d\mathbf{H}_m(Q/Q')$ is the magnetic field at the surface due to the equivalent magnetic source $dP_m(Q')$ at Q'. The infinitesimal dipole moments and the aperture fields are related by

$$d\mathbf{p}_m(Q') = \mathbf{E}_1^a(Q') \times \hat{n}' \, dS_1 \tag{7.3a}$$

$$d\mathbf{p}_m(Q) = \mathbf{E}_2^a(Q) \times \hat{n} \, dS_2 \tag{7.3b}$$

Mutual impedance between linear antennas can be found by extending the dipole moment theory to linear antennas. This is easy for short monopoles where the current is constant and there is only one dominant mode; the mutual impedance is given by

$$Z_{21} = \int_0^{h_1} \int_0^{h_1} \frac{d\mathbf{E}_e(Q/q') \cdot d\mathbf{p}_e(Q)}{I_{11} I_{22}} \tag{7.4}$$

where $d\mathbf{E}_e(Q/Q')$ is the electric field at Q due to the equivalent source $d\mathbf{p}_e(Q')$ at Q.

The plots of isolation of axial slots on a cylinder have been presented in Figures 7.2 and 7.3. Figure 7.2 is an equatorial case where the torsion is

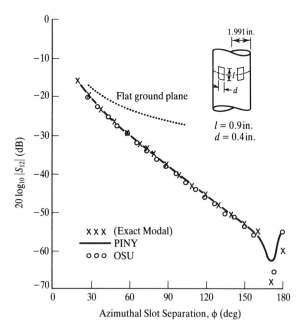

FIGURE 7.2 Isolation between two axial slots in an equatorial plane on a circular PEC cylinder: $a = 1.991$ in.; $z_0 = 0$; $f = 9$ GHz [2].

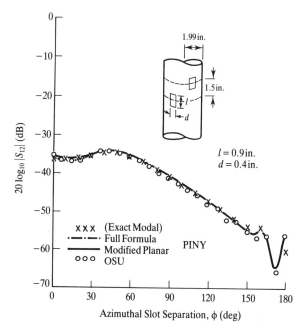

FIGURE 7.3 Isolation between two axial slots in a nonequatorial plane on a circular PEC cylinder: $a = 1.991$ in.; $z_0 = 1.50$ in.; $f = 9$ GHz [2].

neglected and Figure 7.3 shows a case where torsion is included. These results are compared with the exact solution and results from another study [4].

7.2.2 Using Hybrid Techniques

Hybrid methods (Section 1.17) of combining the method of moments with GTD have been employed [6, 7] to estimate coupling between two antennas on a cylindrical surface. The antenna is cast in the MOM format and the format is modified to take care of the presence of the conducting object with the help of high-frequency techniques. The expressions are obtained for the coupling between a transmitting and a receiving antenna situated on the surface of a PEC cylinder. Theoretical results were validated by comparing then with an intrasystem electromagnetic compatibility analysis program (IEMCAP) used by USAF. A brief description of IEMCAP is available in Chapter 10 on high-frequency codes.

We briefly outline the hybrid method [7] to solve the antenna coupling on a cylinder. The hybrid method is discussed in Section 1.17.1 using the curved surface UTD described in Section 1.18. This hybrid method will be used in this section to estimate the coupling. The electric field from a source at Q' as observed at an observation point Q is given by [2, 7]

$$E_n^e(P) = Z_0^2 DG(k_s)\mathbf{p}_e \cdot \hat{n}'\hat{n}\left\{F_n(\xi, p_1, p_2) - \frac{j}{k_s} F_h(\xi, p_1, p_2) + \left(\frac{j}{k_s}\right)^2 F_s(\xi, p_1, p_2)\right\}$$

$$+ T_0^2 \frac{j}{k_s} \{F_s(\xi, p_1, p_2) - F_h(\xi, p_1, p_2)\} \tag{7.5}$$

where Z_0 = the intrinsic impedance of free space
k = the wavenumber of the medium = $\omega\sqrt{\mu\varepsilon}$
δ = the angle that the tangent to the path makes with the cylinder axis
ρ_s = the radius of curvature of the surface at any point on the ray path = $\dfrac{a}{\sin^2 \delta}$
a = the radius of the cylinder
s = the length of the surface along the ray
$G(k_s) = k \dfrac{e^{-jks}}{j2\pi s Z_0}$, the Green's Function
$\xi = \dfrac{ms}{\rho_g}$
$m = \left(k\dfrac{\rho_g}{2}\right)^{1/3}$
$p_1 = m^{-1}kd_1$
$p_2 = m^{-1}kd_2$
d_1 = height of the source point P' above the cylinder surface
d_2 = height of the observation point P above the cylinder axis
T_0 = the torsion factor = $\cos \delta$ for a circular cylinder
D = the spatial factor = 1 for a circular cylinder
$F_h(\xi, p_1, p_2)$ = surface Fock function of hard type
$F_s(\xi, p_1, p_2)$ = surface Fock function of soft type
P_e = strength of a current moment source at P'
$\hat{n}'$ = unit normal vector at source point P'
$\hat{n}$ = unit outward normal vector at observation point P

For $\xi \neq 0$, the Fock function can be expressed as a Taylor series expansion:

$$F_h(\xi, p_1, p_2) = v(\xi) - \frac{j}{4}\xi^{-1}v_1(\xi)[p_1^2 + p_2^2] \tag{7.6a}$$

$$F_s(\xi, p_1, p_2) = u(\xi) + \frac{j}{2}[u'(\xi) - \frac{3}{2}\xi^{-1}u(\xi)][p_1^2 + p_2^2] \tag{7.6b}$$

7.2 MUTUAL COUPLING BETWEEN ANTENNAS ON A CYLINDER

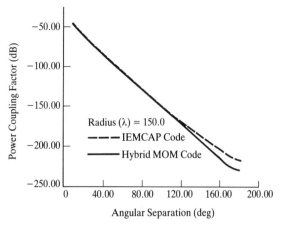

FIGURE 7.4 Power coupling factor between two $\lambda_0/4$ monopoles on a PEC cylinder of radius 150 wavelengths [7].

Numerical computations have been made in [7] for power coupling between two monopole antennas of length $\lambda/4$ with radius varying from λ to 150λ. Figures 7.4 through 7.8 [7] give the power coupling between two $\lambda/4$ monopoles versus the angular separation in degrees with five different values of cylinder radius. The power coupling factor between two $\lambda/4$ monopoles on a cylinder of radius 10λ versus separation along the cylinder axis in wavelengths is shown in Figures 7.9 and 7.10. Figures 7.9 and 7.10 are for angular separations of 90° and 180°.

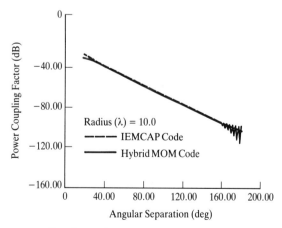

FIGURE 7.5 Power coupling factor between two $\lambda_0/4$ monopoles on a PEC of radius 10 wavelengths [7].

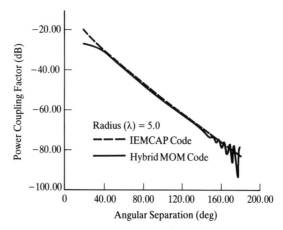

FIGURE 7.6 Power coupling factor between two $\lambda_0/4$ monopoles on a PEC cylinder of radius 5 wavelengths [7].

7.3 SHADING LOSS CALCULATIONS

Shading loss calculations can be done using methods like AAPG, IEMCAP, Keller and Levy's method [8], and the uniform theory of diffraction (UTSD). Section 1.18 surveys the different surface wave diffraction theories but only a few experimental results for such case studies are available [10].

Table 7.1 compares the maximum shading loss in dB for some case studies using AAPG, Keller and Levy's approach, and the uniform theory of surface wave diffraction (UTSD). Some experimental results [10] for monopoles on cylinder are available in the literature.

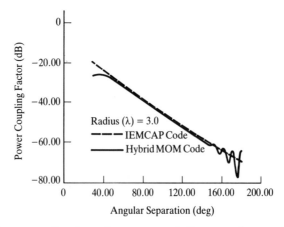

FIGURE 7.7 Power coupling factor between two $\lambda_0/4$ monopoles on a PEC cylinder of radius 3 wavelengths [7].

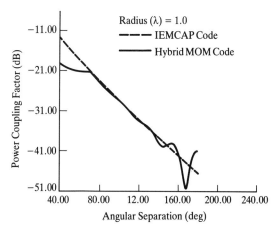

FIGURE 7.8 Power coupling factor between two $\lambda_0/4$ monopoles on a PEC cylinder of radius 1 wavelength [7].

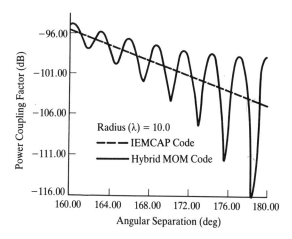

FIGURE 7.9 Power coupling factor between two $\lambda_0/4$ monopoles in dB versus separation along a PEC cylinder axis in wavelengths: angular separation = 90°; radius = 10 wavelengths.

7.4 MUTUAL COUPLING BETWEEN ANTENNAS ON A CONE

The same approach can be used for the mutual coupling calculations of thin slots on a cone and this is available in [2]. Figures 7.11 and 7.12 show the coupling coefficient S_{12} versus frequency. Figure 7.13 is for the slots at the same distance from the tip and Figure 7.14 is for different distances. The coupling shows an oscillatory nature with frequency with a period in the frequency domain. A study of cross-polarization components due to dipole moments has been done in [10].

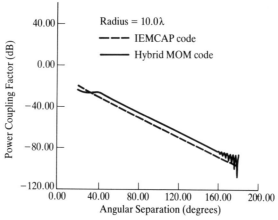

FIGURE 7.10 Power coupling factor between two $\lambda_0/4$ monopoles in dB versus separation along cylinder axis in wavelengths: angular separation = 180°; radius = 10 wavelengths.

TABLE 7.1 Maximum Shading Loss (dB) in Some Case Studies for a CC-144 Challenger Fuselage, Radius $a = 1.3462$ m

			Maximum Shading Loss (dB)		
f (MHz)	$\phi' - \phi$ (deg)	$z' - z$ (m)	AAPG	Keller's Approach	UTSD
118	81.27	−4.8768	6.6	0.5	1.31
229	132.31	−7.28	11.7	1.0	3.36
236	84.12	−2.324	11.9	2.1	5.99
336	84.12	−2.324	14.1	2.3	6.40

7.5 LIMITATIONS OF CURRENT ANTENNA COUPLING MODELS

The antenna-to-antenna coupling models AAPG and IEMCAP do not accurately predict the antenna coupling in the UHF/VHF frequency range for the following reasons [10] and need to be improved in order of importance.

1. The geometric models used in these codes are not very real models of complex structures at the high frequencies of interest. For example, very few modern aircraft can be modeled with a cone on a circular cylinder. Arbitrary geometries should be incorporated in geometric models in these codes.
2. No rigorous curved surface diffraction theory was used in these codes.
3. Basically knife-edge diffraction was used to consider wing edges. This type of approximation is for infinitely electrically thin edges. But at high

7.5 LIMITATIONS OF CURRENT ANTENNA COUPLING MODELS 397

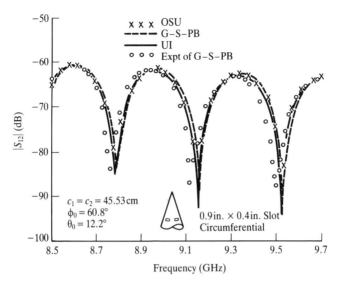

FIGURE 7.11 Coupling coefficient $|S_{12}|$ between two circumferential slots on a PEC cone versus frequency with distance between the slots as parameter, the angular separation $(\phi_0) = 80°$ and $C_1 = C_2 = 45.53$ cm. The half angle of the cone $(\theta_0) = 11°$ [2].

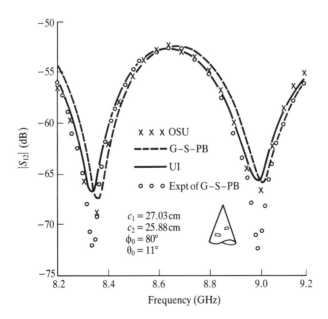

FIGURE 7.12 Coupling coefficient $|S_{12}|$ versus distance between two circumferential slots on a PEC cone with frequency as a parameter. The angular separation $(\phi_0) = 80$; $C_1 = 27.03$ cm, $C_2 = 25.88$ cm; $\theta_0 = 12.2$ [2].

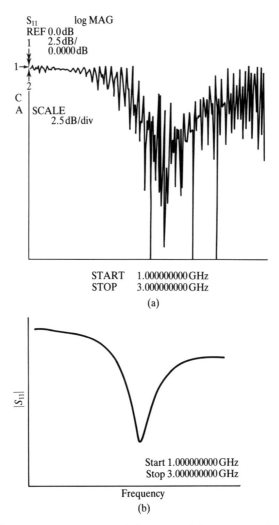

FIGURE 7.13 The frequency response of the transmitting antenna (a monopole on a PEC cylinder) [10]: (a) $|S_{11}|$ versus frequency (without gating); (b) $|S_{11}|$ versus frequency (with gating).

frequencies, some of the edges are electrically thick since the knife-edges may be thin wedges in reality. Therefore, a thick-edge diffraction coefficient (Section 1.13) should be used in these codes.

4. Both codes take into account only the shortest (direct) path between the transmitting and receiving antennas. In many situations, the multipath coupling may be significant compared to the direct path coupling. This is particularly true when the primary source has a wide beamwidth and hence the energy reaches the receiver from a number of secondary sources.

7.5 LIMITATIONS OF CURRENT ANTENNA COUPLING MODELS

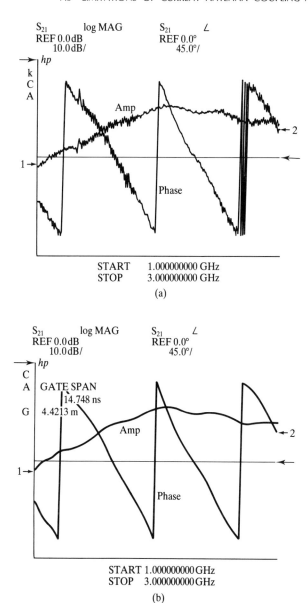

FIGURE 7.14 Amplitude and phase of coupling S_{12} between two monopoles on a cylinder (separation 20 cm and $a = 0.3885$ cm): (a) without gating; (b) direct + edge + greater path between transmitter and receiver; (c) experimental, direct + edge; and (d) only direct contribution. (*continued on next page*)

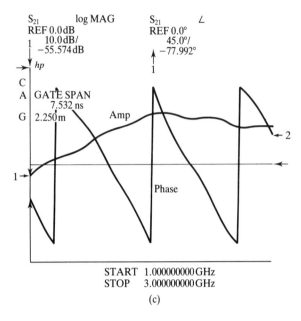

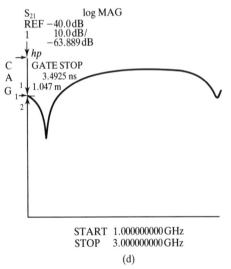

FIGURE 7.14 (*continued*).

5. The reflected fields are taken into account. These may be worthwhile to consider in some examples.
6. More accurate geodesic path calculations (some information is available in Section 1.4) should be included in the codes. The fuselage might be an elliptic cylinder.

7. Diffraction from the end of the cones and cylinders is not included. The curved surface treatment is for an infinite cylinder.
8. The effects of corner diffraction, as described in Section 1.9.2.2, has not been taken into account in these codes. This may be significant if a corner is strongly illuminated in a particular situation..
9. The AAPG code is capable of truncating a cylinder or a cone only with a flat bottom but it does not automatically take into account the diffraction that can occur from such truncations.
10. These codes cannot predict cross-polarized components. Cross-polarization is important in many target detection schemes in modern radars.

7.6 SOME EXPERIMENTAL RESULTS

Some experiments were carried in [8] using an HP8510 network analyzer to find the coupling between $\lambda/4$ monopoles in the equatorial plane of a large cylinder. The contribution to the receiver consists of: (1) the direct coupling through the shortest path between the transmitting and receiving monopoles; (2) the reflections from the edges of the finite cylinder; (3) the coupling between the greater path on the equatorial plane; and (4) numerous other sources in the background. In the experiment, the unwanted reflections were identified in the time domain and time-domain gating was used to eliminate undersired reflections. The time-domain response was then transformed to the frequency domain using time-to-frequency conversion software. It may be noted that the response of the transmitting antenna itself is not flat (Fig. 7.13) and therefore a correction is needed. The cylinder used in the experiment was one with radius $a = 0.3885$ m and $L = 0.9144$ m. The separations between the two monopoles along the arc length were 20 and 60 cm.

Figure 7.14 shows some typical experimental results of the amplitude and phase of S_{12} for the cylinder with a separation of 20 cm in the equatorial plane: (a) is without gating; (b) is the gated response keeping the contributions (1) through (3) above; (c) is gated keeping (1) and (2); and (d) is only the direct contribution. Figure 7.15 shows the variation of coupling between two monopoles in dB versus frequency with a cylinder radius of 0.3885 m and a separation of 20 cm with only the direct contribution.

7.7 CROSS-POLARIZATION COMPONENTS IN INTERANTENNA COUPLING CALCULATIONS

As well as the copolarized field components it is interesting to estimate the cross-polarized component generated by the presence of finite structures. When the geodesic on a cylinder is not restricted to the equatorial plane, it may be possible to generate cross-polarized field components. The existing interantenna

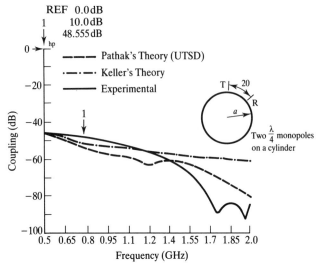

FIGURE 7.15 Coupling between two monopoles in dB versus frequency: $a = 0.3885$ m; separation = 20 cm (only direct contributions).

coupling programs, such as IEMCAP and AAPG, do not consider cross-polarization. Some work on cross-polarization between two small dipole moments on the surface of a cylinder was reported in [13].

The expressions for the fields at any point due to the dipole moments $d\tilde{p}_m$ and $d\tilde{p}_e$ are given by eqn. (1.127).

For the dipole sources on the cylinder, the co- and cross-polarized components are given by

$$H'_{zz'} = -\frac{jk}{2\pi Z_0} d\tilde{p}_m \left[\left(1 - \frac{j}{kt}\right)V(\xi) + t\left(\frac{j}{kt}\right) + \tilde{T}_0^2 \frac{j}{kt}(U(\xi) - V(\xi))\right]\frac{\exp(-jkt)}{\sqrt{t}} \quad (7.7a)$$

$$H'_{\phi\phi'} = -\frac{jk}{2piZ_0} d\tilde{p}_m \left[\frac{j}{k}V(\xi) + \frac{j}{kt}U(\xi) - 2\left(\frac{j}{kt}\right)^2 V(\xi)\right]\frac{\exp(-jkt)}{\sqrt{t}} \quad (7.7b)$$

$$H'_{\phi z} = j\frac{\exp(-jkt)}{2\pi kt} d\tilde{p}_m \tilde{T}_0 [U(\xi) - V(\xi)] \quad (7.7c)$$

$$R'_{zp} = \frac{-jk}{2\pi} d\tilde{p}_m \left[\left(1 - \frac{j}{kt}\right)V(\xi) + \tilde{t}_0^2 \frac{j}{kt}[U(\xi) - V(\xi)]\right]\frac{\exp(-jkt)}{\sqrt{t}} \quad (7.7d)$$

$$E'_{\phi'\rho} = \tilde{T}_0 \frac{d\tilde{p}_m}{4\pi t}[U(\xi) - V(\xi)]\frac{\exp(-jkt)}{\sqrt{t}} \quad (7.7e)$$

7.7 CROSS-POLARIZATION COMPONENTS IN COUPLING CALCULATIONS

Due to the electric dipole moment, the cross-polarized components are

$$H'_{\rho'z} = \tilde{T}_0 \frac{j}{kt} d\tilde{p}_m [U(\xi) - V(\xi)] \frac{\exp(-jkt)}{\sqrt{t}} \quad (7.7\text{f})$$

$$H_{\rho'z} = \frac{-jk}{2\pi} d\tilde{p}^e \left[\left(1 - \frac{6j}{kt}\right) V(\xi) + \tilde{T}_0 \frac{j}{kt} [U(\xi) - V(\xi)] \right] \frac{\exp(-jkt)}{\sqrt{t}} \quad (7.7\text{g})$$

The co- and cross-polarized components and their variations with parameters were computed in [12] for a cylinder both for equatorial and nonequatorial cases. Table 7.1 shows the comparison of shading loss (dB) for equatorial and nonequatorial cases with the same path length. **1** is for $z' = -1$ m, $z = 1$ m (nonequatorial case), 2 is for $z = z' = 1$ m (equatorial case). Figures 7.16 and 7.17 show the normalized field versus azimuthal angle ϕ for typical nonequatorial and equatorial cases. The parameters for Figure 7.16 are $a = 45$ m, $z' = 1$ m, $z = -1$ m, $f = 1000$ MHz and those for Figure 7.17

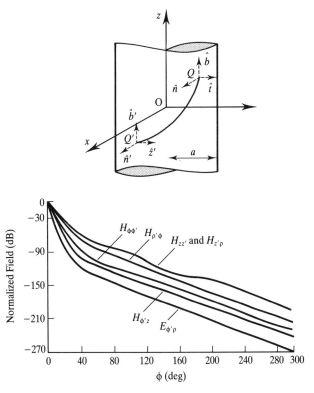

FIGURE 7.16 Normalized field decay versus ϕ for a cylinder with $a = 45$ m, $z' = -1$ m, $z = 1$ m, $f = 1000$ MHz. (From [13] © 1986 IEEE.)

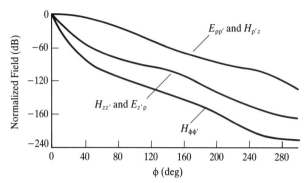

FIGURE 7.17 Normalized field decay versus ϕ for a cylinder with $a = 45$ m, $z' = z = 1$ m; $f = 1000$ MHz. (From [13] © 1986 IEEE.)

are $a = 45$ m, $z' = z = 1$ m, $f = 1000$ MHz. The transmitting dipole moment is assumed to be equal to unity and the receiving antenna is assumed to be a point source. It is found that the cross-polar components $E_{\rho\phi}$ (and $E'_\rho \phi$), $H'_{\phi z}$ (and $H_{\phi z'}$) are weaker than the copolar components $H_{zz'}$ (and $H_{z'z}$) and $H_{\phi\phi'}$ (and $H_{\phi'\phi}$) while $H_{z'\rho}$ is not. In the equatorial case, the cross-polar components do not get excited but by our calculations $E_{z'\rho}$ and $H_{\rho'z}$ are not predicted to assume zero values. The co- and cross-polar components are found to differ considerably in intensity for the same path length in equatorial and non-equatorial cases, particularly for smaller radii of curvature of the cylinder. This is because the effective radius of curvature depends on the angle of entry, even though the path length is the same for the nonequatorial case. The above analytical model is accurate for large radii of curvature of the cylinder. The dipole moment concept can be extended to linear antennas for analytical formulation of mutual coupling between two linear antennas.

7.8 CONCLUSIONS

Keller's model and AAPG/IEMCAP cannot predict cross-polarized fields on a structure when torsion is present and hence can only partially take care of the coupling phenomenon. Existing theories mostly work for large values of kR, where R is the distance of the observation point from the source. In Keller's approach of associating an attenuation constant and a complex surface diffraction coefficient, the first-order mode decays much faster than the zeroth-order mode for large radius of curvature. It is recommended that for intermediate values of radius of curvature, only the zeroth-order mode need be considered. Keller's formulations cannot predict cross-polarized components. The surface ray induced by the tangential component of E^i (soft case) seems to attenuate much faster than the normal component of E^i (hard case). The UTSD formulation can predict co- as well as cross-polarized components. I am still of the opinon that a thorough investigation of surface waves and shading loss mechanisms with practical antennas is necessary to accurately pinpoint EMI/EMC problems of antennas on a smooth conducting surface.

7.9 SUMMARY

This small chapter gave a limited discussion of the application of high-frequency techniques to the problem of mutual coupling between antennas on structures such as ground plane, cylinder, sphere, and cone. The limitations of the current antenna coupling models and a simple example of the effect of cross-polarization on the inter-antenna coupling calculations were also presented.

7.10 REFERENCES

1. *Electromagnetic Interference in New Aircraft*, NASA Tech. Briefs, ARG-12161, Ames Research Center, Moffett Field, CA.
2. P. H. Pathak and N. Wang, "Ray Analysis of Mutual Coupling between Antennas on a Convex Surface," *IEEE Trans. Antennas and Propagat.*, **AP-29**(11), 1981, 911–922.
3. D. A. McNamara, C. W. J. Pistorious, and J. A. G. Malherbe, *Introduction to the Uniform Theory of Diffraction*, Artech House, Boston, MA, 1990, Ch. 8.
4. S. W. Lee and S. Naini, "Approximate Asymptotic Solution of Surface Field due to a Magnetic Dipole on a Cylinder," *IEEE Trans. Antennas and Propagat.*, **AP-26**(4), 1978, 593–597.
5. J. Boersma and S. W. Lee, *Surface Fields due to a Magnetic Dipole on a Cylinder: Asymptotic Expaansion of Exact Solution*, Electromag. Lab., Dept. of Elec. Eng., Univ. of Illinois, Tech. Rep. 784583-7, Oct. 1978.
6. E. P. Ekelman and G. A. Thiele, "A Hybrid Technique for Combining the Moment Method Treatment of Wire Antennas with the GTD for Curved Surfaces," *IEEE Trans. Antennas and Propagat.*, **AP-28**(6), 1980, 831–839.
7. S. A. Davidson, "A Hybrid Method of Moments Technique for Computing Electromagnetic Coupling between Two Monopole Antennas on a Large Cylindrical Surface," MS thesis, Dept. Elec. Eng., Ohio State Univ., 1980.
8. B. R. Levy and J. B. Keller, "Diffraction by a Smooth Object," *Commn. Pure Appl. Math.*, **12**, 1959, 159–209.
9. G. Genello and A. Pasta, "Aircraft Coupling Model Evaluations at SHF/EHF," *IEEE Electromag. Compatibility Symp.*, Wakefield, MA, Aug. 20–22, 1985, pp. 72–74.
10. A. K. Bhattacharyya, *Shading Loss Calculations*, Tech. Note TN-EMC-87-01, Feb. 1987. Prepared for Defence Research Establishment, Canada, Dept. Elec. Eng., Concordia Univ., Montreal.
11. T. E. Durham, *SHE/EHF Antenna-to-Antenna EMI Coupling*, Interim Tech. Rep., Harris Corporation, Melborne, FL, Nov. 1986.

12. G. Genello, "Aircraft Coupling Model Evaluation at SHF/EHF," *IEEE Electromag. Compatibility Symp.*, Wakefield, WA, Aug., 20–22, 1985, pp. 72–74.
13. A. K. Bhattacharyya and S. J. Kubina, "Cross-Polarization Components in Inter-antenna Coupling Calculations," *IEEE Electromag. Compatibility Symp., San Diego, CA*, Session 3B, 1986, pp. 184–187.
14. W. D. Burnside, M. C. Gilreath, R. J. Marhefka, and C. L. Yu, "A Study of KC-135 Aircraft Antenna Patterns," *IEEE Trans. Antennas and Propagat.*, **AP-23**, 1975, 309–316.
15. C. L. Yu, W. D. Burnside, and M. C. Gilreath, "Volumetric Pattern Analysis of Airborne Antennas," *IEEE Trans. Antennas and Propagat.*, **AP-26**, 1978, 636–641.

7.11 ADDITIONAL SOURCES

G. Mazzarella and G. Panariello, "On the Evaluation of Mutual Coupling Between Slots," *IEEE Trans. Antennas and Propagat.*, **AP-35**(11), 1987, 1289–1293.

C. R. Paul, *Introduction to Electromagnetic Compatibility*, John Wiley, 1992.

W. Streifer, "Creeping Wave Propagation Constants for Impedance Boundary Conditions," *IEEE Trans. Antennas and Propagat.*, **AP-12**, 1964, 764–766.

CHAPTER EIGHT

Terrain Scattering and Propagation Modeling Using High-Frequency Techniques

8.1 INTRODUCTION

The role of terrain features is very important in the design and evaluation of ground-to-ground and ground-to-air communication links as well as to low-altitude radar. An accurate and reliable knowledge of the terrain features can help in determining the area of coverage and the potential sources of interference to the link. In this chapter, we deal with terrain modeling, attenuation, and propagation of electromagnetic waves using high-frequency techniques; Chapter 9 deals with radar clutter models.

Terrain modeling may be required to handle a relatively narrow frequency band, or in the case of an ultrawide-band pulsed high-resolution radar a very wide frequency band. The two basic models that are commonly used are the Longley–Rice model and the GTD model. To the best of my knowledge, no book gives a discussion on recent advances in terrain modeling using high-frequency techniques; an early book is reference [1]. It is worthwhile to give here a brief and perhaps incomplete summary of the terrain modeling research. An empirical prediction model for multipath propagation of pulse signals at VHF and UHF on an irregular terrain was described in [2] and results for rural, hilly terrain and a built-up metropolitan area were compared with experiment. An experimental study [3] was made in Stanford University on microwave propagation over mountain-diffraction paths. VHF propagation measurements over hilly and forested terrain at low altitude was reported in [4]. Terrain modeling using half plane geometries was started as early as 1976 where the ground was modeled [5] with a series of strips including the effect of arbitrarily oriented ground planes. Evaluations of Longley–Rice and GTD propagation models were done in [6]. Attenuation due to edge and surface diffraction is important in terrain modeling studies. An attenuation function for multiple knife-edge diffraction has been obtained in [7] and a series solution for the same problem has recently been presented [8]. Finite conductivity uniform GTD and knife-edge diffraction models have been compared in [9] from the point of view of their capability of predicting propagation path loss. Propagation prediction in hilly terrain using GTD wedge diffraction was formulated in [10]. A terrain model of glides slope using high-frequency techniques for instrument landing systems (ILS) is described in [11]. A comparison of lossy wedge diffraction coefficients in the prediction of mixed path propagation loss is available in [12]. A comparison was made in [13] of different lossy wedge diffraction coefficients with applications to mixed path propagation loss prediction. Based on the narrow-band GTD model a comparison was made in [13] of the GTD wide-band propagation model with experiments.

8.2 A SUMMARY OF LONGLEY–RICE AND GTD PROPAGATION MODELS

8.2.1 Longley–Rice Model

The Longley–Rice model [14] is an approximate model which was developed at the Institute of Telecommunication Sciences. It is a point-to-point model for paths for when information describing the terrain is limited. The input terrain profiles were available from the Continental United States (CONUS) data base. The input parameters are the heights of transmitter and receiver, frequency, polarization, ground constants, surface refractivity, and terrain profile. The output parameters are complex reflection coefficients, scatter attenuation as a function of frequency, terrain profile for each propagation path, and atmospheric absorption. The receiving antenna is considered to be in one of the three regions; line of sight, diffraction, and scatterer. For line of sight propagation the path loss is calculated by Schelkunoff and Frii's formula on the basis of the straight path and single reflection from a point on the baseline joining the footprints of the transmitting and receiving antennas. The reflection coefficient is calculated by standard oblique incidence Fresnel formulas using average values of ground parameters. The complex reflection coefficient is modified to incorporate the terrain irregularity. Terrain irregularity is due to the change of terrain elevation with position due to the presence of natural objects like hills, mountains, cliffs, and dense vegetation. When the receiving antenna is far away from the shadow boundaries, no diffraction correction is incorporated; when the receiving antenna comes near the shadow boundary, diffraction attenuation is incorporated in the path loss. When the receiver is positioned below the horizon of the transmitter, the path loss is calculated by using a knife-edge diffraction phenomenon (Section 8.3).

There are problems with this model. It considers the entire region between the two antennas in path loss calculations. This may be incorrect as very often there are obstacles on the line joining the transmitting and receiving antennas. The Longley–Rice model does not incorporate multipath contributions to the receiver given the transmitting antenna and the terrain parameters.

8.2.2 GTD Model

Terrain modeling using uniform GTD is more versatile and realistic [6]. Also, it can be extended to model wide-band communication links and three-dimensional terrain effects. Apart from the direct and reflected rays, it considers all secondary and multipath contributions. The many different rays along the paths between the transmitting and receiving antennas. The first-order contributions are: (1) the direct ray if there is no blocking, (2) and reflected ray(s), and (3) diffracted ray(s). The second-order contributions are: (1) reflected–diffracted, (2) reflected–reflected, (3) diffracted–diffracted, and (4) reflected–diffracted. The third-order contributions are: (1) reflected–reflected–diffracted, (2) reflected–diffracted–diffracted, (3) diffracted–reflected–reflected, (4) reflected–reflected–reflected, (5) diffracted–diffracted–reflected, and (6) diffracted–diffracted–diffracted.

The input parameters in this model are: (1) a piecewise-linear two-dimensional particular terrain profile, (2) location of the transmitting antenna, (3) location of the receiving antenna, (4) frequency, (5) distance, and (6) electrical parameters of the ground. The accuracy of path loss prediction primarily depends on the accuracy with which the particular terrain profile can be obtained. Once the terrain profile is specified, the GTD model generates the many different primary and secondary contributions listed above and calculates the total field at the receiving point. The roughness factor, which is due to the local irregularity, is incorporated in the reflection coefficient and is given by

$$R_g^{\perp,\|} = \delta_s R_0^{\perp,\|} \tag{8.1a}$$

where δ_s is the surface roughness factor given by

$$(\delta_s)^2 = e^{-2(\Delta\phi)} \tag{8.1b}$$

$$\delta\phi = \frac{4\pi\Delta h}{\lambda}\sin\gamma \tag{8.1c}$$

where Δh is the standard deviation of the terrain elevation corresponding to each of the piecewise-linear sections of the terrain, γ is the grazing angle, λ the wavelength, $\Delta\phi$ is the phase difference between the shortest and longest path. $R_s^{\perp,\|}$ is the plane wave reflection coefficient for specular reflection from a rough surface for perpendicular and parallel polarization, and $R_0^{\perp,\|}$ is the above coefficient for a smooth plane surface instead of a rough surface.

410 TERRAIN SCATTERING AND PROPAGATION MODELING

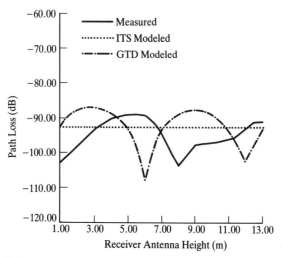

FIGURE 8.1 Path loss versus height of receiving antenna at 1846 MHz for a planar sloping terrain. (From [6] © 1982 IEEE.)

The UTD diffraction coefficients which take care of the continuity at the shadow boundaries and the aspect-dependent reflection coefficients are used.

8.2.3 A Comparison Between Longley–Rice and GTD Propagation Models

In this section we discuss the two models with some examples of practical terrain profiles. Figure 8.1 [6] compares a typical case of path loss versus receiving antenna height at a frequency of 1846 MHz for a simple profile of a planar slopping terrain of path length 0.54 km. A terrain roughness factor (δ_s) of 9 was used. It is found that the GTD model is a better match than the Longley–Rice model. Figure 8.2 shows path loss versus receiver antenna height for a terrain profile; the piecewise-linear approximation is presented in Figure 8.2(a); Figures 8.2(b) through (e) are for frequencies 210, 751, 4595, and 9190 MHz. Again, the GTD model gives a better match with the experiment. With a transmitting antenna height of 6.6 m below 910 MHz and 7.3 m for higher frequencies. The CPU time is greater for the GTD model than the Longley–Rice model and it increases with the number of edges. Figure 8.3(a) shows a terrain profile with the piecewise-linear approximation and Figure 8.3(b) and (c) show the path loss versus receiver antenna height for the profile in Figure 8.3(a). The two sets of measured results are shown and are labeled "open" and "concealed." An open path is an unblocked path and a concealed path is a blocked path. It is again obvious that the GTD model correctly accounts for the reflection and diffraction while the Longley–Rice model cannot. A limitation of the GTD model is that it does not calculate the scatter attenuation.

8.2 A SUMMARY OF LONGLEY–RICE AND GTD PROPAGATION MODELS

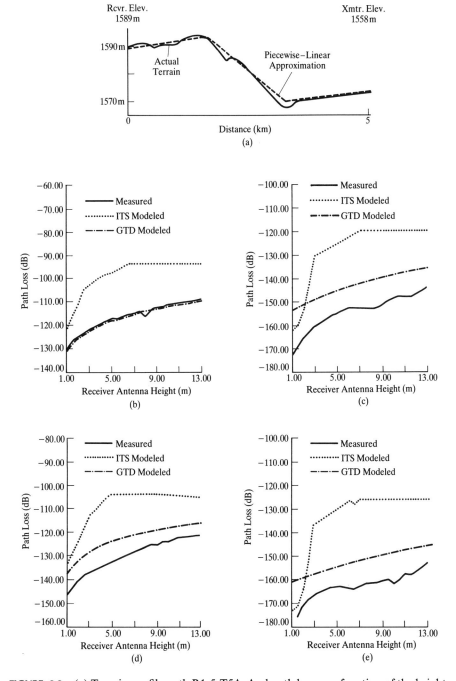

FIGURE 8.2 (a) Terrain profile path R1-5-T5A. And path loss as a function of the height of the antenna at (b) 210 MHz, (c) 751 MHz, (d) 4595 MHz, and (e) 9190 MHz. (From [6] © 1982 IEEE.)

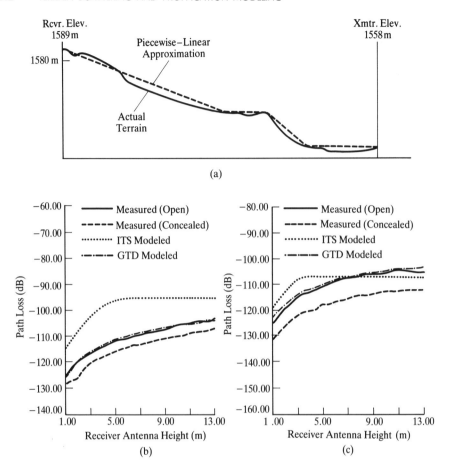

FIGURE 8.3 (a) Terrain profile R1-5-T3. And path loss as a function the height of the receiving antenna at (b) 230 MHz and (c) 910 MHz. (From [6] © 1982 IEEE.)

8.3 KNIFE-EDGE DIFFRACTION AND AN ATTENUATION FUNCTION FOR MULTIPLE KNIFE-EDGE DIFFRACTION

Knife-edge diffraction theory [15] is formulated by postulating a secondary Huyghens' source in the plane of the knife-edge. The edge is considered to be totally absorbing and the transmitting field is calculated from what goes above the edge to the receiver. The major advantage of this approach is that it is very simple. Knife-edge diffraction gives results close to uniform GTD in situations where the knife-edge can be applied. Knife-edge diffraction approximations do not consider important parameters such as polarization, composition, conductivity, edge permeability, edge permittivity, ridge profile, and surface roughness. In many practical situations, multiple edges are encountered. Therefore, a multiple-edge diffraction coefficient is necessary. In this section, we briefly describe a multiple knife-edge diffraction coefficient.

8.3 ATTENUATION FUNCTION FOR MULTIPLE KNIFE-EDGE DIFFRACTION

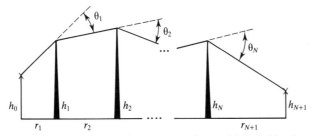

FIGURE 8.4 Geometry and coordinate system for multiple knife-edge diffraction.

Figure 8.4 shows the geometry and coordinate system of the diffraction over N knife-edges. The details of the analysis are available in [16, 17]. The attenuation function A is given by

$$A = (1/2^N) \cdot C_N e^{\sigma N} \sum_{m=0}^{\infty} I_m \tag{8.2}$$

where

$$I_m = 2^m m! \alpha_1^m I(m, \beta_1) I(m, \beta_2) \quad \text{for } N = 2 \tag{8.2a}$$

$$I_m = 2^m \sum_{m_1=0}^{m} \cdots \sum_{m_{N-1}=0}^{m_{N-3}} \prod_{i=1}^{N} \frac{(m_{i-1} - m_{i+1})!}{(m_i - m_{i+1})!} \cdot \alpha_1^{m_i - 1 - m_i} I(n_i, \beta) \quad \text{for } N \geq 3 \tag{8.2b}$$

With

$$n_i = (m_0 - m_1), \quad i = 1 \tag{8.2c}$$

$$n_i = (m_{i-2} - m_i), \quad 2 \leq i \leq N - 1 \tag{8.2d}$$

$$n_i = (m_{N-2} - m_{N-1}), \quad i = N \tag{8.2e}$$

$$\alpha_N \equiv 1 \quad m_0 = m \quad m_k = 0 \tag{8.2f}$$

where $k \geq N - 1$. The constant C_N is given by

$$C_N = 1 \quad N = 1$$

$$C_N = \left[\frac{r_2 r_3 \cdots r_N r_T}{(r_1 + r_2)(r_2 + r_3) \cdots (r_N + r_{N+1})} \right]^{1/2}$$

The α_m and β_m are given by

$$\alpha_m = \left[\frac{r_m r_{m+2}}{(r_m + r_{m+1})(r_{m+1} + r_{m+2})} \right]^{1/2} \quad m = 1, \ldots, N-1$$

$$\beta_m = \theta_m \left[\frac{ikr_m r_{m+1}}{2(r_m + r_{m+1})} \right]^{1/2} \quad m = 1, \ldots, N$$

414   TERRAIN SCATTERING AND PROPAGATION MODELING

In [7], a computer program was prepared which computes the attenuation A for propagation paths involving knife-edges where $0 \leq N \leq 10$. Free-space and single knife-edge paths are for $N = 0$ and $N = 1$.

Example 8.1 [7]: A 30 km propagation path at a frequency $f = 100$ MHz has two fixed knife-edges at distances of 10 km and 20 km from the transmitter (Fig. 8.4). $h_1 = h_3 = 100$ km. $h_0 = h_4 = 0$. A third knife-edge with variable height h_2 is placed in the middle of the path of 15 cm. Figure 8.5 gives the attenuation in dB versus knife-edge height h_2 for the 30 km path. The following observations can be arrived at: (1) If h_2 is well above the reference plane, the path is a fixed double knife-edge and the attenuation is 21.61 dB; (2) with an increase in h_2, the signal oscillates because of interference effects from the two knife-edges.

Example 8.2 [7]: The propagation path at $f = 500$ MHz in which the height h_r of the receiver with respect to a height h_t of the transmitter. Figure 8.6 shows the attenuation in dB versus the height of the receiver h_r for the 50 km path with the number of knife-edges as parameter. It is concluded that an additional knife-edge cause a significant increase in the attenuation when the receiver is below the line-of-sight (LOS) region. The added attenuation A_a is given by

$$A_a = 20 \log(1 + N_0) \tag{8.3}$$

Further discussion on the results of knife-edge diffraction can be found in [7]. In conclusion, knife-edge diffraction theory applied to terrain propagation with multiple edges works for situations where one can make a knife-edge approximation.

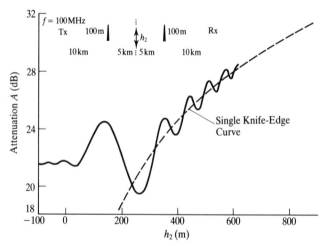

FIGURE 8.5   Attenuation (dB) versus knife-edge height h_2 in Example 8.1 [7].

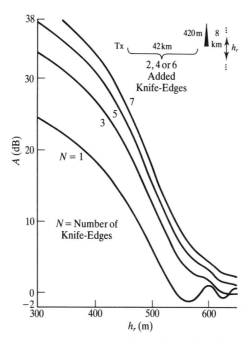

FIGURE 8.6 Attenuation (dB) versus height h_t of the receiver for the 50 km path with number of knife-edges as parameter [7].

8.4 FINITE CONDUCTIVITY GTD

It is well known that the original formulation of uniform GTD was for PEC wedges; heuristic extensions were made in [6, 17]. Now a modified uniform GTD coefficient is available which incorporates finite conductivity and local roughness of wedge surfaces, an edge profile of the wedge, and multiple diffractions. The accuracy of knife-edge diffraction can be improved by including the finite conductivity of the wedge surfaces where knife-edge diffraction is not valid. Figure 8.7 shows the geometry and coordinate system of diffraction from

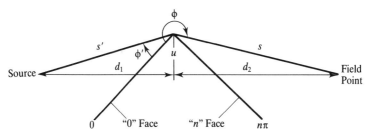

FIGURE 8.7 Geometry and coordinate systems for diffraction from a knife-edge and a wedge [7].

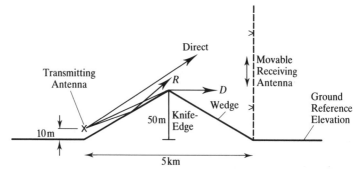

FIGURE 8.8 Illustration of the difference between GTD wedge diffraction and knife-edge diffraction. The direct, reflected, and diffracted rays used in GTD calculations are shown [9].

a knife-edge and a wedge. Figure 8.8 illustrate the difference between GTD wedge diffraction and knife-edge diffraction. Direct, reflected, and diffracted rays are used in this calculation. The following are the expressions of the fields for knife-edge and wedge diffraction:

A. *Knife-Edge Diffraction*

$$E_{KE} = E_0 \frac{e^{-jk(d_1+d_2)}}{d_1+d_2} \cdot \frac{1+j}{2} \cdot \int_v^\infty \exp(-j(\pi/2)\tau^2) \, d\tau \qquad (8.4)$$

where

$$v = u\sqrt{\frac{2(d_1+d_2)}{\lambda d_1 d_2}}$$

and the assumption is that both d_1 and d_2 (Fig. 8.6) are large compared to u and λ.

B. *GTD Formulation*

$$E_{GTD} = E_0 \frac{\exp(-jks')}{s'} D_\parallel^\perp \sqrt{\frac{s'}{s(s'+s)}} \cdot \exp(-jks) \qquad (8.5)$$

where the diffraction coefficient is given by eqns. (A.1) to (A.3) and s and s' are shown in Figure 8.7.

The diffraction coefficient for a field diffracted by a wedge with finite conductivity and local surface roughness is given by [18]

$$D_{\parallel}^{\perp} = \frac{-\exp(-j\pi/4)}{2n\sqrt{2\pi k}} \cot \frac{\pi + (\phi - \phi')}{2n} \cdot F(kLa^+(\phi - \phi')) + \cot \frac{\pi - (\phi - \phi')}{2n}$$

$$\cdot F(kLa^-(\phi - \phi')) + R_0^{\perp,\parallel} \cdot \cot \frac{\pi - (\phi + \phi')}{2n} \cdot F(kLa^-(\phi + \phi'))$$

$$+ R_n^{\perp,\parallel} \cdot \cot \frac{\pi + (\phi + \phi')}{2n} \cdot F(kLa^+(\phi + \phi')) \qquad (8.6)$$

where $R_{0,n}^{\perp,\parallel}$ are the Fresnel reflection coefficients for perpendicular and paarallel polarizations for the 0 and n faces with surface roughnesses included and the surface roughness factors are given by eqn. (8.1).

8.4.1 Comparison of Finite Conductivity Uniform GTD and Knife-Edge Diffraction in Path Loss Prediction

Comparison has been made for several cases of terrain profiles with knife-edge diffraction and wedge diffraction. Figure 8.9 shows the path loss in dB versus height of the antenna for the idealized terrain profile of Figure 8.8 at frequencies 0.1, 1.0, and 10.0 GHz obtained using knife-edge diffraction theory for both polarization. Figure 8.10 shows the same path loss calculated using the direct, reflected, and diffracted rays with the assumption of PEC surface. Figure 8.11 shows the same path loss as in Figure 8.9 assuming terrain parameters $\sigma = 0.012$ S/m, $\varepsilon_r = 15.0$, and standard deviation of surface roughness equal to

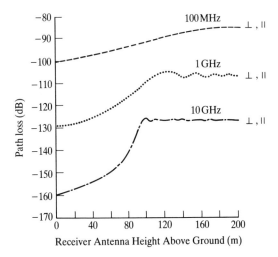

FIGURE 8.9 Path loss (dB) versus antenna height for the terrain profile of Figure 8.8 using knife-edge diffraction at 0.1, 1.0, and 10.0 GHz for both polarizations [9].

418 TERRAIN SCATTERING AND PROPAGATION MODELING

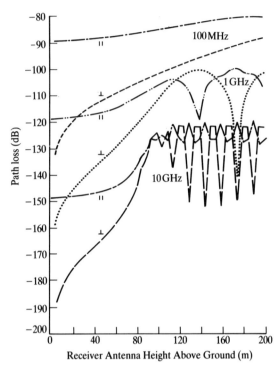

FIGURE 8.10 Path loss (dB) versus antenna height for the idealized terrain profile of Figure 8.8 for both polarizations using the GTD, including direct, reflected, and diffracted rays for smooth PEC surfaces [9].

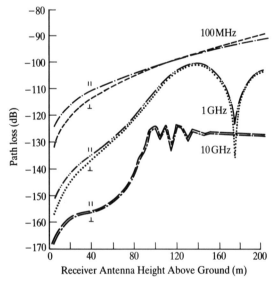

FIGURE 8.11 Same as Figures 8.9 and 8.10 but for the terrain parameters $\sigma = 0.012$ s/m, $\varepsilon_r = 15.0$ and surface roughness standard deviation = 0.5 m [9].

0.5 m. The application of any model to an actual terrain geometry is always a good test. An actual terrain profile with knife-edge and wedge approximations to the dominant diffracting ridge and antenna locations is shown in Figure 8.12. A comparison of calculated and measured path loss versus the height of the receiving antenna above local terrain for the terrain profile of Figure 8.12 at 230 MHz is shown in Figure 8.13 for horizontal polarization and in Figure 8.14 for vertical polarization. The imperfect conductivity GTD results are furthest from experiment for horizontal polarization and closer to the experiment for vertical polarization. The height of the transmitting antenna is 6.6 m, the

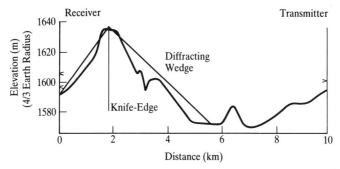

FIGURE 8.12 The terrain profile of path length 28.2 km showing the knife-edge and wedge approximations [9].

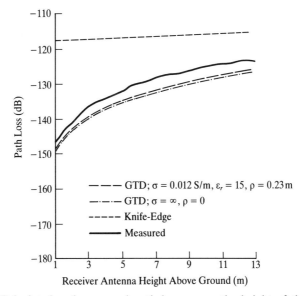

FIGURE 8.13 Calculated and measured path loss versus the height of the antenna at 493 MHz for horizontal ($\perp$) polarization. The terrain profile is shown in Figure 8.12 [9].

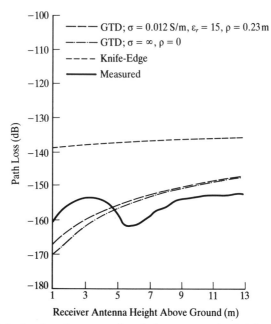

FIGURE 8.14 Calculated and measured path loss versus the height of the antenna at 493 MHz for vertical ($\|$) polarization. The terrain profile is shown in Figure 8.12 [9].

polarization is horizontal and only singly diffracted rays are included. The same case with frequency 1846 MHz and transmitter height 7.3 m is shown in Figure 8.15. Some results are presented for vertical polarization in Figures 8.17 and 8.18 with the terrain profile shown in Figure 8.16. It is concluded that the wedge model is much more accurate than the knife-edge diffraction model, particularly in cases of large wedge angles, grazing incidence, and large diffraction angles.

8.5 HILLY TERRAIN MODELING

Hilly terrain can be modeled by multiple knife-edge and multiple wedge diffraction depending upon the exact terrain profile, each one having distinct advantages and disadvantages. The multiple-wedge model is often accurate but it has problems at or near the shadow boundaries, for example, when the source and the vertices of two wedges are in the same straight line. Multiple-wedge diffraction has been treated in [6, 19] and the problem of shadow boundary fields which are nonoptical can also be treated. Fortunately, since the terrain distribution is rather random, the probability of the edges coming in line with the source is not very high. The geometry for double and triple wedge diffraction using GTD is shown in Figure 8.19. The divergence factor Γ and the distance parameters L used for triply diffracted rays are as follows. Results of multiple-edge modeling compared with experiments are available in [10].

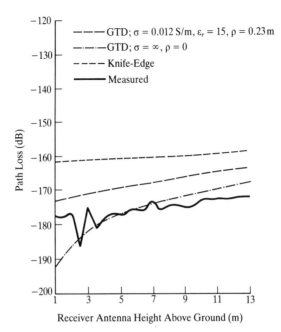

FIGURE 8.15 Calculated and measured path loss versus the height of the antenna at 1846 MHz. The height of the antenna above the local terrain is 7.3 m; horizontal polarization; only singly diffracted rays [9].

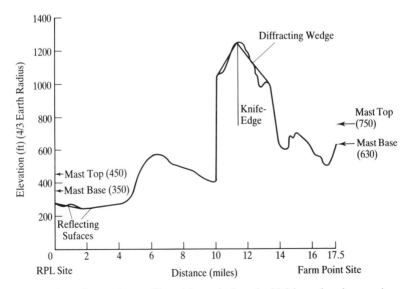

FIGURE 8.16 Actual terrain profile with path length 28.2 km showing wedge and knife-edge approximations to the dominant diffracting edge. Both methods include reflections from indicated surfaces [9].

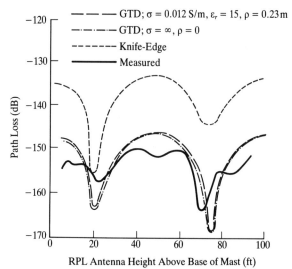

FIGURE 8.17 Calculated and measured path loss versus the height of the antenna for ths terrain profile of Figure 8.16 at 493 MHz; horizontal polarization farm point antenna 112 ft above base [9].

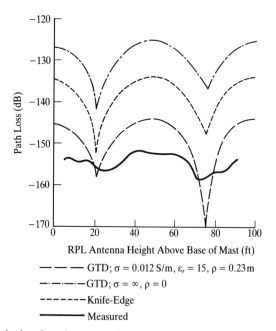

FIGURE 8.18 Calculated and measured path loss versus the height of the antenna for the terrain profile in Figure 8.16 at 493 MHz; vertical polarization: farm point antenna 112 ft above base [9].

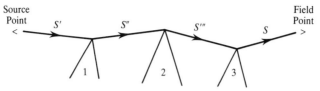

FIGURE 8.19 Geometry and coordinate system for double and triple wedge diffraction using GTD.

For the ray diffracted by wedge 1 to the vertex of wedge 2:

$$\Gamma_{S12} = \sqrt{\frac{S'}{S''(S' + S'')}}$$

$$L_{S12} = \frac{S'' \cdot S''}{S' + S''}$$

For the ray diffracted from wedge 1 to wedge 2 and diffracted by wedge 2:

$$\Gamma_{123} = \sqrt{\frac{S' + S''}{S'''(S' + S'' + S''')}}$$

$$L_{123} = \frac{S'' \cdot S'''}{S'' + S'''}$$

For the ray from the source diffracted by wedge 1 to wedge 2, diffracted by wedge 2 to wedge 3, and then diffracted by wedge 3:

$$\Gamma_{23F} = \sqrt{\frac{S' + S'' + S'''}{S(S + S' + S'' + S''')}}$$

$$L_{23F} = \frac{S \cdot S'''}{S + S'''}$$

The significances of other symbols is indicated on Figure 8.19.

While using the above formulae, the transition region nonray-optical fields should be taken into account.

The results of hilly terrain modeling with multiple edges are available in [10]. Figure 8.20 shows a comparison of GTD/UTD and knife-edge diffraction of path loss with edge separation in meters and with specified parameters. Figure 8.21 compares GTD with Vogler's multiple PEC knife-edge diffraction as a function of the height of the receiver with the second edge not at or around the shadow boundary.

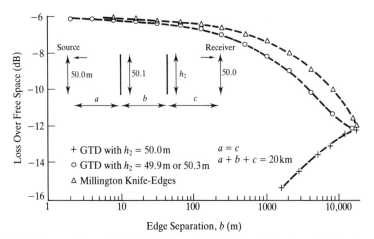

FIGURE 8.20 Comparison of GTD/UTD and knife-edge diffraction. Knife-edge results are for edges completely aligned ($h_1 = h_2 = 50.0$ m). GTD results are for the second edge located to block ($h_2 = 50.1$ m) or not to block ($h_2 = 49.9$, 50.3 m) the SDR. GTD curves take care of DDRs; PEC Case [10].

We will not compare minute details of all propagation models. Additional information on comparison of different methods using the topographic data from a terrain bank can be found in [20, 21].

8.6 IMPEDANCE WEDGE MODELING

The losses in the wedge diffraction process can be taken into account as discussed in Section 8.5 by introducing a reflection coefficient in the diffraction coefficient given by eqn. (8.6). A general way of doing this would be to use a general two-face impedance diffraction coefficient as given by Maliuzhinets. Maliuzhinets' diffraction coefficient is discussed in Sections 1.11.1 and 1.11.2. Another way is to use a heuristic surface impedance diffraction neglecting surface waves. The use of a two-face impedance wedge diffraction coefficient enhances the capability of a model to incorporate mixed paths involving, for example, water, ordinary land, vegetated land, and so on. This includes the interfaces between dissimilar materials. These interfaces are wedges with 90° interior angle and different surface impedances at the two faces.

The Maliuzhinets wedge diffracted field is given by eqn. (1.120a). The two-face impedance plane is shown in Figure 8.22 and the loss over free space with the height of the receiver is plotted [12] in Figure 8.23. The 0 face is aluminum and the n face is earth. Figure 8.23(a) is for a heuristic case and Figure 8.23(b) is for a Maliuzhinets solution. The agreement is good for both cases. In the heuristic case, the plots are smoother than in the Maliuzhinets case.

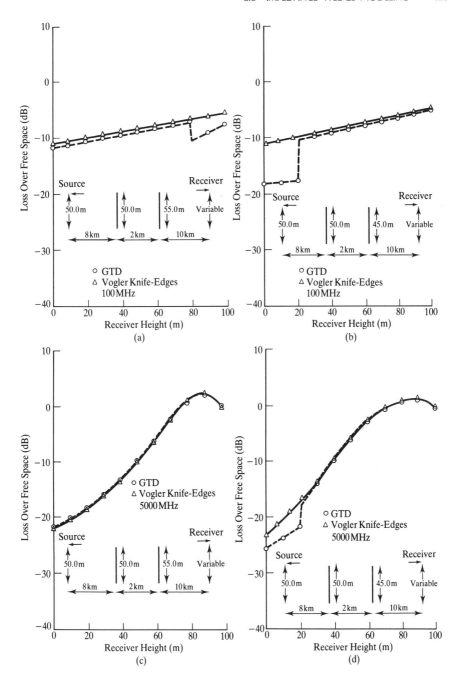

FIGURE 8.21 Comparison of GTD/UTD and multiple knife-edge diffraction as a function of the height of the receiver with the second edge [10]. (a) Above the shadow boundary, $f = 100$ MHz. (b) Below the shadow boundary, $f = 100$ MHz. (c) Above the shadow boundary, $f = 5000$ MHz. (d) Below the shadow boundary, $f = 5000$ MHz.

426 TERRAIN SCATTERING AND PROPAGATION MODELING

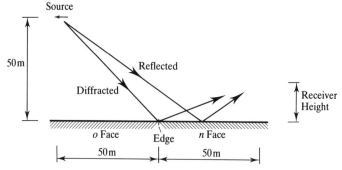

FIGURE 8.22 Diffraction by a two-face impedance half plane.

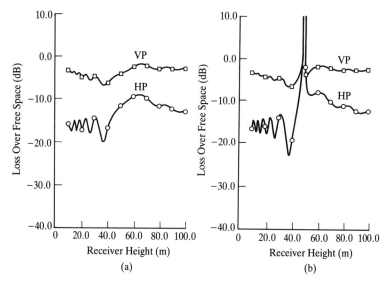

FIGURE 8.23 Loss over free space (dB) with the height of the receiver: (a) using a heuristic approach; (b) using Maliuzhinets' diffraction coefficient. (From [12] © 1988 IEEE.)

8.7 SOME EXPERIMENTAL RESULTS

In this section we summarize some of the experimental results which have been reported in the literature and compare with theoretical models whenever possible. Table 8.1, available in [4], gives a comparison between measurements and models.

In [3], an experimental study of swept-frequency transmission was made over mountain-diffraction paths. The important aspects are polarization dependence, spatial variation, and direct comparison with repeater-relayed signals. A comparison of data from two carefully selected paths shows the uniform transverse obstacle profile must be taken into account to successfully consider

TABLE 8.1 Deviation Between Measurements and Models for Different Paths

Path		RMS Difference (dB)		
Name	Distance (km)	SKEM	DKEM	TKEM
Natty Pond	6.6	1.61	1.23	1.23
West Ware River	12.5	1.25	1.18	1.50
East Ware River	14.8	2.49	1.62	2.18
Forest Hill	7.7	1.58	1.71	1.71
Canesto Brook	8.1	2.45	2.56	2.62
Kendall Cemetery	15.3	3.31	1.41	1.06

SKEM = single knife-edge model
DKEM = double knife-edge model
TKEM = triple knife-edge model

diffraction effects. Tests were done with 48 hours duration and the following parameters were computed every 2 min: the mean power $\bar{P}(f)$ at each sample frequency; the across-the-band mean power $\bar{P}$; the frequency correlation $R(\Delta f)$ of the field strength magnitude; the correlation bandwidth δf_R—the frequency separation at which the frequency correlation drops to a specified value; and the maximum variation of power δP with frequency. The measurements were made with the path profiles shown in Figure 8.24; the transverse profile of the terrain as viewed from the receiver is shown in Figure 8.25. Figures 8.26 and 8.27 show the results of a 48 hour test.

8.8 A THREE-DIMENSIONAL POLARIMETRIC TERRAIN PROPAGATION MODEL

A versatile wave propagation model was proposed in [22], VHF/UHF frequencies considering three-dimensional terrain. The model uses a digital terrain data bank and incorporates multipath effects. The defects [10] of the initial models are: (1) most of them are valid over a narrow frequency band; (2) they are designed for a specific type of terrain, and a generic type; (3) the empirical corrections that can be made cannot often meet the requirements of a complex practical scenario; and (4) they are severely lacking in resolution. With these shortcomings, the early models cannot meet the requirements of modern high-efficiency spectrum management. Of course, the introduction of finite conductivity and two-face impedance wedge diffraction and efficient ray tracing have improved the models quite a lot. A recent wave propagation model [22] has the following features:

1. Three-dimensional digital terrain data with high resolution.
2. More realistic terrain modeling: terrain profile approximation by wedges and convex surfaces and implementation of topography and morphography.

428 TERRAIN SCATTERING AND PROPAGATION MODELING

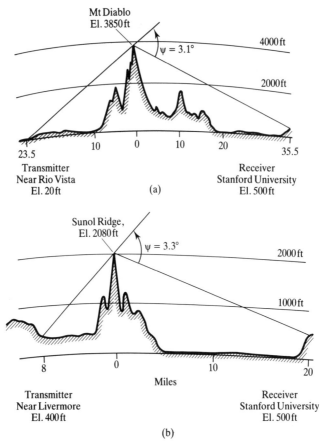

FIGURE 8.24 Path profiles: (a) Mt. Diablo path; (b) Sunol Ridge path. (From [3] © 1966 IEEE.)

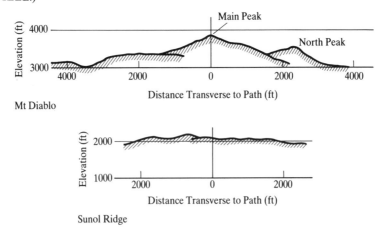

FIGURE 8.25 The transverse obstacle profiles as viewed from the receiver. (From [3] © 1966 IEEE.)

8.8 A THREE-DIMENSIONAL POLARIMETRIC TERRAIN PROPAGATION MODEL 429

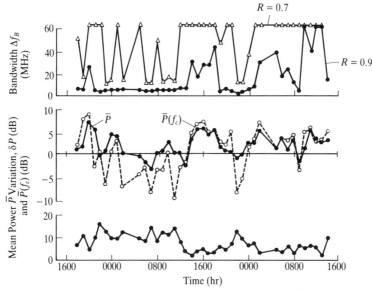

FIGURE 8.26 Bandwidth, mean power, and mean power at each frequency for the Mt. Diablo path, October 7–9, 1962 [22].

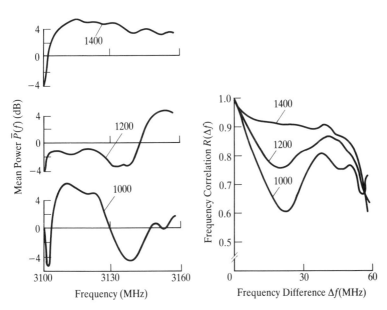

FIGURE 8.27 Mean power and frequency correlation for the Mt. Diablo path, October 8, 1962 [22].

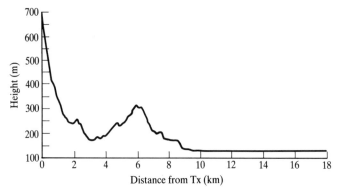

FIGURE 8.28 Terrain profile T × "Merkur" → Rhine. T × height 699 m [22].

3. Propagation calculations include bistatic rough surface scattering apart from PO and GTD.
4. The complex probability density function (PDF) and field strength delay spectrum (FDS) are derived from the multipath signals.

A detailed discussion of the model will not be presented here since it involves many things which are not high-frequency techniques but some results will be presented from [21]. In this recent model, the results are presented in the form of a PDF for the amplitude and phase of the receiver field strength, a tolerance band diagram, a transfer function of the transmitting channel, and the characteristic parameters of the transmitting channel, namely, correlation or coherency bandwidth, multipath spread, and group delay analysis. The predicted field strength for the path (Fig. 8.28) are shown with distance in Figure 8.29 for 2D and 3D models. Differences between the 2D and 3D models are up to 15 dB in some regions where diffraction is predominant, for example, the shadowed region and the hilly region (2.5–7.0 km). For further details, the reader is referred to [22].

8.9 PROPAGATION ALONG BUILDINGS FOR LOW-POWER RADIO SYSTEMS

Existing systems like mobile (cellular) radios and cordless phones have to provide services to densely populated areas having a traffic many times larger than a vehicular mobile system can handle. Eventually, it may be necessary to represent a room or even a floor as a cell. Proper design of future vehicular and personal microcellular mobile radio systems requires a very thorough knowledge of the environment, particularly the location, geometry, and distribution of buildings. This needs to be followed by an effective model of propagation through those buildings. Two references [23, 24] are of special interest in this area. The first one gives a very good review of the published

8.9 PROPAGATION ALONG BUILDINGS FOR LOW-POWER RADIO SYSTEMS

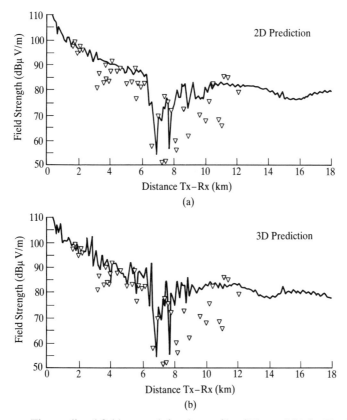

FIGURE 8.29 The predicted field strength for the profile of Figure 8.28 for T × "Merkur"; $f = 94.1$ MHz; transmitted power = 800 W; R × height = 10 m; horizontal polarization; measurement data from [22] for (a) 2D model; (b) 3D model.

information on radiowave propagation and the second and more recent paper describes a more accurate and robust algorithm to estimate diffraction by buildings. The buildings are approximated as rectangular volumes with corners and flat surfaces. The height of this rectangular volume is the height of the building above sea level. Complex buildings are modeled by a series of rectangular shapes of different geometries. A geographic information system [25] is ordinarily used. There are different types of buildings with different wall impedances. We will summarize here the application of high-frequency techniques to calculate diffraction by buildings.

For vertical edges,

$$E_n(P) = -je^{jk\rho} \sqrt{\frac{s}{2\lambda p(s+p)}} \int_{x_1}^{x_2} E_Q(x) e^{j\pi x^2/\lambda p} \, dx \cdot \int_{\eta_1}^{\infty} e^{j\pi\eta^2/2} \, d\eta \quad (8.7a)$$

where

$$\eta_1 = y_1 \sqrt{\frac{2(s+p)}{\lambda sp}} \tag{8.7b}$$

For the single diffraction case, the equation takes the form

$$E_n(P) = \frac{-j}{2} E_{fs} \int_{\xi_1}^{\xi_2} e^{j\pi\xi^2/2} \, d\xi \int_{\eta_1}^{\infty} e^{j\pi\eta^2/2} \, d\eta \tag{8.8a}$$

where the free-space field E_{fs} is given by

$$E_{fs} = e^{jk(x+p)/(s+p)} \tag{8.8b}$$

The total return is a superposition of individual returns and can be expressed in two ways: (1) a vector sum of individual contributions and (2) as the square root of the average power associated with the small cells. Figure 8.30 shows diffraction by the rectangular building. Figure 8.31 [23] shows the diffracted field contributions from the three aperture regions (see Fig. 8.32). The total

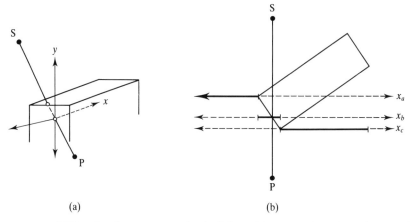

FIGURE 8.30 Diffraction by a rectangular building: (a) perspective view; (b) top view of (a).

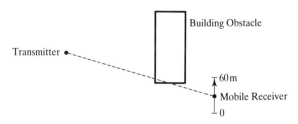

FIGURE 8.31 Top view of the geometry used in example receiver run [24].

8.9 PROPAGATION ALONG BUILDINGS FOR LOW-POWER RADIO SYSTEMS

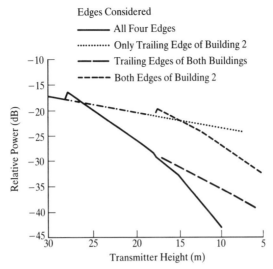

FIGURE 8.32 Diffracted field contributions from a three-aperture region. The total field is an estimate of the local mean diffraction field strength [24].

field represents an estimate of the local mean diffraction field strength. In [23], a sensitivity analysis was also done with computed multiple diffraction by two buildings of heights 15 m and 10 m as a function of height of the transmitter; this is shown in Figure 8.33. The solid line includes diffraction effects from all four edges; in other curves contributions are neglected.

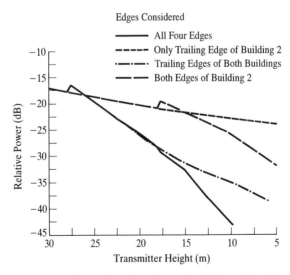

FIGURE 8.33 Computed multiple diffraction by two buildings of heights 15 m and 10 m as a function of the height of the transmitter. The solid line includes all contributions; other lines are with various contributions purposely missing [24].

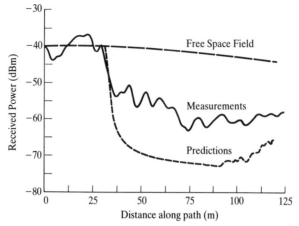

FIGURE 8.34 Comparisons with 914 MHz measurements taken of the diffraction introduced by Patton Hall at the receiver passing in front of it. The transmitter was placed on the roof of another building [24].

Figure 8.34 shows comparisons with 914 MHz measurements, the diffraction produced by a particular hall of residence (Patton Hall) in Virginia Technology, Blacksburg Campus at a receiving point in front of the hall with the transmitter on the roof of another building. Figure 8.35 compares 914 MHz measurements taken adjacent to Davidson Hall. The receiver passed down the sidewalk along one side of the building while the transmitter was positioned on a 15 ft high pole around the corner of the building.

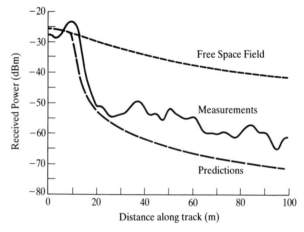

FIGURE 8.35 Comparisons with 914 MHz measurements taken adjacent to Davidson Hall. The receiver passed down the sidewalk along one side of the building while the transmitter radiated from the top of a 15 ft high pole around the corner of the building [24].

8.10 SUMMARY

This chapter dealt with the high-frequency model of propagation of electromagnetic waves. It gave a summary and comparison of Longley–Rice and GTD models, finite conductivity GTD and knife-edge models in path loss prediction. It went on to discuss modeling for hilly terrain and a model using an impedance wedge. The recently developed three-dimensional polarimetric model was also presented. And prediction of diffraction by buildings with applications to low-power radio systems was discussed.

8.11 REFERENCES

1. M. L. Skolnik, *Introduction to Radar Systems*, McGraw-Hill, 1962, Ch. 12, pp. 441–469.
2. H. F. Schmid, "A Prediction Model for Multipath Propagation of Pulsed Signals at VHF and UHF Over Irregular Terrain," *IEEE Trans. Antennas and Propagat.*, **AP-18**(2), 1970, 253–258.
3. A. B. Carlson and A. T. Waterman, Jr., "Microwave Propagation over Mountain-Diffraction Paths," *IEEE Trans. Antennas and Propagat.*, **AP-14**(4), 1966, 489–496.
4. M. L. Meeks, "VHF Propagation over Hilly, Forested Terrain," *IEEE Trans. Antennas and Propagat.*, **AP-31**(3), 1983, 483–489.
5. J. T. Godfrey, H. F. Hartley, G. J. Moussally, and R. A. Moore, "Terrain Modeling using the Half-Plane Geometries with Applications to ILS Glide Slope Antennas," *IEEE Trans. Antennas and Propagat.*, **AP-24**, 1976, 370–378.
6. K. A. Chamberlin and R. J. Luebbers, "An Evaluation of Longley–Rice and GTD Propagation Models," *IEEE Trans. Antennas and Propagat.*, **AP-30**(6), 1982, 1093–1098.
7. L. E. Vogler, "An Attenuation Function for Multiple Knife-Edge Diffraction," *Radio Science*, **17**(6), 1982, 1541–1546.
8. J. H. Whitteker, "A Series Solution for Diffraction over Terrain Modeled as Multiple Bridged Knife Edges," *Radio Science*, **28**(4), 1993, 487–500.
9. R. J. Luebbers, "Finite Conductivity Uniform GTD versus Knife-Edge Diffraction in Prediction of Propagation Path Loss," *Radio Science*, **AP-32**(1), 1984, 760–768.
10. R. J. Luebbers, "Propagation Prediction for Hilly Terrain using GTD Wedge Diffraction," *Radio Science*, **AP-32**(9), 1984, 951–955.
11. M. M. Poulose, P. R. Mahapatra, and N. Balakrishnan, "Terrain Modeling of Glideslope for Instrument Landing System," *IEE Proc., Part H*, **134**(3), 1987, 275–279.
12. R. J. Luebbers, "Comparison of Lossy Wedge Diffraction Coefficients with Applications to Mixed Path Propagation Loss Prediction," *IEEE Trans. Antennas and Propagat.*, **AP-36**(7), 1988, 1031–1034.
13. R. J. Luebbers, W. A. Foose, and G. Reyner, "Comparison of GTD Propagation Model Wide-Band Path Loss Simulation with Measurements," *IEEE Trans. Antennas and Propagat.*, **AP-37**(4), 1989, 499–505.

14. A. G. Longley and P. L. Rice, *Prediction of Tropospheric Radio Transmission Loss over Irregular Terrain*, U.S. Dept. of Commerce, ESSA Rep. ERL-79-ITS-67, 1968.
15. J. L. James, *Geometrical Theory of Diffraction For Electromagnetic Waves*, Peregrinus, London, 1976.
16. L. E. Vogler, "An Attenuation Function for Multiple Knife-Edge Diffraction," *Radio Science*, **17**(6), 1982, 1541–1546.
17. J. H. Whitteker, "A Series Solution for Diffraction over Terrain Modeled as Multiple Bridged Knife Edges," *Radio Science*, **28**(4), 1993, 487–500.
18. W. D. Burnside and K. W. Burgener, "High Frequency Scattering by a Thin Lossless Dielectric Slab," *IEEE Trans. Antennas and Propagat.*, **AP-31**, 1983, 104–110.
19. R. J. Luebbers, "Finite Conductivity Uniform GTD versus Knife Edge Diffraction in Prediction of Propagation Path Loss," *IEEE Trans. Antennas and Propagat.*, **AP-32**(1), 1984, 70–76.
20. R. Luebbers, V. Ungvichian, and L. Mitchell, "GTD Terrain Reflection Model Applied to ILS Glide Slope," *IEEE Trans. Aerosp. Electron. Syst.*, **AES-18**, 1982, 11.
21. R. Grosskopf, "Comparison of Different Methods for the Prediction of the Field Strength in the VHF Range," *IEEE Trans. Antennas and Propagat.*, **AP-35**(7), 1987, 852–859.
22. M. Lebherz, W. Wiesbeck, and W. Krank, "A Versatile Wave Propagation Model for the VHF/UHF Range Considering 3-D Terrain," *IEEE Trans. Antennas and Propagat.*, **AP-40**(10), 1992, 1121–1129.
23. D. Molkdar, "Review on Radio Propagation into and within Buildings," *IEEE Proc., Part H*, **138**(1), 1991, 61–73.
24. T. A. Russell, C. W. Bostian, and T. S. Rappaport, "A Deterministic Approach to Predicting Microwave Diffraction by Buildings for Microcellular Systems," *IEEE Trans. Antennas and Propagat.*, **41**(13), 1993, 1640–1649.
25. T. A. Russell, T. S. Rappaport and C. W. Bostian, "The Use of Building Database in Prediction of Three-Dimensional Diffraction," *42nd Proc. IEEE Vehicular Technology*, Denver, CO, 1992, 943–946.

8.12 ADDITIONAL SOURCES

D. E. Eliades, "Path Integral Analysis of Paraxial Radiowave Propagation over a Building with a Wedged Roof," *J. Electromag. Waves and Applns*, **6**(4), 1992, 513–532.

R. J. Luebbers, "A Semiblind Test of the GTD Propagation Model for Reflective Rolling Terrain," *IEEE Trans. Antennas and Propagat.*, **APS-38**(3), 1990, 403–405.

Y. Ohtkari, Y. Yamaguchi, and T. Abe, "Experimental Study of Propagation Characteristics on Roads on a Snowy Mountain," *IEEE Trans. Electromag. Compatibility*, **30**(2), 1988, 137–144.

M. O. Al-Nuaimi and M. S. Ding, "Prediction Models and Measurements of Microwave Signals Scattered from Buildings," *IEEE Trans. Antennas and Propagat.*, **42**(8), 1994, 1126–1137.

J. H. Whitteker, "Diffraction over a Flat-Topped Terrain Obstacle," *IEEE Proc. Part H*, **137**(2), 1990, 113–116.

CHAPTER NINE

Radar Clutter Modeling Using High-Frequency Techniques

9.1 INTRODUCTION

A major problem encountered by a ground-based or down- or side-looking radar is the clutter environment. It is therefore essential to quantify the radar clutter sources, which are numerous and widely different in kind. A partial list of sources of clutter is given in Section 9.2. The radar range equation for surface clutter power return P_C is given by (Fig. 9.1)

$$P_C = \frac{P_t G A_e \sigma^0 \theta_B (c\tau/2) \sec \phi}{(4\pi)^2 R^3} \tag{9.1}$$

where P_t = transmitted power
G = antenna gain
A_e = effective aperture of the antenna
$\sigma_e = \sigma^0 A_e$
σ^0 = surface clutter cross section per unit area of the surface
θ_B = half-power beamwidth of the antenna

Now, the signal power returned from the target is given by

$$S = \frac{P_t G A_e \sigma_t}{(4\pi)^2 R^4} \tag{9.2}$$

The signal-to-clutter power ratio is given by

$$\frac{S}{C} = \frac{\sigma_t}{\sigma^0 R \theta_B (c\tau/2) \sec \phi} \tag{9.3}$$

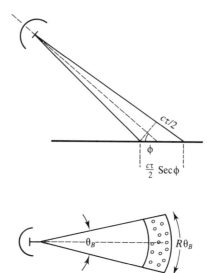

FIGURE 9.1 Surface clutter.

The maximum range with the presence of clutter is found to be

$$R_{max} = \frac{\sigma_t}{(S/C)_{min} \sigma^0 \theta_B \left(\dfrac{c\tau}{2}\right) \sec \phi} \tag{9.4}$$

This maximum range is then different when the signal is corrupted by clutter rather than when the signal is corrupted by system noise alone.

It may be stressed that the clutter return is inversely proportional to the third power of the range R as opposed to the fourth power of R for conventional scattering problem.

An extensive literature has accumulated on radar clutter modeling. Some of the key references are in [1–7]. Broadly, there can be two types of modeling approaches for clutter sources, the analytical model and the statistical model. Depending upon the example, each model has its advantages and disadvantages and applicability. In analytical studies high-frequency techniques have been extensively used: geometrical optics, physical optics, and the geometrical theory of diffraction (GTD) for PEC and non-PEC wedges when applicable. In statistical modeling, which is beyond the scope of this book, many different kinds of statistical distributions are used to model clutter. Often, one statistical distribution is not adequate to represent the clutter situation. For complex scenarios, it will be wise to combine the analytical and statistical modeling techniques to make the clutter model more versatile and realistic.

9.2 SOURCES OF RADAR CLUTTER

There are a large number of artificial and natural clutter sources:

Artificial Clutter Sources

Roads	Land-based vehicles, particularly
Paved roads	moving trucks and other automobiles
Roadside materials	Ships
Parking lots	Aircraft
Buildings	towers
	Antennas

Natural Clutter Sources

Land	Terrain snowpacks
Sea surface, particularly under disturbed conditions	Sea ice
	Tundra
Vegetation	Ice fields
Sparse vegetation	Single bird and flock of birds
Vegetation-covered terrain	Insects and large flocks of insects
Grass	Rainfall
Snow-covered grass	Falling snow
Mountains	Hailstorms
Cliffs	

9.3 PHYSICAL OPTICS METHOD

Usually the clutter surface is divided into a large number of small cells and the total field is a superposition of the returns from each of these cells

$$\mathbf{E}(\mathbf{R}) = -\frac{1}{4\pi} \iint_S d\mathbf{S} [-\omega\mu_0 (\hat{\mathbf{n}} \times \mathbf{H}^s)\Psi + (\hat{\mathbf{n}} \times \mathbf{E}^s) \times \nabla\Psi + (\hat{\mathbf{n}} \cdot \mathbf{E}^s)\nabla\Psi] \quad (9.5a)$$

where $\mathbf{E}^s$ and $\mathbf{H}^s$ are the fields on the surface of the scatterer and

$$R = |\mathbf{R}|: \Psi = \frac{e^{-jkR}}{R} \quad (9.5b)$$

where ∇ = del operator with respect to the coordinates on the surface
$\hat{n}$ = Surface normal unit vector at the point S where it is assumed that $\mathbf{E}^s$ and $\mathbf{H}^s$ are continuous along the surface.

$\Delta\Psi$ is given by

$$\nabla\Psi = \hat{R}\frac{jk}{R}\exp(-jkR)\left(1 + \frac{1}{jkR}\right) \tag{9.5c}$$

The total return is a sum of the N returns from N elementary surface patches and is given by

$$E(R) = \sum_{q=1}^{N} E_q(R) \tag{9.6}$$

The surface integral reduces to

$$E_q(R) = -\frac{1}{4\pi}\iint_{\Delta S_p} dS \frac{e^{-jkr}}{R}$$
$$\times \left\{-j\omega\mu_0(\hat{n} \times H_s) + jk\left(1 + \frac{1}{jkR}\right)((\hat{n} \times E^s) \times R + (\hat{n}\cdot E^s)R)\right\} \tag{9.7}$$

When $R \gg \lambda$, $1 + \dfrac{1}{jkR} \approx 1$.

Using

$$A \times (B \times C) = B(A\cdot C) - C(A\cdot B) \tag{9.8}$$

and $\hat{n} = \hat{z}$

$$E_q = -\frac{j\omega\mu_0}{\pi}\iint_{\nabla S_q} dS \frac{e^{-jkR}}{R}\{(\hat{z} \times H^s) - Y_0((R\cdot\hat{z})E^s - \hat{z}(R\cdot E^s) + R(\hat{z}\cdot E^s))\} \tag{9.9}$$

where, $Y_0 = \sqrt{\dfrac{\varepsilon_0}{\mu_0}}$, the wave admittance of free space.

The free-space Green's function is e^{-jkR}/R with the assumption that the surface element relative to R is given by

$$\frac{e^{-jkR}}{R} \approx \frac{e^{-jkr}}{r}e^{jk\hat{r}_{qp'}\cdot\bar{r}} \tag{9.10}$$

Therefore, $\hat{R} \approx \hat{r}_{qp'}$ and $\bar{R} \approx \bar{r}_{qp'}$.
This gives

$$E_q(\bar{R}_{qp'}) \approx -\frac{jk}{4\pi}Z_0\frac{e^{-jkr_{qp'}}}{r_{qp'}}\iint_{\Delta S_q} dS\, e^{jk\hat{r}_{qp'}\cdot\bar{r}}$$
$$\times \{(\hat{z}\cdot H^s) - Y_0((\hat{z})E^s) - \hat{z}\hat{r}_{qp'}\cdot E^s) + \hat{r}_{qp'}(\hat{z}\cdot E^s))\} \tag{9.11}$$

where the intrinsic wave impedance of free space is denoted by Z_0 and the unit vectors are given by standard coordinate transformation formulae. Look out for two things: (1) the mutual coupling between the adjacent cells and (2) the contributions from multipaths. The details of modeling and software techniques, including the scattering software, are available in [13].

9.4 RADAR CLUTTER FROM DIFFERENT IMPORTANT CLUTTER SOURCES

In this section, we briefly give an account of some common radar clutter sources. These are roads and roadside materials, snow-covered terrain, sea surface scattering using high frequency techniques, airport humped highways.

9.4.1 Roads and Roadside Materials

The University of Kansas, Lawrence, conducted backscattering experiments [14] using the MAS 8-18/35 system on concrete and asphalt taxiways near Johnson County Industrial Airport, Lawrence, Kansas. Measurements were conducted for the following roadside materials: (1) dry asphalt, (2) dry concrete, (3) wet concrete, (4) dry gravel, (5) wet ground, (6) tall grass (80 cm in height), and (7) short grass (10–15 cm in height). The estimated mean value of σ_0 is the average of 30 or more independent samples. Also, this experiment detected the clutter from the boundary of two dissimilar surfaces. Figure 9.2 shows the close-up photograph of the targets investigated. Figure 9.3 shows the angular response of the scattering coefficient of dry asphalt at three different frequencies. Figure 9.4 shows the same for dry concrete. The variation of σ_0 for wet concrete is shown in Figure 9.5. The angular response of dry and wet gravel is shown in Figures 9.6 and 9.7.

It is found from Table 9.1 that the discrimination between grass and dry concrete is more than that between grass and dry asphalt. For dry concrete, the scattering coefficient σ_0 for VV and HH are less at 35.5 GHz than at 17.0 GHz, particularly for angles near normal incidence.

9.4.2 Snow-Covered Terrain

The scattering and emission characteristics of snow, soil, and snow-covered soil with locally and temporally varying parameters like moisture content, surface roughness, and subsurface structure are controlled by absorption, scattering, and volume scattering. Various approaches have been suggested over many years. Some theories are well supported by experimental results. A very good review of high-frequency modeling of snow and soil is available in [15].

One approach to describing the problem of scattering from snow-covered terrain is a two-layer medium scattering model. The formulation of the problem for the general case of a multilayer medium has been discussed at length in [16] using a Green's function technique. Born approximations of first- and second-order are made to extract co- and cross-polarized components of the

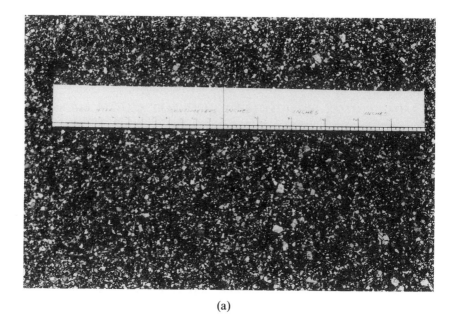

(a)

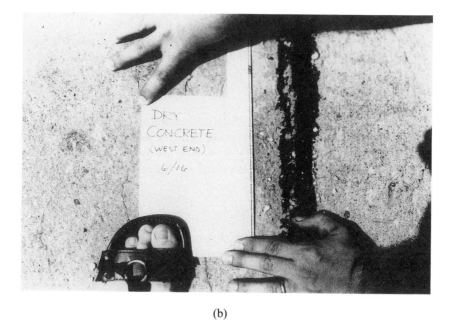

(b)

FIGURE 9.2 Close-up photographs of targets investigated: (a) asphalt taxiway; (b) concrete taxiway; (c) gravel parking; (d) tall grass (80 cm high) (e) short grass (10–15 cm high) [14].

9.4 RADAR CLUTTER FROM DIFFERENT IMPORTANT CLUTTER SOURCES 443

(c)

(d)

FIGURE 9.2 (*continued*).

(e)

FIGURE 9.2 (*continued*).

scattered field. The depolarized backscattering cross-section can also be obtained by solving radiative transfer equations carried to the second order [16, section 2.7] to extract the cross-polarized contributions and also by a random discrete scatter model.

In [17], milimeter wave radar was used to collect data from snow-covered ground at 35 GHz and 94 GHz. It was found in the initial investigation and analysis that so far as the clutter was concerned, the scattering properties of wet and dry snow did not pose a problem in differentiating between the returns from snow-covered ground and returns from a ground target. Once the snow has passed through at least one cycle of melting and refreezing, or has been "metamorphosed," it behaves differently, making it a lot more difficult for target detection in the presence of snow. Some additional conclusions of this SNOWMAN [17] data analysis are: (1) the snow backscatter indicates a strong dependency on depression angle, polarization, frequency, and snow state; (2) free water content appears to be the most significant factor affecting radar reflectivity, and (3) snow clutter decorrelates with frequency agility and with spatial separation of the samples used. One of the early experiments was reported in [18] and shows the angular variation of scattering coefficient at two frequencies for short grass with and without dry powder and wet snow cover.

9.4 RADAR CLUTTER FROM DIFFERENT IMPORTANT CLUTTER SOURCES

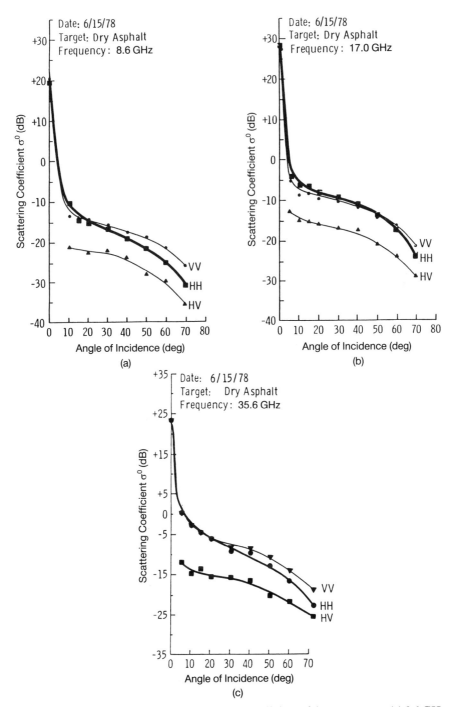

FIGURE 9.3 Angular response of the scattering coefficient of dry concrete at (a) 8.6 GHz, (b) 17 GHz, and (c) 35 GHz [14].

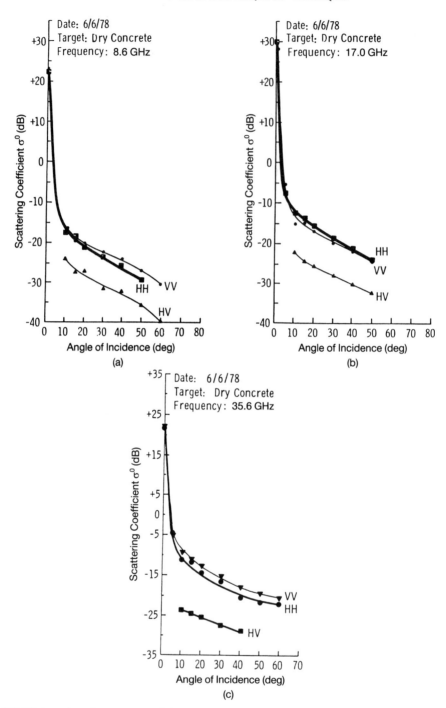

FIGURE 9.4 Angular response of the scattering coefficient of dry concrete at (a) 8.6 GHz, (b) 17 GHz, (c) 35.6 GHz [14].

9.4 RADAR CLUTTER FROM DIFFERENT IMPORTANT CLUTTER SOURCES   447

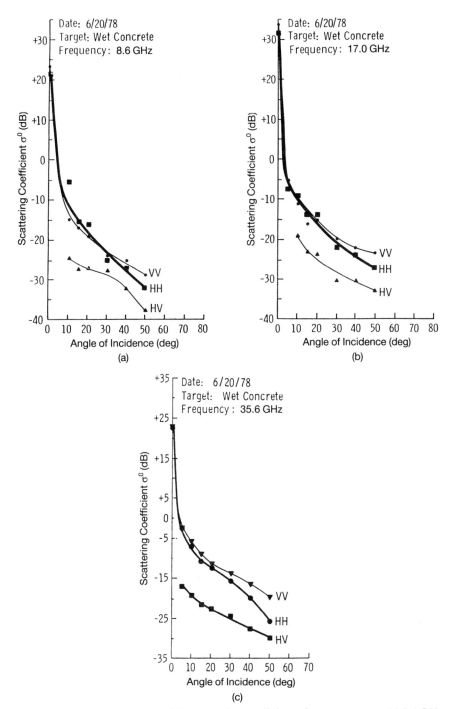

FIGURE 9.5 Angular response of the scattering coefficient of wet concrete at (a) 8.6 GHz, (b) 17 GHz, (c) 35.6 GHz [14].

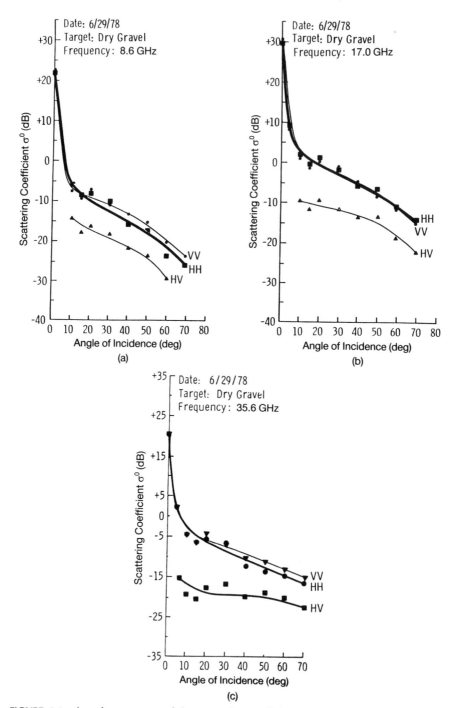

FIGURE 9.6 Angular response of the scattering coefficient of dry gravel at (a) 8.6 GHz, (b) 17 GHz, (c) 35.6 GHz [14].

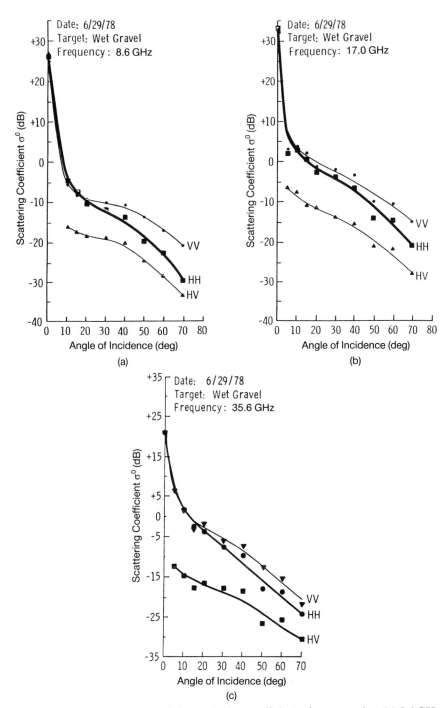

FIGURE 9.7 Angular response of the scattering coefficient of wet gravel at (a) 8.6 GHz, (b) 17 GHz, (c) 35.6 GHz [14].

TABLE 9.1 Discrimination Between Roads and Short Grass Measured in Terms of the Difference in dB, $\theta = 50°$

Frequency (GHz)	Polarization	σ_c^0	σ_a^0
8.6	HH	14	7
8.6	HV	14	8
8.6	VV	11	4
17	HH	11	4
17	HV	10	4
17	VV	10	3
35.6	HH	10	4
35.6	HV	12	3
35.6	VV	7	1
35.6	RR	14	4
35.6	RL	8	2

Subscripts g, c, and a stand for grass, dry concrete, and dry asphalt, respectively.

9.4.3 Sea Surface Scattering Using High-Frequency Techniques

The need for a clear understanding of sea surface scattering relates to the problem of detection, identification, and discrimination of radar targets with the sea in the background. Sea surface conditions may vary significantly throughout the day and during the year, and they depend on the speed and direction of the wind and the presence of sea ice. A vast literature has accumulated in the subject of sea surface scattering. It seems impossible to do justice to the review of this area in a limited space. We summarize in this section some main features of the subject. A balanced discussion on early models is available in [6]. There are two scales of roughness of the sea. Small-scale roughness is caused by *ripple* or *capillary waves* and large-scale roughness is due to *swell* or *gravity waves*. The true picture of a sea surface is a continuous scale of roughness in the time domain. A rigorous analytical model is extremely involved [7–10]. This is the reason why models for the two limiting cases are widely used.

For large-scale roughness Kirchhoff's approximation [10] is used. This approach assumes the return is coming from a mean flat surface held perpendicular to the direction of illumination. This mean surface is assumed over a few wavelengths. The theory is developed in several steps.

The perturbation theory is formulated by expressing the incident and scattered waves in both media as a plane wave spectrum and by applying boundary conditions. The fields are determined to fourth order in the perturbation parameter. The scattered plane waves are integrated from all regions illuminated by the antenna pattern function multiplied by a free-space spherical wave Green's function $\exp(j2kr)/r^2$. This approach has a limitation that it

9.4 RADAR CLUTTER FROM DIFFERENT IMPORTANT CLUTTER SOURCES

accounts for low orders of *local* multiple scattering due to ripple fluctuations but not the *large-scale* multiple scattering. This will cause a large error when there is large swell. The alternative approach [10] using Stratton–Chu or, Kirchhoff–Huyghens integral equations is not a better approach once the same assumptions are made but merely an alternative method of solution.

An early systematic analytical study of sea surface scattering was made in [10]. This report also discusses further extensions from most previous analytical studies. These are: (1) the effect of pulsing or other forms of modulation of the transmitted RF wave, earlier work was mostly based on CW transmission; (2) integration over all incident angles weighted by the antenna pattern shaping functions; and (3) inclusion of the cross-covariance of deviations of two received signal powers from their mean values; and (4) inclusion of higher-order terms which take care of coupling between sea return and target return. This report does not incorporate shadowing effects, which are very important at grazing angles. As mentioned earlier, the return to the receiver is a superposition of returns from different positions weighted by the antenna pattern function $F_{Aa}(\theta'_z, \phi'_z, t)$ and a factor S_{Aa} is obtained from the perturbation solution of the boundary value problem.

The different order returns are given by

$$U_{Aa0}(t) = \int_0^\pi d\theta'_3 \sin \theta'_3 \int_0^{2\pi} d\phi' F_{Aa}(\theta'_3, \phi'_3, t) S_{Aa0}(\hat{\alpha}_0(\theta'_3, \phi'_3, t)) \quad (9.12)$$

$$U_{Aa1}(t) = \int_0^\pi d\theta'_3 \sin \theta'_3 \int_0^{2\pi} d\phi' F_{Aa}(\theta'_3, \phi'_3, t) S_{Aa1}(\hat{\alpha}_0(\theta'_3, \phi'_3, t))$$
$$\cdot Z\left(-\frac{2\omega}{c} \alpha(\theta'_3, \phi'_3, t)\right) \quad (9.13)$$

$$U_{Aa2}(t) = \int_0^\pi d\theta'_4 \sin \theta'_3 \int_0^{2\pi} d\phi' F_{Aa}(\theta'_3, \phi'_3, t) \iint dk S_{Aa2}(\theta'_3, \phi'_3, t), \cdot k) \cdot Z(k)$$
$$\cdot Z\left(-\frac{2\omega}{c} \hat{\alpha}(\theta'_3, \phi'_3, t) - k\right) \quad (9.14)$$

$$U_{Aa3} = \int_0^\pi d\theta'_3 \sin \theta'_3 \int_0^{2\pi} d\phi'_3 F_{Aa}(\theta'_3, \phi'_3, t) \iint dk_1$$
$$\times \iint dk_2 S_{Aa3}(\hat{\alpha}_0(\theta'_3, \phi'_3, t), \cdot k_1 \cdot k_2) Z(k_1) Z(k_2) \ldots$$
$$\times Z\left(-2\frac{\omega}{c} \alpha_0(\theta'_3, \phi'_3, t) - k_1, -k_2\right) \quad (9.15)$$

and

$$U_{Aa4}(t) = \int_0^\pi d\theta'_3 \sin \theta'_3 \int_0^\pi d\phi'_3 F_{Aa}(\theta'_3, \phi'_3, t) \iint dk_1 \iint k_2$$

$$\times \iint dk_3 S_{Aa4} \ldots (\hat{\alpha}_0(\theta'_3, \phi'_3, t), k_1, k_2, k_3) Z(k_1) Z(k_2) Z(k_3) \ldots$$

$$\times Z\left(-2\frac{\omega}{c}\alpha_0(\theta'_3, \phi'_3, t) - k_1, -k_2, -k_3\right) \quad (9.16)$$

where, α_0 is the projection on the xy-plane of the unit vector $\hat{\alpha}_0$ directed from the radar to the observation point. This projection is a function of the angles and time. Time comes in the picture since the sea surface conditions change with time with other parameters remaining the same. The overall weighting function $F_{Aa}(\theta'_3, \phi'_3, t)$ incorporates the possibility that the pulsating function and antenna pattern function may differ for different polarizations and is given by

$$F_{Aa}(\theta'_3, \phi'_3, t) = K \frac{g_{Aa}(r_0(\theta'_3, \phi'_3, t))}{r_0(\theta'_3, \phi'_3, t)^2} \exp\left(j2\frac{\omega}{c} r_0(\theta'_3, \phi'_3, t)\right) \quad (9.17)$$

$f_{Aa}(\theta'_z, \phi'_z)$ is the one-way field pattern of the antenna for the polarization Aa. K is a constant involving radar parameters. Details of the derivations are available in [10]. In such a situation, one has to take an ensemble over surface fluctuations. The cross-covariance functions (CCFs) between all possible combinations of polarizations are available in [10]. Some of the results of this analysis are depicted in Figures 9.8 through 9.11 for $\varepsilon/\varepsilon_r = 70$; $\sigma = 4.5$ mhos/m;

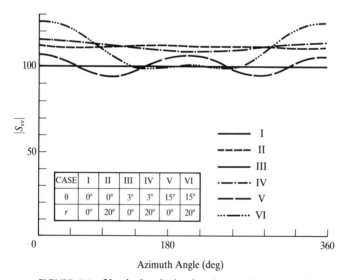

FIGURE 9.8 Vertical polarization (transmit and receive).

9.4 RADAR CLUTTER FROM DIFFERENT IMPORTANT CLUTTER SOURCES   453

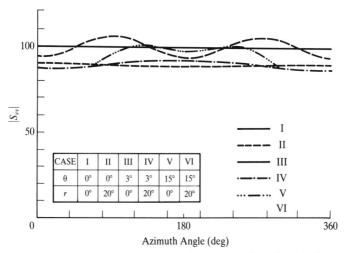

FIGURE 9.9   Horizontal polarization (transmit and receive).

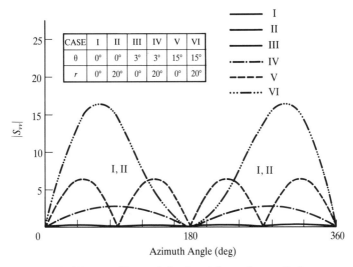

FIGURE 9.10   Transmit horizontal–receive vertical.

$f = 10$ GHz; $r = 100$ m (corresponding to a pulse time interval of 0.067 μsec), $n = 4anf\,\mathbf{K}_0 = 1$. The spatial spectrum $W(|K|)$ of the surface height fluctuations (assumed isotropic) where $|\mathbf{K}|$ is equal to $[(2(\omega/C)\alpha_0]$. Figures 9.8 through 9.11 show the plot of the variation of elements of the polarization matrix with azimuthal angle in relative units for given values of polar angle θ' and the beam angle γ in four cases. Figure 9.12 shows the ratio of the cross-polarized to copolarized component of total received power versus beam angle. The key conclusion of this study is that the cross-polar component is only an order

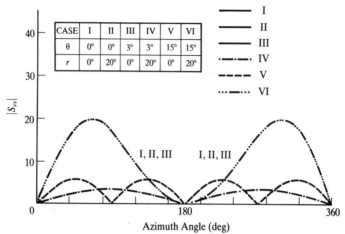

FIGURE 9.11 Transmit vertical–receive horizontal.

stronger than the copolar component, assuming the mean surface to the horizontal and the generation of cross-polarization will be greater for a variable mean surface. Valenzuala [11, 12] studied the depolarization of the scattering of EM waves by a slightly rough surface using Rice's theory. For the sea, it was found that the depolarization increases with wind speed and with the magnitude of the complex dielectric constant of the surface. The depolarized power for backscattering for both polarization is

$$\langle P_{SVH} \rangle^B = \langle P_{SHV} \rangle^B$$

$$= \frac{a^2 \beta^8 \cos^2 \theta}{4\eta_0} \left| \frac{(\varepsilon - 1)^2 \sqrt{\varepsilon - \sin^2 \theta}}{(\varepsilon \cos \theta + \sqrt{\varepsilon - \sin^2 \theta})(\cos \theta + \sqrt{\varepsilon - \sin^2 \theta})} \right|^2$$

$$\cdot \sum \frac{a^4 k^2 l^2}{|D_{k1} d(k, l)|^2} \cdot W(ak - \beta \sin \theta, al) W(ak + \beta \sin \theta, al) \quad (9.18)$$

where k and l vary from $-\infty$ to $+\infty$
  $a = 2\pi/L$, L is the periodicity
  $\beta = 2\pi/\lambda$
  θ = angle of incidence
  η_0 = free-space impedance
  γ = beam angle
  ε = dielectric constant of the material of the rough surface

9.4.4 Airport Humped Runways

According to the requirements of the International Civil Aviation Organization, microwave landing systems (MLSs) should be universally installed. Therefore

9.4 RADAR CLUTTER FROM DIFFERENT IMPORTANT CLUTTER SOURCES

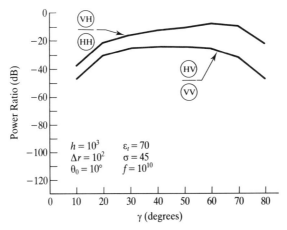

FIGURE 9.12 Ratios of cross-polarized to parallel-polarized component of total received power versus beam angle.

it is necessary to do a very accurate estimate of the power scattered by the airport runway and surroundings, including inverted humps towards the receiver on board the aircraft. This is to ensure the MLS is operational and there is no problem in the initial and final stages of approach, including touchdown and roll-out. In Chapter 8 it was shown that wedge diffraction theory can be used as a powerful tool to predict the propagation losses through terrain. The application of wedge diffraction theory to calculate the power density distribution in and around a humped runway has been done in [18–20]. A typical example in [20] is the centerline profile of Ottawa International Airport Runway 14. The prediction accuracy near the threshold is ± 3 dB for the model in [21].

Runway 14 centerline is shown in Figure 9.13 and the wedge diffraction geometry is shown in Figure 9.14. The signal in the receiver is a sum of direct, reflected, and diffracted signals. It is given by

$$E(\theta) = E_1(\theta) - E_3(\theta) + E_3(\theta - 2\alpha) + E_3(-\theta - 2\beta) - E_3(-\theta - 2\alpha - 2\beta) \quad (9.19)$$

The power density is given by

$$P(\theta) = P\frac{G}{4\pi}(D_1 + D_2)^2 |E^2(\theta)| \quad (9.20a)$$

The power density in the case of flat ground is given by

$$P_{FG}(HT, HR) = PG\frac{4\pi}{\lambda^2}\left(\frac{HT}{D_1 + D_2}\right)^2 \cdot \left(\frac{HR}{D_1 + D_2}\right)^2 \frac{1}{\pi^2}\frac{\sqrt{\lambda/D}}{\alpha + \beta} \quad (9.20b)$$

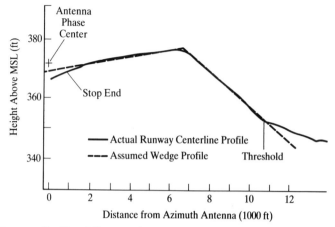

FIGURE 9.13 Profile of Ottawa airport runway 14. (From [21] © 1987 IEEE.)

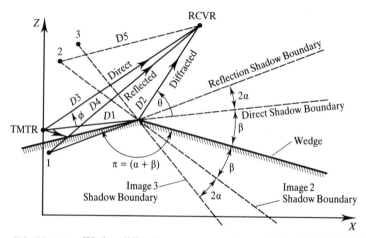

FIGURE 9.14 Wedge diffraction geometry. (From [21] © 1987 IEEE.)

We omit the expressions for different components of the fields and move on to some results [21]. The wedge attenuation versus wedge angle is shown in Figure 9.15. The computed power density in Ottawa Runway 14 versus distance is shown in Figure 9.16.

9.5 BISTATIC CLUTTER MEASUREMENTS

A bistatic clutter measurement program was taken in [21] using a C-46 aircraft. The terrain scattering cross section σ_0 per unit area is defined by

$$\sigma_0 = \frac{4\pi r_s^2 \, dS_a}{S_i \, dA} \tag{9.21}$$

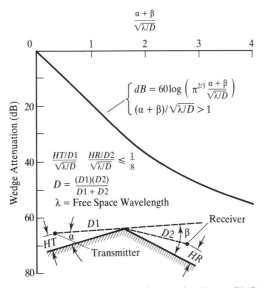

FIGURE 9.15 Wedge attenuation versus wedge angle. (From [21] © 1987 IEEE.)

The incident power density S_i is given by the following expression in terms of the transmitted power P_t, transmitter antenna maximum gain G_T, and normalized power pattern f_T:

$$S_i = \frac{1}{4\pi r_i^2} P_T G_T f_T \tag{9.22}$$

The total power received from the whole of the illuminated surface is

$$P_R = P_T \frac{\lambda^2 G_T G_R L}{(4\pi)^3} \int_{A_{gd}} \frac{f_T f_R \sigma_0 \, dA}{R_i^2 r_s^2} \tag{9.23}$$

Figures 9.17 through 9.21 [22] show the plot of the scattering coefficient versus the scan angle ϕ_s. The bistatic measurement angles are summarized in Table 9.2.

9.6 A TYPICAL RADAR CLUTTER MODEL AND SIMULATION PROGRAM

The development of a comprehensive computer simulation of clutter and multipath returns from a wide variety of artificial and natural clutter sources are described in [13] for various radar scenarios. The investigations are presented in two volumes of a monograph. This monograph includes details of general background theory, discussion of scattering models, and a description

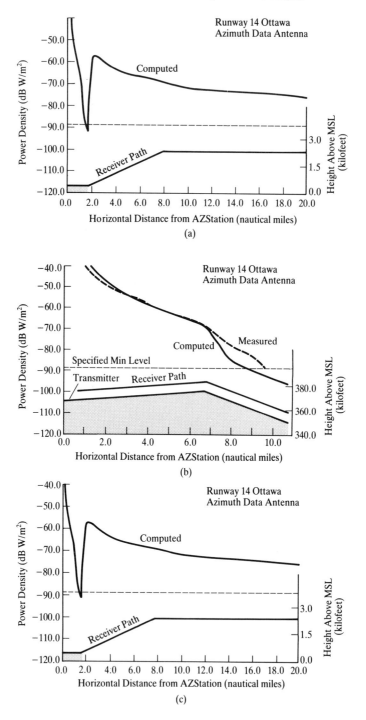

FIGURE 9.16 Computed power density in Ottawa airport runway 14 versus distance. (From [21] © 1987 IEEE.)

9.6 A TYPICAL RADAR CLUTTER MODEL AND SIMULATION PROGRAM   **459**

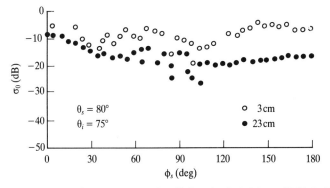

FIGURE 9.17 $\bar{\sigma}_0$ versus ϕ_s for passes 4 and 7, flight 7/17/76. (From [22] © 1978 IEEE.)

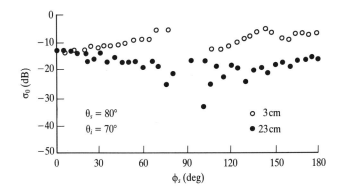

FIGURE 9.18 $\bar{\sigma}_0$ versus ϕ_s for passes 5 and 6, flight 7/17/76. (From [22] © 1978 IEEE.)

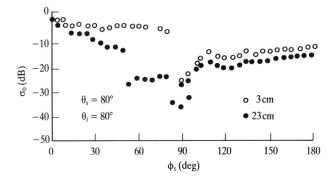

FIGURE 9.19 $\bar{\sigma}_0$ versus ϕ_s for passes 3 and 9, flight 7/17/76. (From [22] © 1978 IEEE.)

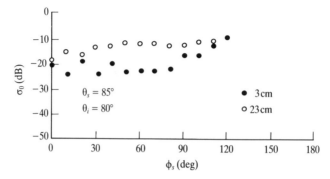

FIGURE 9.20 $\bar{\sigma}_0$ versus ϕ_s for passes 2, 5, and 7, flight 7/9/76 (average of all passes). (From [22] © 1978 IEEE.)

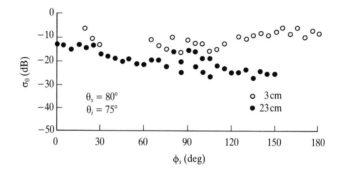

FIGURE 9.21 $\bar{\sigma}_0$ versus ϕ_s for passes 4 and 7, flight 7/17/76. (From [22] © 1978 IEEE.)

of the software program with provision for limiting the number of scattering contributions during a run, which is necessary to keep the CPU time within reasonable limits. The limitations of these scattering models are discussed and results of some simple test scenarios are presented.

9.7 SUMMARY

This chapter summarized radar clutter models often used in communications with background clutter and it emphasized the use of high-frequency techniques. The many different clutter sources were identified. The radar clutter from different clutter sources, such as roads and roadside materials, snow-covered terrain, sea surfaces, and airport humped runways, was discussed. A typical radar clutter model was briefly described.

TABLE 9.2 Summary of Bistatic Measurement Angles [21]

Terrain	Grass and Cement Taxiway, Dry		
	θ_s (deg)	θ_i (deg)	ϕ_s
Site 1 (grass and cement taxiway; dry)	85	80	0.105
		75	0.105
		65	0.105
		50	0.105

	Tall Weeds and Scrub Trees, Dry		
	θ_s (deg)	θ_i (deg)	ϕ_s
	80	80	0.180
	80	80	0.180
		70	0.180
	70	80	0.105
		75	0.105
		70	0.105
		75	0.105
		65	0.105
		50	0.105
Site 2 (tall weeds and scrub trees; dry)	80	80	0.180
	80	80	0.180
		70	0.180
	70	80	0.105
		75	0.105
		70	0.105

9.8 REFERENCES

1. M. I. Skolnik, *Introduction to Radar Systems*, McGraw-Hill, New York, 1980.
2. M. I. Skolnik, *Radar Handbook*, McGraw-Hill, New York, 1970.
3. M. W. Long, *Radar Reflectivity of Land and Sea*, Lexington Books, Lexington, MA, 1975.
4. P. Beckman and A. Spizzichino, *The Scattering of Electromagnetic Waves from Rough Surfaces*, McMillan, New York, 1963.
5. F. T. Ulaby, R. K. Moore, and A. K. Fung, "*Microwave Remote Sensing—Active and Passive, Vol. II Radar Remote Sensing and Surface Scattering and Emission Theory*," Addison-Wesley, Reading, MA, 1982.
6. G. T. Ruck (ed), *Radar Cross Section Handbook, Vol. II*, Plenum Press, New York, 1970, Ch. 9.
7. G. R. Valenzuala, "Scattering of Electromagnetic Waves from a Tilted Slightly Rough Surface," *Radio Science*, **3**(11), 1968, 1057–1066.
8. M. B. Laing, "Sea Clutter Measurement by a Four Frequency Radar," *EASCON Record*, 1969.

9. J. K. De Rosa, "The Characterization of Multipath and Doppler Fading in Earth Scatter Communication, Navigation and Radar," PhD dissertation, Northeastern Univ., Boston, MA, June 1972.
10. H. R. Raemer, *An Analytical Study of Radar Returns in the Presence of a Rough Sea Surface*, NRL Report 8369, N.R.L., Washington D.C., March 12, 1980. (Available from the author at the Northeastern Univ., Boston, MA).
11. [6] p. 50.
12. G. R. Valenzuéla, "Depolarization of EM Waves by Slightly Rough Surfaces," *IEEE Trans. Antennas and Propagation*, **AP-15**(4), 1967, 552–557.
13. H. R. Raemer and A. K. Bhattacharyya, *Analytical Modeling and Computer Simulation of Radar Clutter and Multipath Signals, Vol. I: Background Analysis and Description of Software Excluding Scattering Details, Vol. II: Description of Scattering Software*, Center for Electromagnetic Research, Northeastern Univ., Boston, MA.
14. W. H. Stiles, F. T. Ulaby, and E. Wilson, "Backscatter Response of Roads and Roadside Surfaces," Univ. of Kansas Center for Research, Inc., Lawrence, Kansas, Jan. 1979. (Contract 07-5688, Sandia Lab., Albuquerque, NM 87115.)
15. E. Schanda, "Microwave Modeling of Snow and Soil," *J. Electromag. Waves and Applns*, **1**(1), 1987, 1–24.
16. M. A. Zuniga, "Theoretical Studies for Microwave Remote Sensing," PhD dissertation, Dept. of Physics, MIT, 1980.
17. N. C. Currie et al., "Millimeter-Wave Measurements and Analysis of Snow-Covered Ground," *IEEE Trans. Geoscience and Remote Sensing*, **26**(3), 1988, 307–318.
18. F. T. Ulaby et al., "Experiments on Radar Backscatter of Snow," *IEEE Trans. Geoscience and Remote Sensing*, **GE-15**(4), 1977, 185–189.
19. J. Benjamin and G. E. J. Peake, *Contributions to the U.K. Microwave Landing System Study (Phase I)*, Royal Aircraft Establishment, Tech. Memo RAD 1021, May 1973.
20. J. Capon, *Multipath Parameter Computations for the MLS Simulation Computer Program*, Lincoln Lab. Rep., FAA-RD-76-55, Project Rep. ATC-68, Apr. 1976.
21. A. R. Lopez, "Application of Wedge Diffraction Theory to Estimating Power Density at Airport Humped Runways," *IEEE Trans. Antennas and Propagat.*, **AP-35**(6), 1987, 708–714.
22. R. W. Larsen, A. L. Maffett, R. C. Heimiller, A. F. Fromm, E. L. Johasen, R. F. Rawson, and F. L. Smith, "Bistatic Clutter Measurements," *IEEE Trans. Antennas and Propagat.*, **AP-6**(6), 1978, 801–804.

CHAPTER TEN

High-Frequency Electromagnetic Computer Codes

10.1 INTRODUCTION

Nowadays computer codes are very commonly used in the electromagnetic case studies of Chapters 2 through 9. These codes range in complexity from the very small and simple to the very large and complex. Choosing a code is not trivial; it depends on the requirement of models, the requirement and ready availability of computing facilities, accuracy, resources, cost, and many other factors. Codes can be divided into three categories according to operating frequency: low, high, and hybrid. Hybrid codes use a combination of two or more techniques; some combinations are described in Section 1.17. Many codes are low-frequency codes, mostly involving the method of moments. In this book, since we are interested in high-frequency techniques, we restrict our discussion to high-frequency codes and use MOM codes for comparison. Although [1] is mainly about low-frequency codes, it puts in a nutshell the numerical issues behind all such codes.

10.2 STEPS TO DEVELOP COMPUTER CODES

The major steps [1] in developing a computer code are:

1. Decide the purpose of the code. For example, it may be to study scattering from complex bodies using high-frequency techniques.
2. Decide the mathematical model on which the computer code is based; this may depend upon the problem and the desired results.
3. Complete the mathematical formulation. Lay down the approximations and define the limitations.

4. Formulate the task and make it ready for numerical evaluation.
5. Validate the code. A code can be validated with experimental results corresponding to the same task and by using other analytical or numerical methods applicable to the same problem.

10.3 CHARACTERISTICS OF A GOOD CODE

Since codes are written for a large variety of systems, such as mainframe, massively parallel operation, and PC-based systems, it is very difficult to state precisely the characteristics of a code. For example, when a supercomputer is available for use then the possibility of a memory space problem is not an issue. A reasonably good code should have the following characteristics given the problems to be solved.

1. The code should be user-friendly so it can can be used by someone less conversant with the theory.
2. The documentation should provide a commentary on each section of the code so the job can be understood very easily.
3. Debugging should be easy.
4. The code should be efficient and should require an optimum computer time.
5. The code should make efficient use of limited computer memory.
6. If appropriate, an efficient PC-based version should be available.

10.4 SOME OF THE EXISTING HIGH-FREQUENCY COMPUTER CODES

Various high-frequency codes have been written in the past to do specific jobs; a summary is available in [2] and more details are available in [3–7]. In this section we deal with a few of the codes that are currently available and widely used.

10.4.1 Intrasystem Electromagnetic Compatibility Analysis Program (IEMCAP)

IEMCATP [4] is a computer program in Fortran IV to do EMC analysis of aircraft, space, missile, and ground-based systems. The antenna-to-antenna is modeled using GO. The aircraft is modeled as a cone–cylinder, simulating the fuselage, with attached flat plates or thin wedges, simulating the wings. All parameters for transmitter and receiver along with antennas and dimensions of the model are specified by the user. IEMCAP then determines the magnitude of the power delivered to the receiving antenna of interest based on the input information. It takes into consideration the distance between antennas (i.e. free-space loss), propagation around the fuselage (fuselage shading factor), and

10.4 SOME OF THE EXISTING HIGH-FREQUENCY COMPUTER CODES

any diffraction from the edge of the wing that lies in the direct path between the antennas (using wedge diffraction theory). The fuselage shading factor SF is given by

$$SF_C = -\frac{A}{(\eta A + \xi)} \tag{10.1a}$$

where

$$A = \rho_f \theta_s^2 \sqrt{\frac{2\pi}{\lambda D_C}} \tag{10.1b}$$

$$\eta = \begin{cases} 5.478 \ E-03 & A < 26 \\ 3.34 \ E-03 & A \geq 26 \end{cases} \tag{10.1c}$$

$$\xi = \begin{cases} 0.5083 & \text{for} \quad A < 26 \\ 0.5621 & \text{for} \quad A \geq 26 \end{cases} \tag{10.1d}$$

where ρ_f = the radius of the cylinder in meters
θ_s = the angle around the cylinder due to the propagation path in radians
λ = the wavelength
D_C = the distance of the cylindrical segment of propagation path in meters.

The formulas in eqn. (10.1) are based on the analytical formulation due to Hasserjian and Ishimaru [8] described in Section 1.18.2.2.

10.4.2 Aircraft Interantenna Propagation with Graphics (AAPG)

The AAPG program also uses the same formulations but has a computer graphics capability to display the antenna-to-antenna coupling path and coupled EMI on an aircraft. AAPG is restricted to antenna-to-antenna coupling models on aircraft. It uses the diffraction coefficient for an infinite half plane instead of knife-edge diffraction in cases where IEMCAP gives better results at higher frequencies for curved surface modeling. In this model, the reflected rays are not included. AAPG uses only the direct path between the transmitter and receiver as does IEMCAP. An example of a graphic display is shown in Figure 10.1 and the geometric model used in AAPG code is given in Figure 10.2 consisting of cylinder, cone, and plate models. A new version of AAPG has been developed by Harris Corporation which allows the cylinder and cone to have arbitrary cross sections.

10.4.3 GTD-Based Codes

Numerical Electromagnetic Code–Basic Scattering Code (NEC–BSC). A summary of the basic scattering code, developed in the electroscience laboratory of Ohio State University, is available in [2, 5]. In this code, the fuselage of the complex

466 HIGH-FREQUENCY ELECTROMAGNETIC COMPUTER CODES

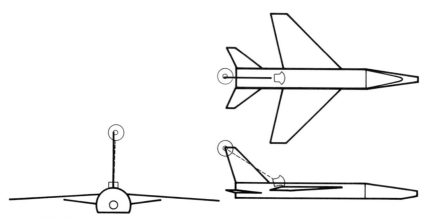

FIGURE 10.1 Example of a graphic display from the AAPG code [2].

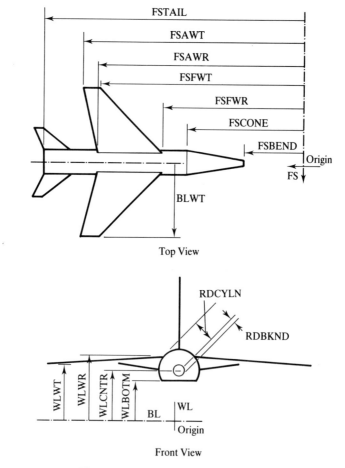

FIGURE 10.2 The geometric model used in the AAPG code [2].

object is modeled by an elliptic cylinder and the wings by flat plates. It is possible to model sources on wings, horizontal and vertical stabilizers, external pods, and sources situated outside the aircraft structure. The interactions between plates and cylinders including the creeping wave contributions are also included. This code has the option of being interfaced with a method of moments approach to compute the mutual coupling effects between various array elements and to model small scattering objects in close proximity to the sources. This code is also capable, on a limited basis, of handing dielectric plates simulating glass windows, absorbing panels, and dielectric-coated structures. It can do power and impedance coupling calculations between two antennas. Antenna current distribution can be chosen from a given set of distributions. Also, there is a provision for placing an infinite ground plane at a suitably chosen point in the geometry. This ground plane can be PEC, of imperfect conductivity, and dielectric. The plates can be polygonal, there is no restriction on their numbers, and they need not be connected to the fuselage. This also includes GTD curved surface diffraction. GTD codes incorporate polarization, unlike AAPG and IEMCAP, which calculate the polarization-independent mutual coupling using the Friis transmission formula. All these aspects make this code very important and attractive.

General Three-Dimensional Aircraft Antenna Radiation Pattern Code (Fuselage Code). This code was also developed at the electroscience laboratory of Ohio State University. It does a very good job of analyzing antennas on aircraft with the restriction that the source is always on the aircraft. The fuselage is modeled by a prolate spheroid and the loss due to creeping waves around the fuselage is included. As usual, plates are used to model wings and stabilizers, both horizontal and vertical. It incorporates second-order effects but has no provision for estimating mutual coupling effects without modifications, although the coupling is directly proportional to the near-zone fields.

It is worthwhile to compare predictions by IEMCAP and BSC codes. The example [2] is a 24 in. cylinder in the frequency range is 2–18 GHz. The results are for antennas in the same plane perpendicular to the cylinder axis and for antennas offset along the z-direction by 36 in. The separation between the two antennas as varied along the circumference. The representative results are available in [2]. The key conclusions of this comparitive study done in [2] are:

1. The TE and TM polarizations for BSC and IEMCAP differ by a maximum of 20 dB, which means the coupling is dependent on polarization but not that strongly.
2. BSC and IEMCAP results differ for large angles of incidence at high frequencies. Of course, experimental results are needed to judge which code gives accurate results for particular examples when there is a discrepancy between results by the two codes.
3. BSC and IEMCAP differ in coupling calculations at high frequencies.

10.4.4 General Electromagnetic Model for Analysis of Complex Systems (GEMACS)

This is a versatile code [9] which uses low-frequency, high-frequency, and hybrid techniques to accommodate targets of arbitrary size. It solves electromagnetic radiation and scattering problems, treating both the exterior and interior of the structure as a total solution. The method of moments (MOM) and the geometrical theory of diffraction (GTD) are used for the exterior analysis. The interior problem is solved using a finite difference (FD) formulation. The MOM, GTD, FD, and their various hybrid combinations can be used to solve almost any real-world problems. The restrictions are the limitations of the geometric models and the inherent limitations of the techniques.

10.4.5 G3F-TUD1

This code [10] was developed at the Technical University of Denmark in the mid-1970s. It is a frequency domain code treating the problems of antennas outside a closed, convex, polyhedral satellite using GTD solutions for the far field as the vector sum of direct, reflected, singly diffracted and doubly diffracted (from edges) rays. The corner and triple are not included. This code can be used to calculate the surface currents on EMP excited objects. As an example [10], for a box shape with 12 edges and 8 vertices, the total CDC 7600 computer time to obtain the far field in 80 directions by G3F-TUD1 is about 4 sec.

10.4.6 GENSCAT

This code has the capability to calculate scattering from the interior of engine inlet ducts and from the exterior surfaces of an aircraft. This code employs all high-frequency methods. It uses a GTD edge integration method. The GTD integration method is better for certain cases, such as a flat-backed cone with a direction of incidence such that the edge is only partially illuminated. The steps of using the GENSCAT program are

1. Construct a computer model of the target by subdividing the complex body into its major components. Each component is mathematically described by one or more surfaces.
2. Select an appropriate scattering technique to be applied to the different parts. For example, the total scattering from a smooth surface is determined by PO and fringe wave contributions.
3. Assemble the user-definable subroutines (UDSs) for each of the major aircraft structures.

10.5 PHYSICAL OPTICS (PO) AND PHYSICAL THEORY OF DIFFRACTION (PTD) CODES

Some of the university of Illinous Urbana Champaign codes [11] have used physical optics (PO) and the physical theory of diffraction (PTD). The PO part consists of two codes, the analytical surface version and the patch version. The analytical surface version can also handle special material. To first order, the effect of any three-dimensional diffraction is taken care of by the code MacWEDGE, either GTD or PTD version. The basic code does not consider multiple bounce and mutual interactions, but along with creeping wave considerations they can be easily incorporated. This code can treat both mono- and bistatic cases. Results of RCS of various targets with and without coating have been computed in [11] using the UI-PTD codes. Some of the results are in Figures 10.3 through 10.6.

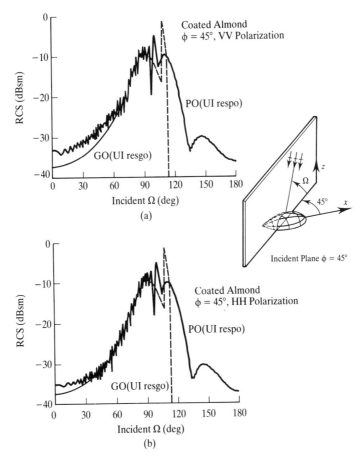

FIGURE 10.3 The RCS in dBsm with incident angle in degrees of a PEC almond structure: (a) for $\phi = 45°$, VV polarization [11]; (b) for $\phi = 45°$, HH polarization [11].

470 HIGH-FREQUENCY ELECTROMAGNETIC COMPUTER CODES

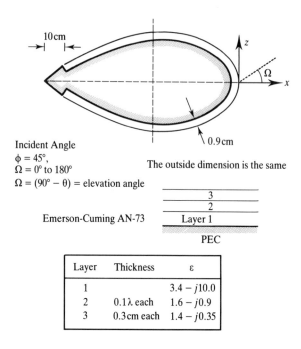

Incident Angle
$\phi = 45°$,
$\Omega = 0°$ to $180°$
$\Omega = (90° - \theta) =$ elevation angle

The outside dimension is the same

Emerson-Cuming AN-73

Layer	Thickness	ε
1		$3.4 - j10.0$
2	0.1λ each	$1.6 - j0.9$
3	0.3 cm each	$1.4 - j0.35$

(a)

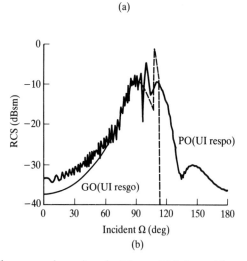

(b)

FIGURE 10.4 (a) The same almond as in Figure 10.3 but with coating. (b) E_ϕ (VV) versus Ω. (c) E_θ versus Ω.

10.5 PHYSICAL OPTICS (PO) AND PHYSICAL THEORY OF DIFFRACTION (PTD) CODES

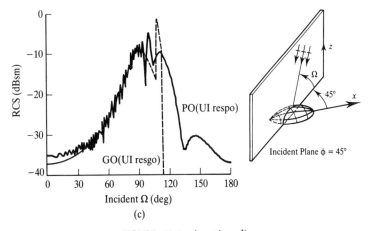

FIGURE 10.4 (*continued*).

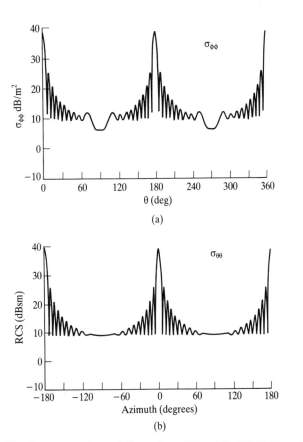

FIGURE 10.5 RCS of a square plate of dimensions ($5\lambda \times 5\lambda$); UI-PTD code compared with MOM code: (a) $\sigma_{\phi\phi}$ and (b) $\sigma_{\theta\theta}$.

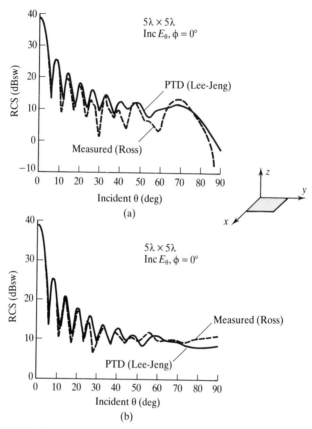

FIGURE 10.6 RCS of a square plate of dimensions $(5\lambda \times 5\lambda)$, $\phi = 0°$; PTD code compared with experiment [Ross, 12]: (a) $\sigma_{\theta\theta}$ and (b) $\sigma_{\phi\phi}$.

10.6 SCOPE FOR FURTHER INVESTIGATIONS

In light of changed priorities for research and development after the cold water, environmental modeling and simulation have now emerged as key areas. Under these changed circumstances, in the opinion of the author, an individual with the background of the contents of the book could deal with the following problems.

1. Modeling of subsurface objects. Some of the technique, though to a limited extent, can be used to model metallic and dielectric objects. This has application in unexploded ordinance of the US Army. The unexploded objects could prove hazardous unless detected and removed.
2. Modified modeling of terrain propagation and scattering with incorporation of the environmental effects, for example, temperature, snow, and dust.

3. In predictive technology, which is concerned with how to keep the weapons in working condition, detect cracks and gaps, effect of snow, dust on the antennas and other radar components.
4. In wideband modeling of radiation and scattering, the high frequency techniques described in this text with applications, along with low frequency methods, show the promise of time-domain modeling of many objects of interest.
5. Development of efficient computer codes and graphics package for each of the above investigations.

10.7 SUMMARY

Starting with steps to develop a code and its desired characteristics, this chapter discussed some of the available high-frequency codes and their limitations. It described key features of the high-frequency codes IEMCAP, AAPG, BSC, fuselage code, GMACS, GENSCAT, UI-PO, and UI-PTD.

10.8 REFERENCES

1. A. J. Poggio and E. K. Miller, "Techniques for Low-Frequency Problems," in *Antenna Handbook*, edited by Y. T. Lo and S. W. Lee, Van Nostrand Reinhold, New York, 1988.
2. T. E. Durham, *SHE/EHF Antenna-to-Antenna EMI Coupling Interim Technical Report*, Harris Corporation, Melbourne, FL. (Contract F-30602-85-C-0114, RADC, Griffis AFB, NY.)
3. H. P. Widmer, *A Technical Description of the AAPG Program*, Rept. ECAC-CR-83-048, Nov. 1983, Annapolis, MD.
4. J. L. Boddanor, R. A. Pearlman, and M. D. Siegel, *Intrasystem Electromagnetic Compatibility Analysis Program*, User Manual Usuage Section, AD A-008-527, Dec. 1974; User's Manual Eng. Section AD-009-526, Dec. 1974; NTIS, U.S. Dept. of Commerce, Springfield, VA.
5. R. J. Marhefka and W. D. Burnside, *Numerical Electromagnetic Code–Basis Scattering Code, NEC-BSC (Version 2), Part I: User's Manual*, Rep. 712242-14, Dec. 1982, Ohio State Univ. Electroscience Lab., Dep. Elec. Eng. Prepared under contract N00123-79-C-1469 for Naval Regional Contracting Office.
6. R. J. Marhefka, *Numerical Electromagnetic Code (NEC)—Basic Scattering Code, Part II: Code Manual (Version 2)*, Rep. 712242-15, Dec. 1982, Ohio State Univ.
7. R. J. Marhefka, *Numerical Electromagnetic Code (NEC)—Basic Scattering Code, Part II: Code Manual (Version 2)*, Rep. 712242-15, Dec. 1982, Ohio State Univ. Electroscience Lab., Dept. Elect. Eng. Prepared under contract N00123-79-C1469 for Naval Regional Procurement Office.
8. G. Hesserjian and A. Ishimaru, "Excitation of a Conducting Cylindrical Surface of Large Radius of Curvature," *IRE Trans.*, **AP-10**, 1962, 264–273.

9. *General Electromagnetic Model For the Analysis of Complex Systems (GEMACS)—Version 5, Vol. I, User Manual: Vol. II, Version 5, Engineering Manual*, Final Tech. Rep. RADC-TR-90-360, Dec. 1990.
10. H. Bach, *Description of and Manual for Polygonal Satellites*, Electromag. Inst., Tech. Univ. Denmark, Sept. 1975.
11. S. W. Lee and R. J. Marhefka (eds.), *Data Book of High-Frequency RCS*, complied by High Frequency Subgroup, Joint Service Electromagnetic Code Consortium, Version 2, Aug. 1, 1989, pp. 33, 42, 53, 69, 83, 95, 145.

APPENDIX A

A Summary of UTD and UAT

Even though UTD and UAT have been described in many easily available references, for the benefit of the readers and for ready reference we briefly summarize the techniques in this appendix.

A.1 UNIFORM THEORY OF DIFFRACTION AND SLOPE DIFFRACTION

The dyadic UTD coefficient (Chapter 1, [10]) for the PEC wedge is given by

$$\bar{D} = -\hat{\beta}_0'\hat{\beta}'D_s - \hat{\phi}\hat{\phi}'D_h \tag{A.1}$$

$$D_{s,h} = \frac{-1}{2n\sqrt{2\pi}} \sum_{m=1}^{4} d_m^{s,h} F(X) \tag{A.2}$$

where

$$d_1^{s,h} = \frac{1}{\sin\beta_0} \cot\left\{\frac{\pi + (\phi - \phi')}{2n}\right\} \tag{A.3}$$

$$d_2^{s,h} = \frac{1}{\sin\beta_0} \cot\left\{\frac{\pi - (\phi - \phi')}{2n}\right\} \tag{A4}$$

$$d_3^{s,h} = \mp \frac{1}{\sin\beta_0} \cot\left\{\frac{\pi + (\phi + \phi')}{2n}\right\} \tag{A.5}$$

$$d_4^{s,h} = \mp \frac{1}{\sin\beta_0} \cot\left\{\frac{\pi - ((\phi + \phi'))}{2n}\right\} \tag{A.6}$$

$$F(X) = i2\sqrt{X}\exp(iX)\int_{\sqrt{(X)}}^{\infty} e^{-i\tau^2}\,d\tau \tag{A.7}$$

$$a^{\pm}(\beta^{pm}) = 2\cos^2\frac{2n\pi N^{\pm} - (\beta^{pm})}{2} \tag{A.8}$$

the integers $N^{\pm}$ (see [8] for details) most nearly satisfy the equations:

$$2v\pi n N^+ - (\beta^+) = \pi \quad \beta^+ = \phi + \phi'$$
$$2\pi n N^- - (\beta^-) = \pi \quad \text{with} \quad \beta^- = \phi - \phi'$$

L is the distance parameter, which depends on the geometry

$$L = \begin{cases} = \dfrac{rr'}{r+r'}, & \text{for cylindrical incident wave incidence} \\ = s^2 \sin \beta_0, & \text{for plane wave incidence} \\ = \dfrac{ss'}{s+s'}, & \text{for conical and spherical wave incidence} \end{cases}$$

The significances of the other symbols are

s = distance of the field point from the wedge along the diffracted ray
s' = distance of the wedge from the source point along the incident ray
β_0 = the smaller angle between the incident ray and the tangent to the edge at the point of incidence.

The modified Fresnell integral may be approximated in the following series:

$$F(X) \simeq \sqrt{\pi X} - 2X \exp(i\pi/4) - \frac{2}{3} X^2 \exp(-i\pi/4) \exp\left\{i\left(\frac{\pi}{4} + X\right)\right\} \quad \text{for} \quad X \ll 1$$

(A.9a)

$$F(X) \simeq 1 + \frac{i}{2X} - \frac{3}{4} \frac{1}{X^2} - i \frac{15}{8} \frac{1}{X^2} + \frac{75}{16} \frac{1}{X^4} \quad \text{for} \quad X \gg 1 \quad \text{(A.9b)}$$

If $X < 10$, $F(X) \approx 1.0 + i0.05$.

$n = 2$ for half plane, $n = 1$ for full plane and $n = 3/2$ for a wedge with 90° interior angle.

The above assumes the edge of the wedge is illuminated uniformly. In situations where this is not satisfied one makes the uniform slope diffraction correction. This correction is given by

$$\begin{aligned}\frac{\partial D_{s,h}(\phi,\phi';\beta_0)}{\partial \phi'} = -\frac{e^{-i\pi/4}}{4n^2\sqrt{2\pi k}\sin\beta_0} &\times \left[\csc\frac{(\pi+(\phi-\phi')}{2n}F_s[kLa^+(\phi-\phi')]\right.\\ &-\csc\left(\frac{(\pi-(\phi-\phi')}{2n}\right)F_s[kLa^+(\phi+\phi')]\\ &\pm\csc\left(\frac{(\pi+(\phi+\phi')}{2n}\right)F_s[kLa^+(\phi+\phi')]\\ &\left.-\csc\left(\frac{\pi-(\phi+\phi')}{2n}\right)F_s[kLa^-(\phi+\phi')]\right]\end{aligned}\quad (A.10)$$

where

$$F_s(X) = i2X[1 - F(X)]$$

Considering the slope diffraction, the complete diffracted field is given by the matrix relation:

$$\begin{bmatrix}E_\parallel^s\\ E_\perp^s\end{bmatrix} = \left\{\begin{bmatrix}D_s & 0\\ 0 & D_h\end{bmatrix} + \frac{\partial}{\partial\phi'}\begin{bmatrix}D_s & 0\\ 0 & D_h\end{bmatrix}\right\}\begin{bmatrix}E_\parallel^i\\ E_\perp^i\end{bmatrix}$$

Very recently a physical theory of slope diffraction has been proposed in a presentation. A summary is available in [A.3] and details can be obtained from the author, who is yet to obtain definite information on the suitability of PTD slope diffraction compared to UTD slope diffraction.

A.2 UNIFORM ASYMPTOTIC THEORY (UAT) OF DIFFRACTION FOR PEC HALF-PLANE DIFFRACTION

UAT [8, 70] states that the incident, reflected, and diffracted fields are expanded asymptotically and the total field satisfies Maxwell's equations asymptotically. The solution is continuous around the incident and reflected shadow boundaries and indentifies with Keller's theory away from the shadow boundaries.

The total field is given by

$$E^t(\mathbf{r}) = E^G + E^d(\mathbf{r}) + O(k^{-1}) \quad (A.12)$$

where, $E^G(\mathbf{r})$ and $E^d(\mathbf{r})$ are the new GO and the diffracted fields.

The UAT expands the GO field in the form

$$\mathbf{E}^G(\mathbf{r}) = \{\mathbf{F}(\xi^i) - \hat{\mathbf{F}}(\xi^i)\}\mathbf{E}(\mathbf{r}) + \{i \to r\} + O(\mathbf{k}^{-1}) \qquad (A.13)$$

$i \to r$ means that the second term is the same as the first term with i replaced by r.

$(\xi^i)^2$ is called the *phase detour parameter*, the difference between the phases of the two paths. One is the straight path from source to observation point; the other is the path from source to observation point via the edge. It is given by

$$\xi^{i,r}(\mathbf{r}) = \varepsilon^{i,r}(\mathbf{r}|\sqrt{k_0[s^d(\mathbf{r}) - s^{i,r}(\mathbf{r})]}|) \qquad (A.14)$$

The Fresnel function dominant term $\hat{F}(X)$ is given by

$$\hat{F}(X) = \frac{1}{2X\sqrt{\pi}} e^{-(X^2 + \pi/4)} \qquad (A.15)$$

UAT seems to work very well in many applications. Some of the examples have been discussed in Chapters 2 and 5.

APPENDIX B

Time-Saving Sampling Methods for Evaluation of Time-Consuming PO Integral

It is well known that a potential difficulty of application of physical optics is the enormous amount of computer (CPU) time required to compute the integral. So there is a need to use integral sampling methods to optimize the CPU time.

Sampling methods are described in [A.1, A.2]. Let us discuss the problem from [1]. The electric far field scattered by an arbitrary reflector antenna is given by

$$\mathbf{E}(\mathbf{r}, \theta, \phi) = -\frac{i\omega\mu}{4\pi R}[\mathbf{I} - \hat{\mathbf{R}}\hat{\mathbf{R}}]I \qquad (A.16)$$

where

$$I = \iint_S \mathbf{J}_s \exp(i\beta\rho \cdot \mathbf{R})\, d\mathbf{S} \qquad (A.17)$$

where $\mathbf{J}_s$ = the induced reflector surface current density
$\mathbf{I}$ = the (3×3) unit matrix
β = propagation constant
$\mathbf{k} = u\hat{\mathbf{x}} + v\hat{\mathbf{y}} + w\hat{\mathbf{z}}$

The radiation integral $I(u, v, w)$ is given by

$$I(u, v, w) = \iint_S \mathbf{J}_s \exp[i(u\hat{\mathbf{x}} + v\hat{\mathbf{y}} + w\hat{\mathbf{z}})]\, d\mathbf{S}$$

$$= \iint_S \mathbf{J}_s \exp(i\mathbf{k} \cdot \boldsymbol{\rho})\, d\mathbf{S} \qquad (A.18)$$

479

APPENDIX B

Since the scattering or antenna structure may be electrically large but finite, the radiation integral is a band-limited function and according to [A.1] is given by

$$I(u, v, w) = \sum_n \sum_m \sum_p \frac{n\pi}{a}, \frac{m\pi}{b}, \frac{p\pi}{c} (ua - n\pi)(vb - m\pi) \operatorname{sinc}(we - p\pi) \quad (A.19)$$

$2a$, $2b$ and $2c$ are the dimensions of the parallelopiped including the reflector surface, and $\operatorname{sinc}(x) = (\sin(x)/x)$. For $u = \beta \sin \theta \cos \Phi$, $v = \beta \sin \theta \sin \Phi$, $w = \beta \cos \theta$.

The equation gives the algorithm to evaluate the integral in (A.19) in terms of samples at the Nyquist rate. An example of how the sampling algorithms will be used in a practical problem of a reflector antenna analysis has been described in Chapter 3.

APPENDIX C

Evaluation of the Function $M_n(\phi, \phi_0, \theta_0, \theta_n)$

The Maliuzhinets function is involved in problems of electromagnetic scattering, radiation and coupling dealing with half planes and wedges. It is defined as

$$M(u) = \exp\left[-\left(\frac{1}{8\pi}\right)\int_0^u (\pi \sin t - 2\sqrt{2\pi} \sin(t/2) + 2t)/\cos t \, dt\right]$$

$$= \exp\left[-\left(\frac{1}{8\pi}\right)\int_0^u \pi \tan t \, dt - 2\sqrt{2\pi} \int_0^u \frac{\sin(t/2)}{\cos t} dt + 2\int_0^u t \sec t \, dt\right]$$

(A.20)

On integration by parts using necessary substitution

$$\int_0^u \pi \tan t = \pi \log(\sec u) \tag{A.21a}$$

$$\int_0^u \frac{\sin(t/2)}{\cos t} dt = \frac{\frac{1}{2}\log 0.1715(\sqrt{2}\cos u/2 + 1)}{\sqrt{2}\cos(u/2) - 1} \tag{A.21b}$$

and

$$I_3 = \int_0^u t \sec t \, dt = \sum_{k=0}^\infty \frac{|E_{2k}|u^{2k+2}}{2k(2k+2)} \tag{A.21c}$$

For $|u| < \pi/2$ where E_{2k} are Euler numbers.
It gives

$$I = \frac{u^2}{2} + \frac{u^4}{8} + \frac{5u^6}{144} + \frac{61}{5760}u^8 + \ldots, |u| < \pi/2 \tag{A.21d}$$

481

For small u

$$I_3 \approx u^2/2 \tag{A.22}$$

Hence, for small u

$$M(u) = \exp\left[-0.125 \log(\sec u) + 0.125\sqrt{2} \log 0.1715 \frac{1.414 \cos u/2 + 1}{1.414 \cos u/2 - 1} - \frac{u^2}{8\pi}\right] \tag{A.23}$$

The function is analytic when $\text{Re}(u) \leq 2$. It obeys the following relationships

$$M(u) = M(-u)$$

$$M(u + \pi/2)M(u - \pi/2) = M_2(\pi/2) \cos(u/4)$$

$$M(u - 2\pi)/M(u + 2\pi) = \tan[0.5(u + \pi/2)]$$

$$M(u \pm 7\pi/2) = \pm \frac{\sin[(\pi \pm u)/4]}{\sin(u/4)} M(3\pi/2 \pm u)$$

$$M(u_1 + iu_2) \sim \exp\left(\frac{u_1 \mp iu_2}{8}\right) \quad \text{as} \quad |u| \to \infty$$

REFERENCES

A.1. O. M. Bucci and G. D. Massa, "Exact Sampling Approach for Reflector Antennas Analysis," *IEEE Trans. Antennas and Propagat.*, **AP-32**(11), 1984, 1259–1262.

A.2. O. M. Bucci, G. Franceschetti, and G. D'Elias, "Fast Analysis of Large Antennas, A New Computational Philosophy," *IEEE Trans. Antennas and Propagat.*, **AP-28**, 1980, 306–310.

Author Index

Ahluwalia, 41
Anderson, 80

Balanis, C.A., 64
Bhattacharyya, A.K., 58
Boersma, J., 41
Booker, 73
Bucci, O., 58
Buyukaksoy, 71

Chakravorty, A., 85
Clemmow, 61, 73
Cramer, P.W., 19

Deschamps, 68

Felsen, 23
Franceschetti, G., 58

Greisser, 64

Hansen, T.B., 50

Karal, 33
Karp, 23
Keller, J.B., 23, 40, 41
Kouyoumjian, R., 64

Lee, S.W., 19, 23, 29, 68, 73
Lewis, 23, 41
Love, A., 29

Maliuzhinets, 58, 66
Marcuvitz, 23
Mittra, R., 19, 73

Pathak, P.H., 41, 64
Pauli, 61

Rusch, W.V.T., 18
Rushdi, A., 19
Rytov, 32

Sanyal, S., 58
Schelkunoff, S. A., 36
Schensted, 23
Seckler, 23
Senior, T.B.A., 51, 84
Serbest, 71
Snyder, 29
Sommerfeld, 68
Sorenson, O., 18

Tiberio, R., 58

Van Waerden, 61, 62
Volakis, J.L., 58, 84

Weiner-Hopf, 53, 64, 71, 81
Weston, 51

Subject Index

AAPG, 396, 465
Airy functions, 114
Angular spectrum method, 73
Antennas. *See* Horn antennas; Reflector antennas
Antennas in complex environment, 355
   microstrip patch on a finite PEC ground plane, 358
   monopole on a cone, 369
   monopole on a finite cylinder, 365
   monopole on a finite PEC and composite ground plane, 355
   monopole on a rocket-shaped body, 369
Antennas on aircraft, 369
   far-field, 385
   near-field, 371
   mutual coupling, 388
Aperture efficiency, 263
Asymptotic solution of Maxwell's equation, 11
ATS F and G spacecraft, 213

Basic scattering code (BSC), 465
Bistatic
   cross-section, 330
   scatterer, 215
Boundary conditions
   absorbing, 34
   approximate, 32
   generalized higher order, 32
   generalized impedance, 35
   hard and soft, 35
   impedance, 32
   nonconventional, 31
Brillouin ray technique, 192
BRL-CAD package, 331
Broad-band flared horns, 197

Cassegrainian reflector, 215
Caustic, 129
   method of equivalent current (MEC), 130
Christoffel symbols, 16
CNI blade antenna, 356
Codes, 463
   AAPG, 396, 465
   BSC, 465
   fuselage code, 467
   G3F-TUD1, 468
   GMACS, 468
   IEMCAP, 96, 126, 391, 464
   McWEDGE, 469
   UI-PO, 469
   UI-PTD, 469
Complex targets, 302
Conical horn, 185
   corrugated, 194
   uncorrugated, 186
Conical wave launcher, 263
Cross-polarization, 252
   in inter-antenna coupling calculations, 401
   of front-fed and offset reflectors, 252
   of offset antennas, 262
Curvature term, 113
Curvature vectors, 15
   geodesic, 15
   normal and tangential, 15

Deep space network (DSN), 213
Dielectric wedge, 86
   diffraction coefficient, 90
   dual integral equations, 87
   results and discussions, 92
Diffraction, 40
   anisotropic half plane, 141

SUBJECT INDEX 485

applications, 135
circular aperture, 54
coefficient, 86, 92
corner, 41
dielectric-coated metal wedge, 144
dielectric half plane, 80
dielectric wedge, 86, 90, 109
differential geometry, 15
double impedance wedge, 97
double PEC wedge, 94
ferrite half plane, 137
incremental length (ILDC), 133
infinite wedge with tensor boundary
   conditions, 144
impedance discontinuity edge, 70
impedance discontinuity in half plane, 70
   applications, 54
   mathematical basis, 53
   matrix of a curvature discontinuity, 51
   spectral theory of, 53
lossy rough wedge, 145
single impedance wedge, 57
single PEC wedge, 41
special wedges, 144
spectral, 54
surface, 115
surface wave, 109
   analytical models, 112
   on coated bodies, 121
   results and discussions, 123
tip, 46
Dihedral corner, 176
Dipole fields, 120

ECAC, 387
Eikonal equation, 11
Elliot type modes, 321
EMI and EMC, 388
Extended spectral ray method (ESRM), 97

Ferrite half plane, 137
Flange, 264
   corrugated, 265
   loaded, 265
Focus-fed reflector, 215
Fresnel reflection coefficients, 86
Fuselage code, 467

Gap between low and high frequency, 151
Gaussian beam, 316
Gaussian beam method, 316
Gaussian curvature, 23, 26
   reflected wavefront, 23
   transmitted wavefront, 26

Generalized scattering matrix formulation
   (GSMF), 73
GENSCAT, 468
Geodesic, 15, 16
   finding a geodesic, 17
Geodesic curvature, 15
Geometric optics, 88
   equivalent line currents, 130
Geometrical theory of diffraction, 40
GMACS, 468
Graphical EM Computing (GRECO), 346
Green's function for impedance cylinder, 122

Half plane, 41
   anisotropic, 141
   dielectric, 138
   ferrite, 137
   impedance, 54
   PEC, 41
Hard boundary, 35
High-frequency techniques, 1, 8
   accuracy testing of, 146
   time-domain (TD), 153
      physical optics (TDPO), 153
      UTD (TDUTD), 157
Horn antennas, 175
   broad-band flared horns, 197
   conical horns, 185
   control of sidelobes, 199
   CONUS, 408
   corrugated, 192
   cross-polarization characteristics, 205
   first order, 176
   multiple diffraction, 178
   pyramidal horns, UAT analysis, 176
      E-plane, 176
      H-plane, 182
   second order, 178
   special horns, 196
   uncorrugated, 186
Hybrid techniques, 103, 321

I-DEAS, 346
IEMCAP, 126, 391, 396, 464
Impedance half planes, 57
Impedance wedge, 57, 61
Impulsive radiating antenna (IRA), 262
Incremental length diffraction coefficient (ILDC),
   133

Knife-edge diffraction, 412

Longley–Rice mode, 408

486 SUBJECT INDEX

Magnetic field integral equation, 39, 108
Maliuzhinets'
  function evaluation, 481
  solution, 59
Method of equivalent current (MEC), 130
  applications, 132
Method of moments combined with GTD, 103
Modified impedance matrix of a body, 106
Monopole fields, 119
  near-zone fields, 119
Monopole on rocket-shaped body, 369
Monopole PEC knife-edge diffraction, 420
Mutual coupling, 395, 396
  limitations of models, 396
  on a cone, 395

NEC-BSC, 465

Paraboloids, 217
  caustic fields, 220
  diffracted fields, 219
  near-field patterns, 217
    E-plane, 220
PEC starship enterprise, 331
Phong local illumination model, 346
Physical optics method, 36, 89
  classical, 39
  polarization-corrected, 39
Primary feed, 216

Radar absorbing material (RAM), 247
Radar cross section (RCS) of, 302
  bistatic, 329
  complex objects, 302
  cylindrical plates, 311
  flat plate using PO and Keller's GTD, 45
  open-ended cavities, 313
  plates at grazing incidence, 306
  polygonal plates, 302
  real-time prediction of complex objects, 345
  strips at grazing incidence, 306
Radius of curvature, 113
Ray-optical methods, 86
Ray path length, 113
Ray tracing, 319
Reflection electromagnetic, 23
  from convex conducting surface, 23
  from dielectric surface, 27
Reflector antennas, 213
  aperture field formulation, 226
  as an EM scattering problem, 215
  axially symmetric reflector, GTD analysis, 222

  cassegrainian, 229
  cross-polarization characteristics, 252
  focus-fed, 216
  high power microwave (HPM) applications, 263
  offset, 224
    GO analysis, 220
  real-time prediction, 345
  rim-loading and sidelobe control, 247
    diffracted fields, 249
    diffraction fields at the junction of annular loaded layers, 250
    focus-fed parabolic and hyperbolic dishes, 247
    incident, 248
    reflected, 248
    total field, 250
  shaping scheme, 243
    GO shaping, 243
  shooting and bouncing ray (SBR), 314
  sidelobe control with loaded and unloaded shrouds, 264
  subreflector main reflector system, 236
    GO and PO analysis, 237
    GTD analysis, 238, doubly diffracted ray (DDR), 241
    with offset feed, 224
Relationship between GTD, PTD, and MEC, 135

Shading loss calculations, 129, 394
Shooting and bouncing rays (SBR), 322
Shroud loaded and unloaded, 264
Slot antenna, 275
  on circular cylinder, 282
  on cones, 292
  on elliptic cylinders, 285
  on ground planes, 276
Slot dipole on convex surface, 113
Soft surface, 35
Source representation, 119
Spectral theory of diffraction, 97
Specular point, 18
Specular point determination, 19
Subaperture expansion, 318
Subreflector, 216

TEM
  coaxial aperture, 275
  feed, 263
  mode coaxial aperture, 278
  structure, 263
  wave launcher, 262
Time-consuming PO integral, 479

SUBJECT INDEX **487**

Time-saving sampling methods, 479
Transport equation, 11

UDS, 465
Uniform asymptotic theory of diffraction, 66, 477
    for impedance half plane, 66
Uniform theory of diffraction (UTD), 61, 475
    for half plane, 62
    for impedance wedge, 61

Uniform theory of slope diffraction (UTSD), 123
Univac, 108, 23

Viking orbiter, 213

Wedges, 41
    double, 92
    multiple, 92
    perfectly electrically conducting (PEC), 41